Complex Manifolds and Geometric Algebraic Analysis

Complex Manifolds and Geometric Algebraic Analysis

Ahmed Lesfari

iSTE

WILEY

First published 2026 in Great Britain and the United States by ISTE Ltd and John Wiley & Sons, Inc.

ISTE Ltd
27-37 St George's Road
London SW19 4EU
UK

www.iste.co.uk

John Wiley & Sons, Inc.
111 River Street
Hoboken, NJ 07030
USA

www.wiley.com

Library of Congress Control Number: 2025945943

British Library Cataloguing-in-Publication Data
A CIP record for this book is available from the British Library
ISBN 978-1-83669-091-7

The manufacturer's authorized representative according to the EU General Product Safety Regulation is Wiley-VCH GmbH, Boschstr. 12, 69469 Weinheim, Germany, e-mail: Product_Safety@wiley.com.

Contents

Preface ix

Chapter 1. Holomorphic Functions of Several Complex Variables 1

1.1. Notations, definitions and properties 1
1.2. Cauchy integral formula and applications 7
1.3. Power series in several variables 9
1.4. Various fundamental properties 20
1.5. Inverse mapping and implicit function theorems 28
1.6. Exercises 32

Chapter 2. Complex Manifolds and Analytic Varieties 37

2.1. Preliminaries 37
2.2. Examples of complex manifolds 42
2.2.1. $\mathbb{C}^n$ and open subsets 42
2.2.2. Complex algebraic curves or compact Riemann surfaces 43
2.2.3. Spheres 43
2.2.4. Projective spaces 47
2.2.5. Grassmannians 50
2.2.6. Tori 51
2.3. Tangent spaces and tangent bundles 52
2.4. Constant rank theorem 53
2.5. Submanifolds, subvarieties and examples 55
2.6. Exercises 58

Chapter 3. Sheaves and Cohomology 65

3.1. Sheaves 65
3.2. Cech cohomology 69
3.3. De Rham cohomology 71
3.4. Dolbeault cohomology 72
3.5. Connections and curvature 74
3.6. Curvature form, first Chern class of line bundles 76
3.7. The Poincaré dual 77
3.8. Exercises 78

Chapter 4. Divisors and Line Bundles 79

4.1. Divisors 79
4.2. Line bundles 81
4.3. Sections of line bundles 83
4.4. Tori and Riemann form 89
4.5. Line bundles on complex tori 91
4.6. Exercises 97

Chapter 5. Some Fundamental Theorems 101

5.1. Preliminaries and various notions 101
5.1.1. Projective variety 101
5.1.2. Tangent and cotangent bundles 103
5.1.3. Dolbeault cohomology group 103
5.1.4. First Chern class 106
5.1.5. Hodge forms, hermitian metrics, harmonic space, Hodge star operator and Hodge theorem 107
5.1.6. Kähler metric, Kähler form, Kähler manifold 109
5.2. The Kodaira–Nakano vanishing theorem 111
5.3. The Lefschetz theorem on hyperplane sections 113
5.4. The Lefschetz theorem on $(1, 1)$-classes 114
5.5. The Kodaira embedding theorem 117
5.6. Exercises 126

Chapter 6. Abelian Varieties 133

6.1. Definitions and examples 133
6.2. Riemann conditions, polarization and Riemann form 138
6.3. Isogenies and reducible Abelian varieties 141
6.4. Dual Abelian varieties 142
6.5. Prym varieties 145
6.6. Projectively normal embedding 154
6.7. Number of even and odd sections of a line bundle 155
6.8. Exercises 156

Chapter 7. Theta Functions and Complex Projective Tori 159

7.1. Meromorphic functions and theta functions 159
7.2. Lefschetz theorem 173
7.3. Exercises 179

Chapter 8. Algebraically Completely Integrable Systems 183

8.1. Preliminaries 183
8.2. The Hénon–Heiles system 190
8.3. Kowalewski's spinning top 204
8.4. Kirchhoff's equations of motion of a solid in an ideal fluid 213
8.5. Exercises 216

Chapter 9. Appendix: Riemann Surfaces and Algebraic Curves 227

Chapter 10. Appendix: Elliptic Functions and Elliptic Integrals 233

10.1. Elliptic functions 233
10.2. Weierstrass functions 235
10.3. Elliptic integrals and Jacobi elliptic functions 240
10.4. Application: simple pendulum 244

References 247

Index 251

Preface

This essential book is intended for graduate students in mathematics, physics and beyond. It may also be of interest to doctoral students wishing to acquire a broad spectrum of mathematical knowledge and combine it with other disciplines. It aims to present several important aspects of complex manifolds with various properties in geometric algebraic analysis. The topics covered in this book are objects of extraordinary richness that appear in many scientific disciplines. The presentation is clear and well organized, and many examples, exercises and problems with solutions are provided throughout the text. This book is divided into 10 chapters, each containing several sections.

Chapter 1 deals with some basic properties of holomorphic functions of several complex variables. In this chapter, we cover fundamental concepts in the theory of holomorphic functions of several complex variables, including complex differentiation, the Cauchy integral formula and its applications, power series expansions, expansion in Reinhardt domains, various fundamental properties, inverse mapping and implicit function theorems.

In Chapter 2, we introduce tools for studying complex manifolds and analytic varieties by illustrating them with numerous examples and applications. Several fundamental exercises and problems are studied in order to elucidate the specificities of these manifolds. We explore in particular open subsets, complex algebraic curves or compact Riemann surfaces, spheres, projective spaces, grassmannians and tori, among others. We also study tangent spaces and tangent bundles in the case of complex manifolds. It is well known that we can treat a complex manifold M of dimension n as a real manifold of dimension $2n$. We can therefore define tangent vectors by identifying them with the derivations of the algebra of real differential functions on M with values in $\mathbb{R}$. We can also define tangent vectors by identifying them with the complex-valued derivations of the $\mathbb{C}$-algebra of holomorphic functions. We prove the constant rank theorem, which states, up to a change of coordinates, that an application whose rank is locally constant is equivalent to a linear application. Next, we present submanifolds in different ways, starting by reviewing some definitions of a submanifold and then demonstrating some results that bring together the different ways of locally describing submanifolds: by equations, by the graph of an application and by a parameterization. Each of these characterizations has its own interest and often comes up when solving exercises, and each plays a crucial role.

In Chapter 3, we sketch the foundational material from sheaves and cohomology. First, we recall the definition of sheaves and we examine many examples. Then, we give an overview of Cech, de Rham and Dolbeault cohomologies, connections, curvature form, the first Chern class of line bundles and the Poincaré dual.

Chapter 4 concerns the study of divisors and line bundles on complex manifolds. We are interested in divisors, line bundles, sections of lines bundles, tori, Riemann form and line bundles on complex tori.

Chapter 5 is devoted to some fundamental theorems. We first introduce various notions concerning projective variety, tangent and cotangent bundles (Dolbeault), cohomology groups, the first Chern class, Hodge forms, hermitian metrics, harmonic space, the Hodge star operator and Hodge theorem, the Kähler metric, the Kähler form, the Kähler manifold and Moishezon variety. Next, we prove the Kodaira–Nakano vanishing theorem, the Lefschetz vanishing theorem on hyperplane sections and the Lefschetz vanishing theorem for classes of type (1,1). Particular attention is devoted to the study of the Kodaira embedding theorem and other versions of this theorem. This asserts that a compact complex analytic variety is algebraically projective if and only if it is a Hodge variety, in other words if and only if it admits a closed, entire and positive two-differential form of type (1,1), or again if and only if it admits a holomorphic line bundle whose first Chern class is positive. The connection with the theory of Kähler manifolds will be explained. A Kähler manifold is a complex manifold equipped with a hermitian metric whose imaginary part, which is a two-form of type (1,1) relative to the complex structure, is closed. Such a metric is called a Kähler metric, and the form is called a Kähler form. In the language of Kähler varieties, the Kodaira embedding theorem means that a compact complex variety is projective if and only if it admits a Kähler form whose cohomology class is entire. We discuss another interesting result concerning Kähler varieties obtained by Moishezon. A Moishezon variety is a compact complex analytic variety that becomes projective after a finite number of blow-ups of smooth centers and has the maximum number of algebraically independent meromorphic functions. More precisely, a compact Kähler variety of dimension n is projective if and only if it has n algebraically independent meromorphic functions.

In Chapter 6, we first deal with some definitions and examples of Abelian varieties. We also discuss Chow's theorem, which states that any closed analytic subvariety of the projective space $\mathbb{P}^N(\mathbb{C})$ is a projective variety. He states that such a variety is defined by homogeneous polynomial equations and can therefore be studied by either analytic or algebraic methods. Then, one of the objectives is to see how to characterize complex tori that have an embedding in a complex projective space and are therefore considered as Abelian varieties. The so-called Riemann conditions or Riemann bilinear relations, as well as the Riemann form, play a crucial role in this study. We also introduce isogenies, reducible Abelian varieties, dual Abelian varieties and projectively normal embedding. We provide sufficient conditions, which guarantee that a complex n-dimensional manifold is analytically isomorphic to an n-dimensional complex torus and a Kähler manifold. We discuss the relation with the Hodge theory and an immediate consequence is that a complex manifold will complete to Abelian variety by adjoining some divisors. Several examples are given. Moreover, the classical definition of Prym varieties deals with the unramified covers of curves. The aim here is to give explicit algebraic descriptions of the Prym varieties associated with ramified double covers of algebraic curves. In addition, we provide two explicit formulas for determining the number of even and odd sections of a line bundle.

Chapter 7 deals with theta functions and complex projective tori. We begin by reviewing some results concerning theta functions which play a major role in much current research and explore some of the numerous consequences, a subject of renewed interest in recent years. The purpose of this chapter is to provide a short and quick exposition of some important aspects of meromorphic theta functions for compact Riemann surfaces. The study of theta functions is done via an analytical approach using meromorphic functions in the framework of Mumford. Some interesting examples are given. We show how the meromorphic functions on the Riemann surfaces of an arbitrary genus can be constructed explicitly in terms of the multi-dimensional theta functions. We discuss the important role of the zeros of theta functions and the Jacobi inversion problem, which askes whether we can find a divisor that is the preimage for an arbitrary point in the Jacobian. The Lefschetz theorem on projective embeddings over the complex numbers is of utmost importance in the complex geometric theory of compact manifolds. The next step is to provide an analytic proof of this theorem on the projectivity of complex tori using theta functions. This theorem shows how these theta functions provide an embedding of the complex torus into a projective space. We explain how positive line bundles on Abelian varieties can be explicitly described in terms of multipliers and how their sections can be described by theta functions.

The aim of Chapter 8 is to discuss an interesting interaction between complex algebraic geometry and dynamical systems. We briefly explain how Abelian varieties and in particular Prym varieties are involved in the modern study and active area of algebraically completely integrable hamiltonian dynamical systems. We first consider some notions and properties concerning these systems and explain how the solutions are expressed in the form of a Laurent series of nonlinear differential equations in the complex domain. Such meromorphic solutions depending on a sufficient number of free parameters play a crucial role in the study of so-called algebraically integrable differential equations. This means that we require that the invariants of the differential system be polynomial (in adequate coordinates) and that, moreover, the complex varieties obtained by equating these polynomial invariants to generic constants form the affine part of an algebraic complex torus (Abelian variety) in such a way that the complex flows generated by the invariants are linear on these complex tori. These ideas are applied to the study of three important algebraically completely integrable hamiltonian systems, namely the Hénon–Heiles differential system, Kowalewski's spinning top and Kirchhoff's equations of motion of a solid in an ideal fluid.

The book is supplemented with two appendixes (Chapters 8 and 9), one on Riemann surfaces and algebraic curves and the other on elliptic functions and elliptic integrals. In addition, numerous examples, exercises and problems with solutions are scattered throughout the text.

I have included a bibliography of some key works and a detailed index allowing the reader to quickly locate any item discussed in the book. We sincerely thank ISTE and John Wiley & Sons, Inc. for the quality of their work, their thoroughness and their professionalism. I dedicate this book to all those who have kindly contributed to it in one way or another.

A. LESFARI
October 2025

1

Holomorphic Functions of Several Complex Variables

1.1. Notations, definitions and properties

We consider the space of complex numbers of dimension n,

$$\mathbb{C}^n = \{z = (z_1, ..., z_n) : z_j \in \mathbb{C}, j = 1, ..., n\},$$

the Cartesian product of n copies of $\mathbb{C}$. We set $x_j = \text{Re } z_j$, $y_j = \text{Im } z_j$, $j = 1, ..., n$. Then, $z_j = x_j + iy_j$, $i = \sqrt{-1}$, $(x_1, y_1, x_2, y_2, ..., x_n, y_n) \in \mathbb{R}^{2n}$. We identify $\mathbb{C}^n$ with $\mathbb{R}^{2n}$ by means of the isomorphism of $\mathbb{R}$-vector space,

$$\sigma : \mathbb{C}^n \longrightarrow \mathbb{R}^{2n}, \quad (z_1, ..., z_n) \longmapsto (x_1, y_1, x_2, y_2, ..., x_n, y_n).$$

Since $z_j = x_j + iy_j$ and its conjugate being $\overline{z}_j = x_j - iy_j$, then

$$x_j = \frac{z_j + \overline{z}_j}{2}, \qquad y_j = \frac{z_j - \overline{z}_j}{2i}.$$

Recall that $\mathbb{R}$ is equipped with the usual topology defined by the absolute value, $\mathbb{R}^{2n}$ is equipped with the product topology and $\mathbb{C}^n$ is equipped with the topology carried by σ^{-1}. Therefore, by using this isomorphism, we introduce a norm and therefore a topology on the space $\mathbb{C}^n$, and since all norms on $\mathbb{R}^{2n}$ are equivalent, then all norms on $\mathbb{C}^n$ define the same topology. Among these norms, we can use this one, $\|z\| = \sqrt{|z_1|^2 + \cdots + |z_n|^2}$, which makes $\mathbb{C}^n$ a topological vector space. As above, having identified z with $(x_1, ..., x_n, y_1, ..., y_n)$, $\mathbb{C}^n$ with $\mathbb{R}^{2n}$, by this identification this norm in $\mathbb{C}^n$ coincides with the euclidean norm in $\mathbb{R}^{2n}$. We can also use the norm, $\max\{|z_j|; 1 \leq j \leq n\}$.

DEFINITION 1.1.– *a) A function $f : \mathbb{C}^n \longrightarrow \mathbb{C}$, is said to be $\mathbb{C}$-linear (resp. $\mathbb{R}$-linear) if*

(i) $f(z_1 + z_2) = f(z_1) + f(z_2)$, for all $z_1, z_2 \in \mathbb{C}^n$.

(ii) $f(\lambda z) = \lambda f(z)$ for all $z \in \mathbb{C}^n$ and all $\lambda \in \mathbb{C}$ (respectively $\mathbb{R}$).

b) The function f is said to be $\mathbb{C}$-antilinear if

(i)' $f(z_1 + z_2) = f(z_1) + f(z_2)$, for all $z_1, z_2 \in \mathbb{C}^n$.

(ii)' $f(\lambda z) = \overline{\lambda} f(z)$ for all $z \in \mathbb{C}^n$ and all $\lambda \in \mathbb{C}$.

Any $\mathbb{C}$-linear function is written in the form

$$f(z) = \sum_{j=1}^{n} a_j z_j, \quad a_j \in \mathbb{C},$$

and any $\mathbb{R}$-linear function on $\mathbb{C}^n$ is written as

$$f(z) = \sum_{j=1}^{n} (a_j z_j + b_j \overline{z}_j), \quad (a_j, b_j \in \mathbb{C}).$$

Using the definition above (with $\lambda = i$), we obtain the following property.

PROPERTY 1.1.– *The function $f : \mathbb{C}^n \longrightarrow \mathbb{C}$ is $\mathbb{C}$-linear if and only if f is $\mathbb{R}$-linear and $f(iz) = if(z)$ for all $z \in \mathbb{C}^n$. The function f is $\mathbb{C}$-antilinear if and only if f is $\mathbb{R}$-linear and $f(iz) = -if(z)$ for all $z \in \mathbb{C}^n$.*

We call the subset of $\mathbb{C}^n$ defined by $D(a,r) = \{(z_1, ..., z_n) \in \mathbb{C}^n : |z_j - a_j| < r_j, j = 1, ..., n\}$ the polydisk (open) of $\mathbb{C}^n$ with center a and poly-radius r, where $a = (a_1, ..., a_n) \in \mathbb{C}^n$ and $r = (r_1, ..., r_n) \in (\mathbb{R}_+^*)^n$. In other words, it is the product $D(a,r) = \prod_{j=1}^{n} D(a_j, rj)$ of n disks D_j with center a_j and radius r_j. The point $a = (a_1, ..., a_n)$ is called the center of the polydisk, and the number $r = (r_1, ..., r_n)$ is sometimes called the multi-radius (or as above the poly-radius) of the polydisk. (In fact, the analogues of open balls in dimension n are here the polydisks.) The set

$$\partial_0 D(a,r) = \{(z_1, ..., z_n) \in \mathbb{C}^n : |z_j - a_j| = r_j, \ j = 1, ..., n\},$$

is called the distinguished boundary of $D(a,r)$. Note that $\partial_0 D(a,r)$ is not the boundary (in the topological sense) of the polydisk. It is the product of the boundaries

$$\partial_0 D = \{(z_1, ..., z_n) : |z_1 - a_1| = r_1, ..., |z_n - a_n| = r_n\} = \partial D_1 \times \cdots \times \partial D_n.$$

In the sense of varieties, the boundary ∂D is of dimension $n+1$, while $\partial_0 D$ is of dimension n. (We sometimes say that $\partial_0 D$ is the boundary of Šilov or the distinguished boundary but it should be noted that $\partial_0 D$ is not equal to the boundary of D as soon as $n > 1$.)

A multi-index of size n is a vector $\alpha = (\alpha_1, ..., \alpha_n)$ with coefficients α_i (positive integers). We denote by $|\alpha| = \alpha_1 + \cdots + \alpha_n$ the order or length of α, $\alpha! = \alpha_1! \cdots \alpha_n!$. For multi-indices α, β, we define

$$\alpha + \beta = (\alpha_1 + \beta_1, ..., \alpha_n + \beta_n), \qquad \beta \leq \alpha \Longleftrightarrow \beta_k \leq \alpha_k, \forall 1 \leq k \leq n,$$
$$\alpha - \beta = (\alpha_1 - \beta_1, ..., \alpha_n - \beta_n) \text{ si } \beta \leq \alpha.$$

Let $z = (z_1, ..., z_n) \in \mathbb{C}^n$ and α be a multi-index. Let $z^\alpha = z_1^{\alpha_1} \cdots z_n^{\alpha_n}$ and $\partial^\alpha = \partial_1^{\alpha_1} \cdots \partial_n^{\alpha_n}$ (differential operator) where $\partial_k^j = \dfrac{\partial^j}{\partial x_k^j}$, and write

$$\partial^\alpha = D^\alpha = \frac{\partial^{|\alpha|}}{\partial z_1^{\alpha_1} ... z_1^{\alpha_n}}.$$

(We will use the notations D^α or ∂^α interchangeably.)

Let Ω be an open of $\mathbb{C}^n$ and $f : \Omega \subset \mathbb{C}^n \longrightarrow \mathbb{C}$, a function of class C^1 (as a function $f : \Omega \subset \mathbb{R}^{2n} \longrightarrow \mathbb{R}^2$ because $\mathbb{C}^n \sim \mathbb{R}^{2n}$).

DEFINITION 1.2.– *We say that f is $\mathbb{C}$-differentiable at $z \in \mathbb{C}^n$ if*

$$f(z+h) = f(z) + L(h) + o(h), \qquad \lim_{h \to 0} \frac{o(h)}{|h|} = 0,$$

where L is a $\mathbb{C}$-linear function. The function L is called the differential of f at z and is denoted as df.

The differential of f is decomposed into a $\mathbb{C}$-linear part denoted as ∂f and a $\mathbb{C}$-antilinear part denoted as $\overline{\partial} f$,

$$df = \sum_{j=1}^{n} \frac{\partial f}{\partial x_j} dx_j + \sum_{j=1}^{n} \frac{\partial f}{\partial y_j} dy_j = \underbrace{\sum_{j=1}^{n} \frac{\partial f}{\partial z_j} dz_j}_{\partial f} + \underbrace{\sum_{j=1}^{n} \frac{\partial f}{\partial \overline{z}_j} d\overline{z}_j}_{\overline{\partial} f},$$

where $dz_j = dx_j + i dy_j$, $d\overline{z}_j = dx_j - i dy_j$, and

$$\frac{\partial}{\partial z_j} = \frac{1}{2}\left(\frac{\partial}{\partial x_j} - i \frac{\partial}{\partial y_j}\right), \qquad \frac{\partial}{\partial \overline{z}_j} = \frac{1}{2}\left(\frac{\partial}{\partial x_j} + i \frac{\partial}{\partial y_j}\right),$$

are differential operators on $\mathbb{C}^n$ (the so-called Wirtinger operators). (Obviously it should be noted that $\dfrac{\partial}{\partial z_j}$ and $\dfrac{\partial}{\partial \overline{z}_j}$ are the differential operators defined above, and these notations do not here mean partial derivatives with respect to the variables z_j, $\overline{z}_j$.)

Let $f : \Omega \longrightarrow \mathbb{C}$, $(z_1, ..., z_n) \longmapsto f(z_1, ..., z_n)$, be a function defined on an open subset Ω of $\mathbb{C}^n$. There are several ways to define a holomorphic function on Ω, and the results that will be demonstrated later can be used as a definition. We will first state the following definition, and we will find it in the form of a result (corollary 1.3) that we demonstrate.

DEFINITION 1.3.– *Let Ω be an open set of $\mathbb{C}^n$. A function*

$$f : \Omega \longrightarrow \mathbb{C}, \quad (z_1, ..., z_n) \longmapsto f(z_1, ..., z_n),$$

of class C^1 (with respect to real coordinates) is said to be holomorphic in Ω if

$$\overline{\partial} f = \sum_{j=1}^{n} \frac{\partial f}{\partial \overline{z}_j} d\overline{z}_j = 0,$$

in Ω (Cauchy–Riemann conditions or Cauchy–Riemann equations). In other words, if the differential df is

$$df = \partial f = \sum_{j=1}^{n} \frac{\partial f}{\partial z_j} dz_j.$$

The Cauchy–Riemann conditions or Cauchy–Riemann equations above are equivalent to the system of n (complex) equations:

$$\frac{\partial f}{\partial \overline{z}_j} = \frac{1}{2}\left(\frac{\partial f}{\partial x_j} + i\frac{\partial f}{\partial y_j}\right) = 0, \quad j = 1, ..., n$$

or the system of $2n$ (real) equations:

$$\frac{\partial u}{\partial x_j} = \frac{\partial v}{\partial y_j}, \qquad \frac{\partial u}{\partial y_j} = -\frac{\partial v}{\partial x_j}, \quad j = 1, ..., n$$

where $u = \operatorname{Re} f$ and $v = \operatorname{Im} f$.

PROPOSITION 1.1.– *Let Ω be an open set of $\mathbb{C}^n$. A function $f : \Omega \longrightarrow \mathbb{C}$ of class C^1 is holomorphic if and only if its differential at every point is $\mathbb{C}$-linear.*

PROOF.– Indeed, if the function f is holomorphic on Ω, then by definition $\overline{\partial} f = 0$ on Ω. Therefore, the differential of f is

$$df = \partial f = \sum_{j=1}^{n} \frac{\partial f}{\partial z_j} dz_j,$$

and it is $\mathbb{C}$-linear because for all $z, z' \in \mathbb{C}^n$, we have

$$\partial f(z + z') = \sum_{j=1}^{n} \frac{\partial f}{\partial z_j} d(z_j + z'_j) = \partial f(z) + \partial f(z'),$$

and for all $\alpha \in \mathbb{C}$, we have

$$\partial f(\alpha z) = \sum_{j=1}^{n} \frac{\partial f}{\partial z_j} d(\alpha z_j) = \alpha \partial f(z).$$

Let us show that if *df* is $\mathbb{C}$-linear, then f is holomorphic or, equivalently, if f is not holomorphic, then *df* is not $\mathbb{C}$-linear. If f is not holomorphic on Ω, then there exists $a \in \Omega$ such that for some k, we have $\dfrac{\partial f}{\partial \overline{z}_k}(a) \neq 0$. At this point a, we have

$$df(0, ..., 0, 1, 0, ..., 0) = \frac{\partial f}{\partial z_k}(a) + \frac{\partial f}{\partial \overline{z}_k}(a),$$

and

$$df(0, ..., 0, i, 0, ..., 0) = i \left(\frac{\partial f}{\partial z_k}(a) - \frac{\partial f}{\partial \overline{z}_k}(a) \right),$$

hence,

$$df(0, ..., 0, 1, 0, ..., 0) \neq df(0, ..., 0, i, 0, ..., 0),$$

and consequently *df* is not $\mathbb{C}$-linear. ■

DEFINITION 1.4.– *A continuous function $f : \Omega \longrightarrow \mathbb{C}$ on an open set $\Omega \subset \mathbb{C}^n$ is said to be holomorphic on Ω if it is holomorphic with respect to each variable (i.e. if $f(z_1, ..., z_n)$ is holomorphic in each variable separately).*

This definition means that for each $z_1^0, ..., z_{j-1}^0, z_{j+1}^0, ..., z_n^0$, the function

$$\varphi : \{z_j \in \mathbb{C} : (z_1^0, ..., z_{j-1}^0, z_j, z_{j+1}^0, ..., z_n^0) \in \Omega\} \longrightarrow \mathbb{C},$$

defined by

$$\varphi(z_j) = f(z_1^0, ..., z_{j-1}^0, z_j, z_{j+1}^0, ..., z_n^0),$$

is holomorphic.

REMARK 1.1.– *In fact, we will see later (corollary 1.2 (Osgood) and more generally Hartog's theorem 1.10) and that if $\dfrac{\partial f}{\partial \overline{z}_j} = 0$, $\forall j = 1, ..., n$, then f is holomorphic in Ω. In other words, the function f is holomorphic if it is in each variable z_j.*

Let Ω be an open set of $\mathbb{C}^n$. We denote by $\mathcal{O}(\Omega)$ the set of holomorphic functions on Ω.

PROPOSITION 1.2.– *a) The set $\mathcal{O}(\Omega)$ is a $\mathbb{C}$-algebra for addition, multiplication of functions and multiplication by complex constants.*

b) If $f \in \mathcal{O}(\Omega)$ and if $f(z) \neq 0$, for all $z \in \Omega$, then $\frac{1}{f} \in \mathcal{O}(\Omega)$.

PROOF.– *a)* A direct calculation shows that:

$$\overline{\partial}(f+g) = \overline{\partial}f + \overline{\partial}g, \qquad \overline{\partial}(fg) = g\overline{\partial}f + f\overline{\partial}g,$$

and the result follows.

b) By replacing $g = \frac{1}{f}$ in the expression

$$\overline{\partial}(fg) = g\overline{\partial}f + f\overline{\partial}g,$$

we obtain

$$0 = f\overline{\partial}\left(\frac{1}{f}\right),$$

because $\overline{\partial}f = 0$ since f is holomorphic on Ω. ■

PROPOSITION 1.3.– *Let Ω be an (open connected) domain of $\mathbb{C}^n$. If $f \in \mathcal{O}(\Omega)$ and if f is real-valued or if $|f|$ is constant, then f is constant.*

PROOF.– If f is real-valued, then $\frac{\partial f}{\partial x_k}, \frac{\partial f}{\partial y_k}$, $k = 1, ..., n$, are real. The Cauchy–Riemann equations are written as

$$\frac{\partial f}{\partial x_k} + i\frac{\partial f}{\partial y_k} = 0,$$

so $\frac{\partial f}{\partial x_k} = \frac{\partial f}{\partial y_k} = 0$ and consequently, the function f is constant on Ω since the latter is connected. Finally, if $|f| = constant \equiv \rho$, then

$$f = \rho e^{i\theta(z)}, \qquad \overline{\partial}f = \rho e^{i\theta(z)} i\overline{\partial}\theta(z) = 0.$$

Since θ is holomorphic and real-valued on Ω, then by the above, $\theta = constant$ on Ω. ■

1.2. Cauchy integral formula and applications

THEOREM 1.1.– *Let $D = D_1 \times \cdots \times D_n$ be a polydisk of $\mathbb{C}^n$. Let f be a continuous function on $\overline{D}$ and holomorphic separately with respect to each variable $z_j \in D$. Then, for all $z \in D$, we have the Cauchy integral formula*

$$f(z) = \frac{1}{(2\pi i)^n} \int_{\partial_0 D} \frac{f(\zeta)}{(\zeta_1 - z_1)...(\zeta_n - z_n)} d\zeta_1...d\zeta_n.$$

Moreover, the function f is of class C^∞ on D and it is holomorphic on D.

PROOF.– The result is obtained by repeated use of the Cauchy integral formula. We fix $n-1$ variables, apply the Cauchy integral formula with respect to each variable and the rest consists of reasoning by induction. For $n = 1$, the function $z_1 \longmapsto f(z_1, ..., z_n)$ being holomorphic in D_1, then

$$f(z) = \frac{1}{2\pi i} \int_{\partial_0 D_1} \frac{f(\zeta_1, z_2, ..., z_n)}{\zeta_1 - z_1} d\zeta_1.$$

Let ζ_1 be fixed and consider the function $z_2 \longmapsto f(\zeta_1, z_2, ..., z_n)$ holomorphic in D_2. We have

$$f(\zeta_1, z_2, ..., z_n) = \frac{1}{2\pi i} \int_{\partial_0 D_2} \frac{f(\zeta_1, \zeta_2, z_3, ..., z_n)}{\zeta_2 - z_2} d\zeta_2,$$

hence,

$$f(z) = \frac{1}{(2\pi i)^2} \int_{D_1} \frac{d\zeta_1}{\zeta_1 - z_1} \int_{\partial_0 D_2} \frac{f(\zeta_1, \zeta_2, z_3, ..., z_n)}{\zeta_2 - z_2} d\zeta_2.$$

Continuing in this way, we obtain

$$f(z) = \frac{1}{(2\pi i)^n} \int_{D_1} \frac{d\zeta_1}{\zeta_1 - z_1} \int_{D_2} \frac{d\zeta_2}{\zeta_2 - z_2} \cdots \int_{\partial_0 D_n} \frac{f(\zeta_1, ..., \zeta_n)}{\zeta_n - z_n} d\zeta_n.$$

The function

$$(\zeta_1, ..., \zeta_n) \longmapsto \frac{f(\zeta_1, ..., \zeta_n)}{\zeta_n - z_n},$$

being continuous on $\overline{D}$, therefore on $\partial_0 D$, and according to Fubini's theorem, we can invert the integrals and replace them with a single integral on $\overline{D}$,

$$f(z) = \frac{1}{(2\pi i)^n} \int_{\partial_0 D} \frac{f(\zeta)}{\zeta - z} d\zeta,$$

where

$$z = (z_1, ..., z_n), \qquad \frac{1}{\zeta - z} = \frac{1}{(\zeta_1 - z_1)...(\zeta_n - z_n)}, \qquad d\zeta = d\zeta_1...d\zeta_n,$$

and the demonstration ends. ■

The preceding theorem has immediate consequences.

COROLLARY 1.1.– *Let Ω be an open set of $\mathbb{C}^n$. Then, any function f holomorphic on Ω is of class $\mathcal{C}^\infty$ on Ω and its partial derivatives of any order are holomorphic on Ω.*

PROOF.– Let $\Omega \subset \mathbb{C}^n$ be an open set and D be a polydisk with center $(a_1, ..., a_n) \in \Omega$ such that $\overline{D} \subset \Omega$. By theorem 1.1, the function f is of class $\mathcal{C}^\infty$ on D, therefore in Ω. Note that for all $\alpha = (\alpha_1, ..., \alpha_n) \in \mathbb{N}^n$, the function

$$(z_1, ..., z_n) \longmapsto \frac{1}{(\zeta_1 - z_1)^{\alpha_1}} \cdots \frac{1}{(\zeta_n - z_n)^{\alpha_n}},$$

where $(\zeta_1, ..., \zeta_n) \in \partial_0 D$, is holomorphic with respect to each variable $z_j \in D_j$. Therefore, the partial derivatives of f with respect to $(z_1, ..., z_n)$ that can be obtained by differentiation under the integral sign (in the Cauchy integral formula) are holomorphic in D and the result follows. ■

COROLLARY 1.2 (OSGOOD).– *Let Ω be an open set of $\mathbb{C}^n$. Then, any continuous and holomorphic function with respect to each variable is holomorphic on Ω.*

PROOF.– Let Ω be an open set of $\mathbb{C}^n$ and D be a polydisk with center $(z_1^0, ..., z_n^0) \in \Omega$ such that: $\overline{D} \subset \Omega$. Hence, f is holomorphic on D by theorem 1.1, and consequently, f is holomorphic on Ω. ■

We will see later (Hartog's theorem) that the continuity assumption in this corollary is useless.

COROLLARY 1.3.– *A function $f : \Omega \longrightarrow \mathbb{C}$ defined on an open set Ω of $\mathbb{C}^n$ is holomorphic if and only if it is of class $\mathcal{C}^1$ (with respect to the real coordinates) and satisfies the Cauchy–Riemann equations:*

$$\frac{\partial u}{\partial x_j} = \frac{\partial v}{\partial y_j}, \qquad \frac{\partial u}{\partial y_j} = -\frac{\partial v}{\partial x_j}, \qquad j = 1, ..., n$$

where $z_j = x_j + iy_j$ and $f(z) = u(z) + iv(z)$.

PROOF.– If the function f is of class $\mathcal{C}^1$ and satisfies the Cauchy–Riemann equations, then it is continuous and according to the theory of functions of a complex variable, f is $\mathbb{C}$-differentiable with respect to each variable and therefore it is also holomorphic.

Conversely, if the function f is holomorphic, it is continuous and $\mathbb{C}$-differentiable with respect to each variable at each point of Ω. Since f is holomorphic in each variable separately, it satisfies the Cauchy–Riemann equations with respect to each variable. Let $D(a,r)$ be the polydisk of $\mathbb{C}^n$ with center a and the same radius $r > 0$ such that $\overline{D}(a,r) \subset \Omega$ (we denote by $\overline{D}(a,r)$ the closed polydisk). Let $z \in D(a,r)$. For the same reasons as before, we can use the Cauchy integral formula several times. We have

$$f(z) = f(z_1, ..., z_n) = \frac{1}{2\pi i} \int_{|\zeta_n - a_n| = r} \frac{f(z_1, ..., z_{n-1}, \zeta_n)}{\zeta_n - z_n} d\zeta_n,$$

and reasoning as in the proof of theorem 1.1, we obtain

$$f(z) = \frac{1}{(2\pi i)^n} \int_{|\zeta_n - a_n| = r} \cdots \int_{|\zeta_1 - a_1| = r} \frac{f(\zeta_1, ..., \zeta_n)}{(\zeta_n - z_n) \cdots (\zeta_1 - z_1)} d\zeta_1 \cdots d\zeta_n.$$

Since the domain of integration is compact and the function under the integral sign is continuous in all variables and indefinitely differentiable as a function of $z_1, ..., z_n$ (or of the real and imaginary parts), we can differentiate under the integral sign as often as we like with respect to x_j and y_j and deduce that f is indefinitely differentiable. ■

Other consequences and applications of the previous theorem will be studied further in other sections.

1.3. Power series in several variables

A series (or more precisely a multi-series or even a multiple series),

$$\sum_{\alpha \in \mathbb{N}^n} c_\alpha = \sum_{\alpha_1, \ldots, \alpha_n = 0}^{\infty} c_{\alpha_1, \ldots, \alpha_n},$$

where $\alpha = (\alpha_1, ..., \alpha_n) \in \mathbb{N}^n$, $c_\alpha = c_{\alpha_1, \ldots, \alpha_n} \in \mathbb{C}$, converges to a complex number S if

$$\forall \varepsilon > 0, \exists N \in \mathbb{N} : \forall (N_1, ..., N_n) \in \mathbb{N}^n, N_j > N, \left| \sum_{\alpha_1 = 0}^{N_1} \cdots \sum_{\alpha_n = 0}^{N_n} c_\alpha - S \right| < \varepsilon.$$

In this case, we write $S = \sum_{\alpha \in \mathbb{N}^n} c_\alpha$ and S is called the sum of the series. If the series converges, then its sum is unique. We say that this series converges absolutely if the series $\sum |c_\alpha|$ converges. In other words, if the set

$$\left\{ \sum_{\alpha \in L} |c_\alpha| : L \subset \mathbb{N}^n, \text{ finite} \right\},$$

is bounded, or if

$$\sum_{\alpha \in \mathbb{N}^n} |c_\alpha| = \sup \left\{ \sum_{\alpha \in L} |c_\alpha| : L \subset \mathbb{N}^n \text{ fini} \right\} < +\infty.$$

Any absolutely convergent series converges and furthermore, if σ is a permutation of $\mathbb{N}^n$, the series $\sum_{\alpha \in \mathbb{N}^n} c_{\sigma(\alpha)}$ converges for S and its sum is independent of the order of the permutation.

EXAMPLE 1.1.– *The series*

$$\sum_{\alpha \in \mathbb{N}^n} z^\alpha = \sum_{\alpha_1=0}^{\infty} \cdots \sum_{\alpha_n=0}^{\infty} z_1^{\alpha_1} \cdots z_n^{\alpha_n}, \quad z = (z_1, ..., z_n) \in \mathbb{C}^n, \quad 0 \neq |z_j| < 1,$$

converges and we have,

$$\sum_{\alpha \in \mathbb{N}^n} z^\alpha = \frac{1}{1 - z_1} \cdots \frac{1}{1 - z_n},$$

for all z such that: $|z_j| < 1$.

We now consider series whose terms are functions. A series $\sum_{\alpha \in \mathbb{N}^n} f_\alpha(z)$ of continuous complex-valued functions on a compact K converges normally if

$$\sum_{\alpha \in \mathbb{N}^n} \|f_\alpha\|_K < +\infty.$$

If the functions $f_\alpha(z)$ are defined on an open $\Omega \subset \mathbb{C}^n$, then the series $\sum f_\alpha(z)$ converges normally if it converges normally on every compact of Ω. If the series $\sum f_\alpha(z)$ converges normally on Ω, then it converges absolutely for all $z \in \Omega$ to a function F on Ω,

$$F(z) = \sum_{\alpha \in \mathbb{N}^n} f_\alpha(z).$$

Such a series that converges normally converges uniformly on Ω,

$$\forall \varepsilon > 0, \exists N \in \mathbb{N} : \forall (N_1, ..., N_n) \in \mathbb{N}^n, N_j > N, \left| \sum_{\alpha_1=0}^{N_1} \cdots \sum_{\alpha_n=0}^{N_n} f_\alpha(z) - F(z) \right| < \varepsilon.$$

(N does not depend on z.) Let us consider more particularly in $\mathbb{C}^n$ equipped with the coordinates $(z_1, ..., z_n)$, the case of a multiple power series with complex coefficients, i.e. the case where $f_\alpha(z) = c_\alpha(z - a)^\alpha$, with $z = (z_1, ..., z_n)$, $c_\alpha = c_{\alpha_1,...,\alpha_n} \in \mathbb{C}$ and $(z - a)^\alpha = (z_1 - a_1)^{\alpha_1} \cdots (z_1 - a_n)^{\alpha_n}$.

DEFINITION 1.5.– *Let Ω be an open set of $\mathbb{C}^n$. A function $f : \Omega \longrightarrow \mathbb{C}$ is called analytic in Ω if it is developable in a power series in the neighborhood of each point $a = (a_1, ..., a_n)$ of Ω,*

$$f(z) = \sum_{\alpha \in \mathbb{N}^n} c_\alpha (z-a)^\alpha = \sum_{\alpha_1, \dots, \alpha_n = 0}^{\infty} c_{\alpha_1, \dots, \alpha_n} (z_1 - a_1)^{\alpha_1} ... (z_n - a_n)^{\alpha_n},$$

where $c_{\alpha_1, \dots, \alpha_n} \in \mathbb{C}$. The domain of convergence of this multiple series is the set of $z \in \mathbb{C}^n$ such that: for all $w \in \mathbb{C}^n$ in a neighborhood of $z - a$, we have the absolute convergence,

$$\sum_{\alpha \in \mathbb{N}^n} |c_\alpha (w-a)^\alpha| < \infty.$$

THEOREM 1.2.– *Let f be a holomorphic function on the polydisk $D(a, r)$ and continuous on $\overline{D}(a, r)$. Then, f is developable as a convergent power series at $(z - a)$ in $D(a, r)$,*

$$\forall z \in D(a, r), \quad f(z) = \sum_{\alpha \in \mathbb{N}^n} c_\alpha (z-a)^\alpha,$$

where the coefficients are determined by $c_\alpha = \dfrac{\partial^\alpha f(a)}{\alpha!}$. Moreover, this series converges normally on any polydisk $\overline{D}(a, \rho)$, $\rho < r$.

PROOF.– Without restricting the generality, we can always return by translation to the case where $a = 0$. We develop $\dfrac{1}{\zeta - z}$ in multiple geometric series:

$$\frac{1}{\zeta - z} = \frac{1}{(\zeta_1 - z_1)...(\zeta_n - z_n)} = \frac{1}{\zeta_1 ... \zeta_n} \left(\frac{\zeta_1 ... \zeta_n}{(\zeta_1 - z_1)...(\zeta_n - z_n)} \right),$$

or

$$\frac{1}{\zeta - z} = \frac{1}{\zeta} \left(\frac{1}{\left(1 - \frac{z_1}{\zeta_1}\right) \cdots \left(1 - \frac{z_n}{\zeta_n}\right)} \right) = \frac{1}{\zeta} \sum_{\alpha_1 + \dots + \alpha_n = 0}^{\infty} \left(\frac{z_1}{\zeta_1}\right)^{\alpha_1} \cdots \left(\frac{z_n}{\zeta_n}\right)^{\alpha_n},$$

or

$$\frac{1}{\zeta - z} = \frac{1}{\zeta} \sum_{|\alpha|=0}^{\infty} \left(\frac{z}{\zeta}\right)^\alpha = \sum_{|\alpha|=0}^{\infty} \frac{z^\alpha}{\zeta^{\alpha+1}}.$$

We multiply both sides by $\dfrac{1}{(2\pi i)^n} f(\zeta)$, hence

$$\frac{1}{(2\pi i)^n} \frac{f(\zeta)}{\zeta - z} = \frac{1}{(2\pi i)^n} \sum_{|\alpha|=0}^{\infty} \frac{f(\zeta)}{\zeta^{\alpha+1}} z^\alpha.$$

The function f is continuous, therefore bounded on $\partial_0 D$. It is easily shown (Weierstrass criterion) that the above series converges normally, therefore absolutely and uniformly, on $\partial_0 D$. We can therefore integrate term by term on $\partial_0 D$ (in the Cauchy integral formula). Therefore,

$$f(z) = \frac{1}{(2\pi i)^n} \int_{\partial_0 D} \frac{f(\zeta)}{\zeta - z} d\zeta = \frac{1}{(2\pi i)^n} \int_{\partial_0 D} \sum_{|\alpha|=0}^{\infty} \frac{f(\zeta)}{\zeta^{\alpha+1}} (z)^\alpha d\zeta,$$

thus,

$$f(z) = \sum_{|\alpha|=0}^{\infty} \left(\frac{1}{(2\pi i)^n} \int_{\partial_0 D} \frac{f(\zeta)}{\zeta^{\alpha+1}} d\zeta \right) z^\alpha.$$

Deriving under the integral sign in the Cauchy integral formula, we obtain

$$\partial^\alpha f(0) = \frac{\alpha!}{(2\pi i)^n} \int_{\partial_0 D} \frac{f(\zeta)}{\zeta^{\alpha+1}} d\zeta,$$

and consequently,

$$f(z) = \sum_{|\alpha|=0}^{\infty} \frac{\partial^\alpha f(0)}{\alpha!} z^\alpha,$$

which completes the proof. ■

Let us examine other versions concerning the study of the normal convergence of such power series.

PROPOSITION 1.4.– *Let* $\sum_{\alpha \in \mathbb{N}^n} c_\alpha (z-a)^\alpha$, $c_\alpha \in \mathbb{C}$, *be a power series that converges absolutely for* $z = \zeta$. *Let* $D(a, r)$ *be an open polydisk of* $\mathbb{C}^n$ *with center* a *and radius* r *with* $r_j < |w_j - a_j|$. *Then, this series converges normally on* $\overline{D}(a, r)$.

PROOF.– Without restricting the generality, we set $a = (0, ..., 0)$ and let r_j be with $r_j < |w_j|$. By hypothesis, the series in question converges absolutely for $z = \zeta$, i.e. $\sum |c_\alpha \zeta^\alpha|$ converges. Therefore, there exists for all α, a constant K such that: $|c_\alpha \zeta^\alpha| \leq K$. Let $z \in \overline{D}(0, r)$, hence

$$|c_\alpha z^\alpha| \leq |c_\alpha| r^\alpha \leq K q^\alpha, \quad q = (q_1, ..., q_n), \quad 0 < q_j < 1, \quad j = 1, ..., n.$$

Sine the series $\sum q^\alpha$ converges and

$$\begin{aligned} \sum_{\alpha \in \mathbb{N}^n} |c_\alpha \zeta^\alpha| \leq \sum_{\alpha \in \mathbb{N}^n} K q^\alpha &= K \sum_{\alpha_1=0}^{\infty} \cdots \sum_{\alpha_n=0}^{\infty} q_1^{\alpha_1} \cdots q_n^{\alpha_n}, \\ &= K \frac{1}{1-q_1} \cdots \frac{1}{1-q_n} < \infty, \end{aligned}$$

then the series in question converges normally on $\overline{D}(a, r)$. ■

PROPOSITION 1.5.– *Let*

$$z = (z_1, ..., z_n) \in \Delta = \left\{ z \in \mathbb{C}^n : \sup_{\alpha \in \mathbb{N}^n} |c_\alpha (z-a)^\alpha| < \infty \right\}.$$

Then, the power series $\sum_{\alpha \in \mathbb{N}^n} c_\alpha (z-a)^\alpha$, $c_\alpha \in \mathbb{C}$, *converges normally for all* $w = (w_1, ..., w_n)$ *with* $|w_j| < |z_j - a_j|$, $j = 1, ..., n$.

PROOF.– Note that Δ is the set of points $z \in \mathbb{C}^n$ for which all terms of the series are bounded. We note that the domain (which is a connected open set) of convergence of this series is the interior of Δ. To avoid making the notations cumbersome, it suffices to set $a = (0, ..., 0)$. By hypothesis, $z \in \Delta$, so there exists a constant A such that for all $\alpha \in \mathbb{N}^*$, $|c_\alpha z^\alpha| \leq A$. Let w with $|w| \leq q|z|$, $q = (q_1, ..., q_n)$, $0 \leq q \leq 1$, $j = 1, ..., n$ (i.e. $|w_j| \leq q_j |z_j|$). Therefore,

$$|c_\alpha w^\alpha| \leq |c_\alpha q^\alpha z^\alpha| \leq A q^\alpha.$$

Since $\sum_{\alpha \in \mathbb{N}^n} q^\alpha$ is a convergent series with

$$\sum_{\alpha \in \mathbb{N}^n} q^\alpha = \prod_{j=1}^{n} \frac{1}{1 - q_j},$$

then we deduce (as in the proof of the previous proposition) the result in question. ■

PROPOSITION 1.6.– *Any function f on an open set $\Omega \subset \mathbb{C}^n$ is holomorphic if and only if it is analytic.*

PROOF.– Let us show that any holomorphic function f on a polydisk $D \subset \Omega$ is analytic (i.e. developable in a neighborhood of each point a in a convergent power series). Let D' be a polydisk such that: $\overline{D'} \subset D$. For all $z \in D$, we have $z \in D$ and the result follows from theorem 1.2 applied to $\mathcal{D}'$. Conversely, let f be an analytic function on Ω. Let $a \in \Omega$ be a fixed point and consider a polydisk $D(a, r) \subset \Omega$ with center a and radius r. Since the function f is analytic, it is therefore developable in a neighborhood of each point a in a power series converging in $(z - a)$ in $D(a, r)$ in the form,

$$f(z) = \sum_{\alpha \in \mathbb{N}^n} c_\alpha (z-a)^\alpha, \qquad c_\alpha = \frac{\partial^\alpha f(a)}{\alpha!}.$$

By theorem 1.2 or proposition 1.4, this series converges normally, hence uniformly on $\overline{D}(a, r)$ or on a compact K of $D(a, r)$ and since the terms of this series are polynomials, then they are holomorphic. Therefore, the function f is holomorphic (as a uniform limit of holomorphic functions on K). The function f is holomorphic on a neighborhood of each point of Ω, and consequently, it is holomorphic on Ω. ■

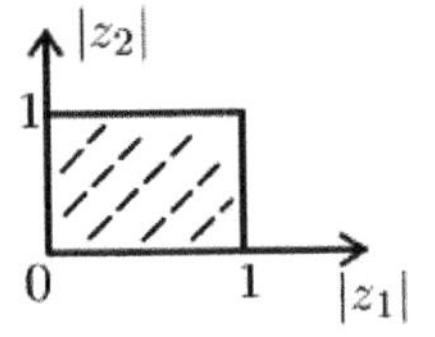

Figure 1.1. *Domain of convergence*

REMARK 1.2.– *Multiple power series have similar properties to power series in one complex variable. However, the domain of convergence of the power series obtained in the previous theorem is a little more complicated. For example, consider a series of the form* $\sum c_{k,l}(z_1 - a_1)^k (z_2 - a_2)^l$. *If we fix the component* z_1, *we obtain a single-variable power series* z_2 *whose coefficients depend on* z_1. *Its radius of convergence (at* z_2*) therefore varies with* z_1. *Similarly, we can fix* z_2 *and obtain a power series at* z_1.

EXAMPLE 1.2.– *The convergence domain of the series* $\dfrac{1}{1 - z_1 z_2} = \sum_{s=0}^{\infty} z_1^s z_2^s$ *is* $\{|z_1 z_2| < 1\}$ *(see Figure 1.1).*

EXAMPLE 1.3.– *The convergence domain of the series*

$$\frac{z_1}{(1 - z_1)(1 - z_2)} = \sum_{|k|=0}^{\infty} z_1^{k_1+1} z_2^{k_2}.$$

is the "bidisk" $\{|z_1| < 1, |z_2| < 1\}$ *completed by the complex line* $\{z_1 = 0\}$.

For the study of the convergence problem, other notions concerning the so-called Reinhardt domains are provided by the following.

DEFINITION 1.6.– *An open* $\Omega \subset \mathbb{C}^n$ *is said to be a Reinhardt domain of center* $a = (a_1, ..., a_n) \in \mathbb{C}^n$ *if* $(a_1 + (z_1 - a_1)e^{i\theta_1}, ..., a_n + (z_n - a_n)e^{i\theta_n}) \in \Omega$, *for each* $(z_1, ..., z_n) \in \Omega$ *and* $0 < \theta_1, ..., \theta_n < 2\pi$. *In other words, if it contains any point* $z = \{a_j + (z_j - a_j)e^{i\theta_j}\}$, $0 < \theta_j < 2\pi$, $j = 1, ..., n$. *A Reinhardt domain with center* $a \in \mathbb{C}^n$ *is said to be complete if for each point* $(z_1, ..., z_n) \in \Omega$, *it contains the polydisk* $\{(w_1, ..., w_n) \in \Omega : |w_j - a_j| \leq |z_j - a_j|, j = 1, ..., n\}$.

EXAMPLE 1.4.– *The polydisks and balls in* $\mathbb{C}^n$ *are Reinhardt domains. The disks* $\{z : |z - a| < r\}$ *are complete Reinhardt domains. However, the crowns* $\{z : r_1 < |z - a| < r_2\}$ *are non-complete Reinhardt domains.*

In terms of the language of Reinhardt domains, we express some results as follows.

PROPOSITION 1.7.– *Let $\Omega \subset \mathbb{C}^n$ be a Reinhardt domain and f be a holomorphic function on Ω and continuous on $\overline{D}(a,r)$. Then, f admits a unique expansion in power series*

$$f(z) = \sum_{\alpha \in \mathbb{N}^n} c_\alpha (z-a)^\alpha, \quad z \in \Omega, \quad c_\alpha = \frac{f^{(\alpha)}(a)}{\alpha!},$$

and this series converges absolutely and uniformly on any compact set of Ω.

PROOF.– Consider a polydisk

$$D(a,r) = \{(z_1, ..., z_n) \subset \mathbb{C}^n : |z_j \quad a_j| < r_j, j = 1, ..., n\},$$

of $\mathbb{C}^n$ with center $a = (a_1, ..., a_n) \in \mathbb{C}^n$ and radius $r = (r_1, ..., r_n) \in \left(\mathbb{R}_+^*\right)^n$, such that: $\overline{D} \subset \Omega$. According to the Cauchy integral formula (theorem 1.1), we have

$$f(z) = \frac{1}{(2\pi i)^n} \int_{\gamma_1} \cdots \int_{\gamma_n} \frac{f(\zeta)}{(\zeta_1 - z_1)...(\zeta_n - z_n)} d\zeta_1 ... d\zeta_n,$$

where, $\gamma_j = \{z : |z_j - a_j| = r_j, j = 1, ..., n\}$. Taking into account that the series,

$$\frac{1}{\zeta_j - z_j} = \frac{1}{\zeta_j - a_j} \sum_{k=0}^{\infty} \left(\frac{z_j - a_j}{\zeta_j - a_j}\right)^k, \quad \zeta_j \in \gamma_j,$$

converges uniformly with respect to ζ_j, z_j and that $|z_j - a_j| \leq \rho_j < r_j$, then

$$f(z) = \sum_{\alpha \in \mathbb{N}^n} c_\alpha (z-a)^\alpha,$$

where

$$c_\alpha = \frac{1}{(2\pi i)^n} \int_{\gamma_1} \cdots \int_{\gamma_n} \frac{f(\zeta)}{(\zeta_1 - a_1)^{\alpha_1+1} ... (\zeta_n - a_n)^{\alpha_n+1}} d\zeta_1 ... d\zeta_n = \frac{f^{(\alpha)}(a)}{\alpha!}.$$

The convergence of the series is uniform on $\{z : |z_j - a_j| \leq \rho_j, j = 1, ..., n\}$. The proof ends. ■

DEFINITION 1.7.– *Let $\varphi : z \longmapsto \varphi(z) = (\ln|z_1|, ..., \ln|z_n|)$, be a mapping from the set $\{z \in \mathbb{C}^n : z_1...z_n \neq 0\}$ into $\mathbb{R}^n$. We say that an open set $\Omega \subset \mathbb{C}^n$ is logarithmically convex if the image $\varphi(\Omega^*)$ of the set $\{z \in \Omega : z_1...z_n \neq 0\}$ is a convex set in $\mathbb{R}^n$ (i.e. if $\{(u_1, ..., u_n) \in \mathbb{R}^n : (e^{u_1}, ..., e^{u_n}) \in \Omega\}$ is a convex set in $\mathbb{R}^n$).*

PROPOSITION 1.8.– *Any complete and logarithmically convex Reinhardt domain is the domain of convergence of a power series.*

PROOF.– This is a consequence of the above. ■

THEOREM 1.3.– *Let f be a holomorphic function on a Reinhardt domain (the product of n crowns), $\Omega = \{z \in \mathbb{C}^n : 0 \leq r_j \leq |z_j - a_j| \leq R_j, j = 1, ..., n\}$. Then, f admits a unique expansion in a (multiple) Laurent series*

$$f(z) = \sum_{\alpha \in \mathbb{Z}^n} c_\alpha (z-a)^\alpha, \qquad c_\alpha = \frac{1}{(2\pi i)^n} \int_\Gamma \frac{f(\zeta)}{(\zeta - a)^{\alpha+1}} d\zeta,$$

where $z = (z_1, ..., z_n)$, $a = (a_1, ..., a_n)$, $\alpha+1 = (\alpha_1+1, ..., \alpha_n+1)$ and Γ denotes the product of n circles defined using the equations: $\zeta_j = c_j + \rho_j e^{i\theta}$, $j = 1, ..., n$, $0 \leq \theta \leq 2\pi$, where the constants ρ_j are between r_j and R_j. Moreover, this series converges absolutely and uniformly in Ω. (The domains of convergence of these Laurent series are Reinhardt domains.)

PROOF.– We use reasoning by recurrence or by induction on n.

– For $n = 1$, the result is known but we will repeat it here because it will give an idea for dealing with the case $n \geq 2$ (and, for this case, we pose $\alpha = k$). Here, for this particular case, $\Gamma = \gamma$ denotes a circle (or a closed path) surrounding the point a and contained in the crown. Let $K = \{z \in \mathbb{C} : r_1 \leq |z - a_1| \leq r_2\}$ be the closed crown contained in Ω where r_1 and r_2 are two numbers such that: $r_1 < r_1^- < R_1^- < R_1$. Consider the circles γ_1 and γ_2 with center a and radii $r_1 + \varepsilon$ and $R_1 - \varepsilon$, $\varepsilon > 0$ where $r_1 < r_1 + \varepsilon < r_1^- < R_1^- < R_1 - \varepsilon < R_1$. We make Ω simply connected by means of a cut pq while trying to avoid the point $z \in \Omega$ as shown in Figure 1.2.

Since f is holomorphic in Ω, we have according to the Cauchy integral formula,

$$\begin{aligned} f(z) &= \frac{1}{2\pi i}\left(\int_{\gamma_2} \frac{f(\zeta)}{\zeta - z} d\zeta + \int_q^p \frac{f(\zeta)}{\zeta - z} d\zeta + \int_{\gamma_1^-} \frac{f(\zeta)}{\zeta - z} d\zeta + \int_p^q \frac{f(\zeta)}{\zeta - z} d\zeta\right), \\ &= \frac{1}{2\pi i}\int_{\gamma_2} \frac{f(\zeta)}{\zeta - z} d\zeta - \frac{1}{2\pi i}\int_{\gamma_1} \frac{f(\zeta)}{\zeta - z} d\zeta. \end{aligned}$$

Let us calculate the first integral, i.e. $\frac{1}{2\pi i}\int_{\gamma_2} \frac{f(\zeta)}{\zeta - z} d\zeta$. We have,

$$\frac{1}{\zeta - z} = \frac{1}{(\zeta - a)\left(1 - \frac{z-a}{\zeta-a}\right)} = \frac{1}{\zeta - a}\sum_{k=0}^{\infty}\left(\frac{z-a}{\zeta-a}\right)^k,$$

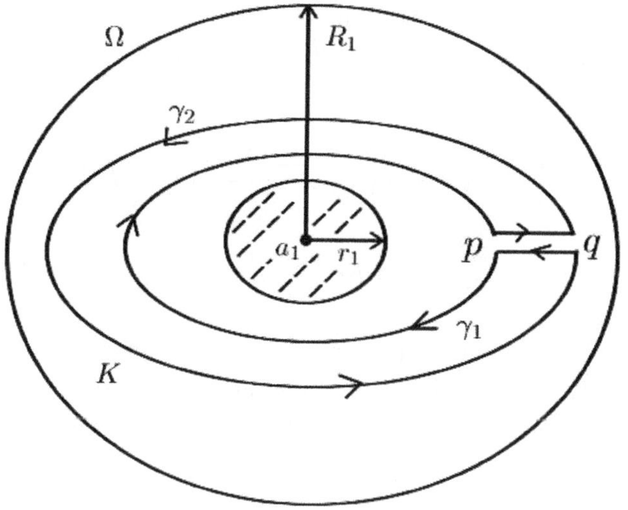

Figure 1.2. *Laurent series*

because for $\zeta \in \gamma_2$, z fixed, we have $\left|\dfrac{z-a}{\zeta-a}\right| \leq \dfrac{R_1^-}{R_1-\varepsilon} < 1$. Therefore, as $\left|\dfrac{z-a}{\zeta-a}\right|^k \leq \left(\dfrac{R_1^-}{R_1-\varepsilon}\right)^k$, and the geometric series $\sum\left(\dfrac{R_1^-}{R_1-\varepsilon}\right)^k$ converges (because $\dfrac{R_1^-}{R_1-\varepsilon} < 1$), according to the Weierstrass criterion of convergence, the series $\sum\left(\dfrac{z-a}{\zeta-a}\right)^k$ converges normally, therefore absolutely and uniformly on K. We can therefore integrate this series term by term:

$$\begin{aligned}\frac{1}{2\pi i}\int_{\gamma_2}\frac{f(\zeta)}{\zeta-z}d\zeta &= \frac{1}{2\pi i}\int_{\gamma_2}\frac{f(\zeta)}{\zeta-a}\sum_{k=0}^{\infty}\left(\frac{z-a}{\zeta-a}\right)^k d\zeta,\\ &= \sum_{k=0}^{\infty}\left(\frac{1}{2\pi i}\int_{\gamma_2}\frac{f(\zeta)}{(\zeta-a)^{k+1}}d\zeta\right)(z-a)^k,\\ &= \sum_{k=0}^{\infty}\left(\frac{1}{2\pi i}\int_{\gamma}\frac{f(\zeta)}{(\zeta-a)^{k+1}}d\zeta\right)(z-a)^k, \qquad [1.1]\end{aligned}$$

by Cauchy's theorem because f is holomorphic and γ_2, γ are homotopic. Let us now calculate the second integral, i.e. $\frac{1}{2\pi i}\int_{\gamma_1}\frac{f(\zeta)}{\zeta - z}d\zeta$. We have

$$-\frac{1}{\zeta - z} = \frac{1}{(z-a)\left(1 - \frac{\zeta - a}{z-a}\right)} = \frac{1}{z-a}\sum_{k=0}^{\infty}\left(\frac{\zeta - a}{z-a}\right)^k,$$

because for $\zeta \in \gamma_1$, z fixed, we have $\left|\frac{\zeta - a}{z-a}\right| \le \frac{R_1 + \varepsilon}{r_1} < 1$. Since we have $\left|\frac{\zeta - a}{z-a}\right|^k \le \left(\frac{R_1+\varepsilon}{r_1}\right)^k$, and $\sum\left(\frac{R_1+\varepsilon}{r_1}\right)^k$ converges because $\frac{R_1+\varepsilon}{r_1} < 1$, then according to the Weierstrass criterion, the series in question converges absolutely and uniformly in K. We can therefore integrate this series term by term,

$$\begin{aligned}\frac{1}{2\pi i}\int_{\gamma_1}\frac{f(\zeta)}{\zeta - z}d\zeta &= \frac{1}{2\pi i}\int_{\gamma_1}\frac{f(\zeta)}{z-a}\sum_{k=0}^{\infty}\left(\frac{\zeta - a}{z-a}\right)^k d\zeta,\\ &= \sum_{k=0}^{\infty}\frac{1}{2\pi i}\int_{\gamma_1}\frac{f(\zeta)}{(\zeta - a)^{-k}}(z-a)^{-k-1}d\zeta,\\ &= \sum_{k=-1}^{-\infty}\left(\frac{1}{2\pi i}\int_{\gamma_1}\frac{f(\zeta)}{(\zeta - a)^{k+1}})d\zeta\right)(z-a)^k,\\ &= \sum_{k=-1}^{-\infty}\left(\frac{1}{2\pi i}\int_{\gamma}\frac{f(\zeta)}{(\zeta - a)^{k+1}}d\zeta\right)(z-a)^k, \end{aligned} \qquad [1.2]$$

according to Cauchy's theorem since f is holomorphic and γ_2, γ are homotopic. Combining [1.1] and [1.2], we obtain the following series:

$$f(z) = \sum_{k=-\infty}^{\infty} c_k (z-a)^k,$$

with the coefficients given by

$$c_k = \frac{1}{2\pi i}\int_{\gamma}\frac{f(\zeta)}{(\zeta - a)^{k+1}}d\zeta.$$

Let us show that the series expansion we obtained is unique. Suppose that in K, we have

$$f(z) = \sum_{k=-\infty}^{\infty} d_k (z-a)^k,$$

and show that $d_k = c_k$. Indeed, we have

$$\frac{f(z)}{(z-a)^{m+1}} = \sum_{k=-\infty}^{\infty} d_k (z-a)^{-k-m-1}.$$

This series is uniformly convergent and we can integrate term by term:

$$\int_\gamma \frac{f(z)}{(z-a)^{m+1}} dz = \sum_{k=-\infty}^{\infty} d_k \int_\gamma (z-a)^{-k-m-1} dz.$$

Or

$$\int_\gamma (z-a)^\nu dz = \begin{cases} 2\pi i \text{ if } & \nu = -1 \\ 0 \text{ otherwise} \end{cases}$$

so

$$\int_\gamma \frac{f(z)}{(z-a)^{m+1}} dz = d_m . 2\pi i,$$

and consequently,

$$d_m = \frac{1}{2\pi i} \int_\gamma \frac{f(z)}{(z-a)^{m+1}} dz = c_m.$$

– Now suppose that $n \geq 2$. Here (as stated in the theorem), Γ denotes the product $\gamma_1 \times \cdots \times \gamma_n$ of n circles defined using the equations: $\zeta_j = c_j + \rho_j e^{i\theta}, j = 1, ..., n, 0 \leq \theta \leq 2\pi$, where the ρ_j are between r_j and R_j. Let $z = (z_1, z_2, ..., z_n) = (z_1, z^*) \in \mathbb{C}^n$ where $z^* \equiv (z_2, ..., z_n) \in \mathbb{C}^{n-1}$. Let us express Ω as a product of the form $\Omega_1 \times \cdots \times \Omega_n$ (they can be visualized as a kind of rings or crowns). By fixing ζ_1 where $|\zeta_1 - a_1| = \rho_1$, we note that the function $f(\zeta_1, .)$ is holomorphic on the product $\Omega_2 \times \cdots \times \Omega_n$. By induction hypothesis, we have

$$\begin{aligned} f(\zeta_1, z^*) &= \sum_{\alpha^* \in \mathbb{Z}^{n-1}} c_{\alpha^*} (z^* - a^*)^{\alpha^*}, \\ &= \frac{1}{(2\pi i)^{n-1}} \sum_{\alpha^* \in \mathbb{Z}^{n-1}} (z^* - a^*)^{\alpha^*} \int_{\gamma_2 \times \cdots \times \gamma_n} \frac{f(\zeta_1, \zeta^*)}{(\zeta^* - a^*)^{\alpha^* + 1}} d\zeta^*. \end{aligned}$$

By fixing z^*, we also have the expression

$$f(z_1, z^*) = \frac{1}{2\pi i} \sum_{\alpha_1 \in \mathbb{Z}} (z_1 - a_1)^{\alpha_1} \int_{\gamma_1} \frac{f(\zeta_1, z^*)}{(\zeta_1 - a_1)^{\alpha_1 + 1}} d\zeta_1,$$

where recall that γ_1 is the circle ($|\zeta_1 - a_1| = \rho_1$) with center a_1 and radius ρ_1. For any z such that: $r_j < |z_j - a_j| < \rho_j$, we deduce from the above expressions of f the following relation punctually valid for any j on Ω:

$$f(z) = \frac{1}{(2\pi i)^n} \sum_{\alpha \in \mathbb{Z}^n} (z-a)^\alpha \int_\Gamma \frac{f(\zeta)}{(\zeta - a)^{\alpha+1}} d\zeta.$$

Let $m_j = \min_{z \in K} |z_j - a_j|$, $M_j = \max_{z \in K} |z_j - a_j|$, and choose r_j^- and R_j^- such that: $r_j < r_j^- < m_j \leq M_j < R_j^- < R_j$. Taking into account the Cauchy integral formula, we express the coefficients c_α as follows:

$$c_\alpha = \frac{1}{(2\pi i)^n} \int_{\Gamma^\varepsilon} \frac{f(\zeta)}{(\zeta - a)^{\alpha+1}} d\zeta,$$

where $\Gamma^\varepsilon = \gamma_1^{\varepsilon_1} \times \cdots \times \gamma_n^{\varepsilon_n}$ is the product of n circles: $\gamma_j^{\varepsilon_j} = \{|\zeta_j - a_j| = \rho_j^{\varepsilon_j}\}$, with $\zeta_j = a_j + r_j^{\varepsilon_j} e^{i\theta}$, $r_j^{\varepsilon_j} = r_j^+$ if $\alpha_j \geq 0$ and $r_j^{\varepsilon_j} = r_j^-$ if $\alpha_j < 0$. Therefore, for all $\alpha \in \mathbb{Z}^n$ and all $z \in K$, we have $|c_\alpha (z-a)^\alpha| \leq C_K A_j B_j$, where

$$C_K = \sup_{\zeta \in \Delta} |f(\zeta)|, \quad A_j = \prod_{j=1, \alpha_j \geq 0}^{n} \left(\frac{M_j}{R_j^-} \right)^{\alpha_j}, \quad B_j = \prod_{j=1, \alpha_j < 0}^{n} \left(\frac{r_j^-}{m_j} \right)^{|\alpha_j|},$$

and

$$\Delta = \left\{ z \in \mathbb{C}^n : 0 \leq r_j^- \leq |z_j - a_j| \leq R_j^-, j = 1, ..., n \right\}.$$

By denoting by q the maximum of the numbers $\dfrac{M_j}{R_j^-}, \dfrac{r_j^-}{m_j}, j = 1, ..., n$, we finally obtain the following expression:

$$\sum_{\alpha \in \mathbb{Z}^n} |c_\alpha (z-a)^\alpha| \leq C_K \sum_{\alpha \in \mathbb{Z}^n} q^{|\alpha|} = \frac{C_K}{(1-q)^n} < \infty,$$

on K. ■

1.4. Various fundamental properties

PROPOSITION 1.9 (CAUCHY INEQUALITIES).– *Let $D(a, r)$ be a polydisk of $\mathbb{C}^n$, with center a and radius r. If f is a holomorphic function on $D(a, r)$ and $|f| \leq M \equiv \sup_{z \in D(a,r)} |f(z)|$ on $D(a, r)$, then for all $\alpha \in \mathbb{N}^n$, we have*

$$|\partial^\alpha f(a)| \leq \frac{M\alpha!}{r^\alpha}.$$

PROOF.– We apply the Cauchy integral formula to the function f on the polydisk $D'(a,\rho)$ with center a and radius ρ such that: $0<\rho<r$. By differentiating as before under the integral sign in this formula, we obtain

$$\partial^\alpha f(a) = \frac{\alpha!}{(2\pi i)^n}\int_{\partial_0 D'} \frac{f(\zeta)}{(\zeta-a)^{\alpha+1}}d\zeta.$$

Therefore, by setting $\zeta = a+\rho e^{i\theta}$, $\theta=(\theta_1,...,\theta_n)\in[0,2\pi]^n$, we obtain

$$|\partial^\alpha f(a)| \le \frac{\alpha!}{(2\pi)^n}\int_{[0,2\pi]^n}\left|\frac{f(a+\rho e^{i\theta})}{(\rho e^{i\theta})^{\alpha+1}}\right|\rho d\theta \le \frac{\alpha! M}{\rho^\alpha},$$

and by making ρ tend towards r, we obtain the inequality in question. ■

THEOREM 1.4 (LIOUVILLE).– *Any holomorphic and bounded function on $\mathbb{C}^n$ is a constant.*

PROOF.– It is enough to use Cauchy's inequality. By hypothesis, the function f is bounded, so there exists a constant $M>0$ such that for all $z\in\mathbb{C}^n$, we have: $|f(z)|\le M$ and the Cauchy inequality (previous proposition) is written as

$$|\partial^\alpha f(a)| \le \frac{M\alpha!}{r^\alpha}.$$

Moreover, f being holomorphic, we have

$$f(z) - \sum_{\alpha\in\mathbb{N}^n} c_\alpha (z-a)^\alpha, \qquad |c_\alpha| - \left|\frac{\partial^\alpha f(a)}{\alpha!}\right| \le \frac{M}{\rho^{|\alpha|}},$$

where $\rho = r_1 = \cdots = r_n$. By making $\rho\to\infty$ tend where $|\alpha|\ge 1$, we obtain $c_\alpha = 0$ and the result follows. ■

We will prove the following identity theorem for holomorphic functions, and we will see in exercise 1.9 a version of this theorem for power series.

THEOREM 1.5 (PRINCIPLE OF ANALYTIC CONTINUATION OR IDENTITY PRINCIPLE).– *Let Ω be a domain (i.e. a connected open set) of $\mathbb{C}^n$ and f, g be two holomorphic functions on Ω. If f and g coincide in the neighborhood of a point of Ω, then they coincide on Ω.*

PROOF.– Let $\Delta = \{z\in\Omega : f(z) = g(z)$ on a neighborhood of $z\}$. Note that Δ is an open subset of Ω and it is nonempty. Furthermore, Δ is closed in Ω since the derivatives of f are all continuous. Since the open set Ω is connected, then $\Delta=\Omega$ or $\Delta=\emptyset$. Now $\Delta\ne\emptyset$, so the only possible case is $\Delta=\Omega$, hence the result. ■

The following result (due to Hartog) is false in dimension 1. Any holomorphic function on $\Omega\backslash K$ where K is a compact subset of an open Ω of $\mathbb{C}^n$ with $n\ge 2$ and

$\Omega\backslash K$ assumed to be connected extends to a holomorphic function on Ω. This can be formulated in another way:

THEOREM 1.6 (HARTOG'S EXTENSION THEOREM).– *Let*

$$D_n(0,r_1) = \{z \in \mathbb{C}^n : |z_j| < r_1\}, \quad r_1 > 0, \quad n \geq 2,$$

and

$$D_n(0,r_2) = \{z \in \mathbb{C}^n : |z_j| < r_2\}, \quad 0 < r_2 < r_1, \quad n \geq 2.$$

Then, any holomorphic function in $D_n(0,r_1)\backslash\overline{D}_n(0,r_2)$ *is the restriction to* $D_n(0,r_1)\backslash\overline{D}_n(0,r_2)$ *of a holomorphic function in* $D_n(0,r_1)$*. In other words, any holomorphic function on* $D_n(0,r_1)\backslash\overline{D_n}(0,r_2)$ *admits a (unique) extension to* $D_n(0,r_1)$ *which is itself holomorphic (or the canonical restriction homomorphim:* $\mathcal{O}\left(D_n(0,r_1)\right) \longrightarrow \mathcal{O}\left(D_n(0,r_1)\backslash\overline{D_n}(0,r_2)\right)$*, is bijective).*

PROOF.– Let $f : D_n(0,r_1)\backslash\overline{D_n}(0,r_2) \longrightarrow \mathbb{C}$, be a holomorphic function,

$$D_n(0,r_3) = \{z \in \mathbb{C}^n : |z_j| < r_3\}, \quad 0 < r_2 < r_3 < r_1,$$

and let U denote the locus defined by

$$U = \{z = (z_1, ..., z_n) \in D_n(0,r_3)\backslash\overline{D_n}(0,r_2) : z = c\},$$

with $c = (c_2, ..., c_n) \in \mathbb{C}^{n-1}$. Note that $U = \emptyset$ if $c \notin D_{n-1}(0,r_3)$, $U = D_1(0,r_3)$ if $c \in D_{n-1}(0,r_3)\backslash D_{n-1}(0,r_2)$ and $U = D_1(0,r_3)\backslash D_1(0,r_2)$ if $c \in D_{n-1}(0,r_2)$. The function g defined on $D_n(0,r_3)\backslash\overline{D_n}(0,r_2)$ by

$$g(z) = \int_{|\zeta_1|=r_1} \frac{f(\zeta_1, z_2, ..., z_n)}{\zeta_1 - z_1} d\zeta_1,$$

is holomorphic and is suitable for f on $D_{n-1}(0,r_3)\backslash D_{n-1}(0,r_2)$ as well as on $D_n(0,r_3)\backslash D_n(0,r_2)$. ■

COROLLARY 1.4.– *For* $n \geq 2$*, the restriction homomorphim:*

$$\mathcal{O}\left(D_n(0,r_1)\right) \longrightarrow \mathcal{O}\left(D_n(0,r_1)\backslash\overline{D_n}(0,r_2)\right),$$

is an isomorphism and in particular there are no isolated "singularities" for holomorphic functions of two or more variables, and therefore no isolated zeros either.

As we pointed out above, this theorem is false in the special case $n = 1$. Indeed, it suffices to consider the function with an isolated singularity, for example, $f(z) = 1/z$. This function is clearly holomorphic in $\mathbb{C}\backslash\{0\}$, but has no holomorphic extensions passing through the singular point 0, and therefore no holomorphic extensions on $\mathbb{C}$.

Hartog's theorem implies that the singularities of holomorphic functions of two or more variables are never isolated. For example, the function $1/z_1z_2$ is singular not only at $(0,0)$ but at every point of the form $(z_1,0)$ or $(0,z_2)$. In addition, it also provides information about the zeros of holomorphic functions. In the case of a complex variable, the zeros of holomorphic functions of one variable are always isolated. But if a holomorphic function f had an isolated zero at $p \in \mathbb{C}^n$ with $n \geq 2$, then $1/f$ would have an isolated singularity, which is impossible. Thus, the zeros of holomorphic functions of more than one variable are never isolated either. Hartog's theorem is therefore a phenomenon that highlights the difference between the theory of functions of one complex variable and that of functions of several complex variables.

THEOREM 1.7 (MAXIMUM MODULUS PRINCIPLE).– *Let f be a holomorphic function on a domain Ω of $\mathbb{C}^n$. If there exists a point $a \in \Omega$ such that for all z belonging to a neighborhood of a, we have $|f(z)| \leq |f(a)|$, then $f(z) = g(a)$ for all $z \in \Omega$. (In other words, f is constant.)*

PROOF.– *1st step*. Let D be a polydisk with center w and radius r, relatively compact in Ω (i.e. $\overline{D} \subset \Omega$). We denote by

$$V(D) = \int_D dV(\zeta),$$

the volume of D where dV is the volume element of $\mathbb{C}^n$. One of the consequences of the Cauchy integral formula is

$$V(D)f(a) = \int_D f(\zeta)dV(\zeta).$$

We reason by induction. For $n = 1$, $D = D_1$ is a disk of $\mathbb{C}$ and we have $V(D_1) = \pi r^2$. By hypothesis, the function f is holomorphic in Ω and according to the mean property for $\rho \in [0,r]$, we have

$$f(a) = \frac{1}{2\pi}\int_0^{2\pi} f(a+\rho e^{i\theta})d\theta.$$

Therefore,

$$\int_{D_1} f(\zeta)dV(\zeta) = \int_0^r \left(\int_0^{2\pi} f(a+\rho e^{i\theta})d\theta\right)\rho d\rho = \pi r^2 f(a).$$

Let us assume that the formula in question is true to order n and show that it is true to order $n+1$. By hypothesis, we have

$$V(D_n)f(a) = \int_{D_n} f(\zeta)dV(\zeta), \quad D_n \subset \mathbb{C}^n$$

Hence, noting $D \equiv D_{n+1} = D_n \times D_0$,

$$V(D)f(a) = V(D_n)V(D_0)f(a) = \int_{D_n} f(\zeta)dV(\zeta) \int_{D_0} dV(\zeta) = \int_D f(\zeta)dV(\zeta).$$

It is inferred that:

$$V(D)|f(a)| \leq \int_D |f(\zeta)|dV(\zeta).$$

2nd step. Let D be a polydisk with center a and radius r such that for all $z \in D$, we have: $|f(a)| - |f(z)| \geq 0$. Hence,

$$0 \leq \int_D (|f(a)| - |f(z)|)dV(\zeta) = V(D)|f(a)| - \int_D |f(\zeta)|dV(\zeta) \leq 0,$$

according to the above. Therefore, $|f(a)| = |f(z)|$ for all $z \in D$ and according to proposition 1.3, the function f is constant on D. Taking into account the principle of analytic continuation, we deduce that the function f is constant on Ω. ■

THEOREM 1.8 (OPEN MAPPING THEOREM).– *Let f be a non-constant holomorphic function on a connected open set $\Omega \subset \mathbb{C}^n$. Then, the mapping $f : \Omega \longrightarrow \mathbb{C}$ is open, i.e. it maps open sets to open sets.*

PROOF.– Let $a \in \Omega$ be a convex open set $U \subset \Omega$ (or an open ball $B(a,r) \subset \Omega$ with center a and radius r). According to the principle of analytic continuation or the identity principle (theorem 1.5), there exists $b \in U$ such that: $f(b) \neq f(a)$. We define a function φ of a single variable by setting $\varphi(z) = f(a + z(b - a))$, where $z \in V \equiv \{z \in \mathbb{C} : a + z(b - a) \in U\}$. The open set V is connected and contains the points 0 and 1. Furthermore, we have $\varphi(0) = f(a) \neq f(b) = \varphi(1)$. The function φ is holomorphic; it is not constant and according to the open map theorem in dimension 1, $\varphi(V)$ is a neighborhood of $f(a)$ included in $f(\Omega)$. This completes the proof of the theorem. ■

REMARK 1.3.– *The principle of the maximum can be easily deduced from the open mapping theorem. Indeed, suppose that f is not constant. Therefore, according to the open mapping theorem, $f(\Omega)$ would be an open set contained in the closed disk $\overline{D}(a, |f(a)|) = \{w \in \mathbb{C} : |w| \leq |f(a)|\}$ with center a and radius $|f(a)|$. Therefore, this disk is necessarily contained in the open disk $D(a, |f(a)|) = \{w \in \mathbb{C} : |w| < |f(a)|\}$, which is absurd.*

THEOREM 1.9 (WEIERSTRASS).– *Let Ω be an open set of $\mathbb{C}^n$ and (f_n) be a sequence of functions holomorphic in Ω. If (f_n) converges uniformly on every compact set of Ω to a function f, then f is holomorphic on Ω and for all $\alpha \in \mathbb{N}^n$, the sequence $(\partial^\alpha f_n)$ converges uniformly on every compact set of Ω to $\partial^\alpha f$.*

PROOF.– By hypothesis, (f_n) converges uniformly on any compact set of Ω to f and since f_n are continuous on Ω, we deduce that f is continuous on Ω. Furthermore, the function f is holomorphic with respect to each variable at every point $z \in \Omega$ and by Osgood's lemma (corollary 1.2), it is holomorphic on Ω. Let D be a polydisk with center a and radius r such that for every $z \in D$, $\overline{D} \subset \Omega$. We know (see proof of theorem 1.2 or that of proposition 1.9) that for $\alpha \in \mathbb{N}^n$,

$$\partial^\alpha f_n(z) = \frac{\alpha!}{(2\pi i)^n} \int_{\partial_0 D} \frac{f_n(\zeta)}{(\zeta - z)^{\alpha+1}} d\zeta, \quad \partial^\alpha f(z) = \frac{\alpha!}{(2\pi i)^n} \int_{\partial_0 D} \frac{f(\zeta)}{(\zeta - z)^{\alpha+1}} d\zeta.$$

Hence,

$$\partial^\alpha (f_n - f)(z) = \frac{\alpha!}{(2\pi i)^n} \int_{\partial_0 D} \frac{f_n(\zeta)}{(\zeta - z)^{\alpha+1}} d\zeta.$$

By setting $\zeta = z + re^{i\theta}$, we obtain

$$|\partial^\alpha (f_n - f)(z)| \leq \frac{\alpha!}{(2\pi)^n} \cdot \frac{(2\pi)^n}{r^\alpha} \sup_{z \in \partial_0 D} |f_n(z) - f(z)| = \frac{\alpha!}{r^\alpha} \sup_{z \in \partial_0 D} |f_n(z) - f(z)|.$$

It follows that the sequence $(\partial^\alpha f_n)$ converges uniformly on $\partial_0 D$ (compact). Since every compact is covered by a finite number of disks, we therefore have a uniform convergence on every compact of Ω. ■

Among the consequences of this theorem, we show that the space $\mathcal{O}(\Omega)$ is closed in the space of continuous functions $\mathcal{C}(\Omega)$ equipped with uniform convergence on compact sets, and it is complete.

The following Hartog's theorem (whose proof is lengthy) ensures that if a function of several complex variables is holomorphic with respect to each variable separately, then it will be holomorphic with respect to all variables.

THEOREM 1.10 (HARTOG'S).– *Let $\Omega \subset \mathbb{C}^n$ be an open set and $f : \Omega \longrightarrow \mathbb{C}$ be a holomorphic function with respect to each variable. Then, f is holomorphic in the open set Ω.*

Hint. This result was proven in corollary 1.2 (Osgood's lemma) for continuous functions. So, to prove this theorem (see, for example, Hormander 1990), we can prove that if a function is holomorphic with respect to each variable, then it is continuous with respect to all variables. Moreover, as the statement is local, the idea of the proof of this theorem (see Silverman 2009) consists of proving it when the open Ω is a polydisk and it suffices to show that the function f is holomorphic at any point. We reason by induction on the dimension n. The result is trivial for $n = 1$, i.e. for a single variable. We assume that the theorem is true in $\mathbb{C}^{n-1}$, i.e. for a function of $n - 1$ variables and we must prove it in $\mathbb{C}^n$. ■

REMARK 1.4.– *This theorem allows us to give a characterization of holomorphic functions in several variables by reducing to one variable. Thus, we can give a new definition of holomorphic function on an open* $\Omega \subset \mathbb{C}^n$*; a complex function* $f : \Omega \longrightarrow \mathbb{C}$ *is said to be holomorphic in* Ω *if it is holomorphic with respect to each variable separately. Note that the real analogue of this result is false, and this is a big difference between real analysis and complex analysis.*

PROPOSITION 1.10 (SCHWARZ LEMMA).– *Let* $D = D(0, r)$, $r = (r, ..., r)$, *be a polydisk with center* 0 *and radius* r *and* f *be a holomorphic function in a neighborhood of* $\overline{D}$. *We assume that the total order at* 0 *of* f *is* k *and that for all* $z \in \overline{D}$, $|f(z)| \leq M$. *Then,*

$$\forall z \in \overline{D}, \quad |f(z)| \leq M \left| \frac{z}{r} \right|^k .$$

PROOF.– We write $f(z)$ in the form $f(z) = P_k(z) + P_{k+1}(z) + \cdots$ where $P_k \neq 0$ are polynomials. Let $z \in D$, $z \neq 0$ and set

$$g(t) = \frac{f\left(\frac{tz}{|z|}\right)}{t^k}, \quad t \in \mathbb{C}, |t| \leq r$$

Hence, the following Taylor expansion:

$$g(t) = f(z) = P_k\left(\frac{z}{|z|}\right) + P_{k+1}\left(\frac{z}{|z|}\right) t + \cdots$$

and we have

$$|g(t)| \leq \frac{M}{r^k}, \quad |t| = r,$$

because by hypothesis, $f\left(\frac{tz}{|z|}\right) \leq M$. According to the maximum modulus principle in $\mathbb{C}$, we have the inequality

$$|g(t)| \leq \frac{M}{r^k}, \quad |t| \leq r.$$

Hence,

$$\left| \frac{f(z)}{|z|^k} \right| = |g(z)| \leq \frac{M}{r^k},$$

and therefore

$$|f(z)| \leq M \left| \frac{z}{r} \right|^k .$$

The proof ends. ■

Let X be a topological space, $a \in X$ and S the set of functions, each defined in a neighborhood of a.

DEFINITION 1.8.– *We say that two functions $f, g \in S$ are equivalent in a if there exists a neighborhood U of a such that: $f|_U = g|_U$. The set of elements equivalent in $a \in X$ to $f \in S$ is called the germ of f.*

Let f be a holomorphic function on a domain Ω of $\mathbb{C}$. The following conditions are equivalent: *(i)* the function f is identically zero; *(ii)* there exists a point of Ω where the germ of f is zero; and *(iii)* there exists a point of Ω where all the derivatives of f are zero. In particular, for any point of $z \in \mathbb{C}$, the ring of germs at point z of holomorphic functions is integral.

We will focus now on functions with a value in $\mathbb{C}^p$, which we denote by $f = (f_1, ..., f_p)$. Let $f : \Omega_1 \subset \mathbb{C}^n \longrightarrow \Omega_2 \subset \mathbb{C}^p$, $(z_1, ..., z_n) \longmapsto (w_1, ..., w_p)$, be a map where $w_1 = f_1(z_1, ..., z_n), ..., w_p = f_p(z_1, ..., z_n)$.

DEFINITION 1.9.– *We say that f is holomorphic if the m functions $f_1, ..., f_p$ are holomorphic in Ω_1.*

Note that if

$$g : \Omega_2 \subset \mathbb{C}^p \longrightarrow \mathbb{C}^q, \quad (w_1, ..., w_p) \longmapsto g(w_1, ..., w_p) = g(w),$$

is a function defined in Ω_2, then the composite

$$gof : \Omega_1 \longrightarrow \mathbb{C}^l, \quad z \longmapsto gof(z) = g(f(z)),$$

is a well-defined function on Ω_1. We will see that the composite of two holomorphic maps is holomorphic.

PROPOSITION 1.11.– *If the functions f and g defined above are holomorphic, then the composite gof is also holomorphic and for all $z \in \Omega_1$, we have*

$$d(gof)(z) = dg(f(z))odf(z).$$

PROOF.– Let $z = (z_1, ..., z_n) \in \Omega_1 \subset \mathbb{C}^n$ and $(w_1, ..., w_p) \in \Omega_2 \subset \mathbb{C}^p$. Using the formula for differentiating composite functions (or chain rule), we obtain

$$\frac{\partial(gof)}{\partial \overline{z}_j}(z) = \sum_{k=1}^{p} \frac{\partial g}{\partial w_k}(f(z))\frac{\partial f_k}{\partial \overline{z}_j}(z) + \sum_{k=1}^{p} \frac{\partial g}{\partial \overline{w}_k}(f(z))\frac{\partial \overline{f}_k}{\partial \overline{z}_j}(z) = 0,$$

by virtue of the Cauchy–Riemann equations, and since gof is of class C^1, the result follows. ■

PROPOSITION 1.12.– *Expressed (matrix-wise) in terms of a Jacobian matrix, the formula obtained in the previous proposition is written for all* $a \in \Omega_1$, *in the form* $J_{gof}(a) = J_g(f(a))J_f(a)$, *where the Jacobian matrix of* f *at point* $a \in \Omega_1$ *is defined by*

$$J_f(a) = \begin{pmatrix} \frac{\partial f_1}{\partial z_1}(a) \ ... \ \frac{\partial f_1}{\partial z_n}(a) \\ \vdots \qquad\qquad \vdots \\ \frac{\partial f_p}{\partial z_1}(a) \ ... \ \frac{\partial f_p}{\partial z_n}(a) \end{pmatrix} = \left(\frac{\partial f_j}{\partial z_k}(a)\right)_{\substack{1 \leq j \leq p \\ 1 \leq k \leq n}}.$$

PROOF.– According to the previous proposition, we have

$$d(gof)(a) = dg(f(a))odf(a),$$

or explicitly,

$$\frac{\partial gof(a)}{\partial \overline{z}_j} = \sum_{k=1}^{p} \left(\frac{\partial g}{\partial u_k}\frac{\partial u_k}{\partial \overline{z}_j} + \frac{\partial g}{\partial v_k}\frac{\partial v_k}{\partial \overline{z}_j}\right).$$

By decomposing f_j into real part and imaginary part: $f_j = u_j + iv_j$, where $j = 1, ..., n$, we obtain

$$\frac{\partial gof(a)}{\partial \overline{z}_j} = \sum_{k=1}^{p} \frac{1}{2}\left(\frac{\partial g}{\partial u_k} - i\frac{\partial g}{\partial u_k}\right)\frac{\partial f_k}{\partial \overline{z}_j} + \sum_{k=1}^{p} \frac{1}{2}\left(\frac{\partial g}{\partial u_k} + i\frac{\partial g}{\partial u_k}\right)\frac{\partial \overline{f}_k}{\partial \overline{z}_j},$$

$$= \sum_{k=1}^{p} \left(\frac{\partial g}{\partial w_k}\frac{\partial f_k}{\partial \overline{z}_j} + \frac{\partial g}{\partial \overline{w}_k}\frac{\partial \overline{f}_k}{\partial \overline{z}_j}\right).$$

Since f and g are holomorphic, then $\frac{\partial f_k}{\partial \overline{z}_j} = \frac{\partial g}{\partial \overline{w}_k} = 0$, for all k. Therefore, for all j, we have $\frac{\partial gof(a)}{\partial \overline{z}_j} = 0$ and from the Cauchy–Riemann equations, we deduce that the function gof is holomorphic. To obtain the result in question, it suffices to express the expression in terms of a Jacobian matrix $d(gof)(a) = dg(f(a))odf(a)$. ■

1.5. Inverse mapping and implicit function theorems

DEFINITION 1.10.– *Let* $f : \Omega \longrightarrow \mathbb{C}^n$ *be a holomorphic mapping on an open set* Ω *of* $\mathbb{C}^n$. *We say that* f *is biholomorphic if it is holomorphic and if it admits a reciprocal mapping* $g \equiv f^{-1} : f(\Omega) \longrightarrow \Omega$ *which is also holomorphic. In other words, if* f *and its inverse are holomorphic.*

REMARK 1.5.– *Obviously, biholomorphic applications can only exist between equidimensional spaces. Note that we have* $gof = id_\Omega$ *and* $fog = id_{f(\Omega)}$.

As in real analysis, we have complex versions of the (holomorphic) local inversion theorems and implicit functions, which we will see below.

THEOREM 1.11 (LOCAL INVERSION THEOREM).– *Let* $f : \Omega \longrightarrow \mathbb{C}^n$ *be a holomorphic map on an open subset* Ω *of* $\mathbb{C}^n$. *Let* $a \in \Omega$, $b = f(a)$. *If* $\det J_f(a) \neq 0$, *then there exists an open neighborhood* $U(a) \subset \Omega$ *of* a *and an open neighborhood* $V(b) \subset \mathbb{C}^n$ *of* b *such that:* $f(U(a)) \subset V(b)$ *and the restriction* $f|_{U(a)} : U(a) \longrightarrow V(b)$ *is biholomorphic.*

PROOF.– Taking into account that $\mathbb{C}^n \simeq \mathbb{R}^{2n}$ while considering f as a map of $\Omega \subset \mathbb{R}^{2n}$ in $\mathbb{R}^{2n}$, we obtain by decomposing f_j into a real part and imaginary part: $f_j = u_j + iv_j$, $j = 1, ..., n$,

$$\det J_f^{\mathbb{R}} = \det \frac{\partial(u_1 v_1, ..., u_n, v_n)}{\partial(x_1, y_1, ..., x_n, y_n)} = \left| \det \frac{\partial(f_1, ..., f_n)}{\partial(z_1, ..., z_n)} \right|^2 = |\det J_f(a)|^2,$$

where $J_f^{\mathbb{R}}$ (respectively J_f) denotes the real (respectively complex) Jacobian matrix. By hypothesis $\det J_f(a) \neq 0$, therefore the Jacobian matrix $\left(\frac{\partial f_j}{\partial z_k}(a)\right)_{1 \le j,k \le n}$ is invertible. By the real local inversion theorem, there exists a neighborhood $U(a)$ of a and a neighborhood $V(b)$ of $b = f(a)$ such that the restriction of f to $U(a)$ is a bijection of $U(a)$ onto $V(b)$ and the converse (real) map $g = f^{-1} : V(b) \longrightarrow U(a)$ is of class C^1, $gof = id_{U(a)}$, i.e. $g(f(a)) = a$, $fog = id_{V(b)}$. Let us show that the map g is holomorphic. Since $fog = id_{V(b)}$ is holomorphic, i.e. according to the Cauchy–Riemann equations, we have $\dfrac{\partial(fog)}{\partial \overline{w}} = 0$, or, according to the differentiation formula of a composite function, $\left(\dfrac{\partial f}{\partial z} og\right) o \dfrac{\partial g}{\partial \overline{w}} = 0$. Note (still taking into account the real local inversion theorem) that $\dfrac{\partial f}{\partial z}$ is invertible around $a = g(f(a))$ and therefore $\dfrac{\partial g}{\partial \overline{w}} = 0$. Consequently, the function g is holomorphic around $b = f(a)$. ■

The (holomorphic) implicit function theorem ensures that under certain conditions, an equation can define a function implicitly. More precisely, let $z = (z_1, ..., z_k, ..., z_n) \in \mathbb{C}^n$ and $g_1, ..., g_k$ (where $k < n$) be holomorphic functions in the neighborhood of a point $z = (z_1^0, ..., z_k^0, ..., z_n^0) \in \mathbb{C}^n$ and suppose that: $\det \left(\frac{\partial g_j}{\partial z_l}\right)_{1 \le j,l \le k} \neq 0$. Then, the system of equations:

$$g_1(z_1, ..., z_n) = 0, ..., g_k(z_1, ..., z_n) = 0,$$

locally admits a solution with respect to $z_1, ..., z_k$ and furthermore, the solution $z_j = f_j(z_{k+1}, ..., z_n)$, $j = 1, ..., k$, is holomorphic in the neighborhood of the point $(z^0_{k+1}, ..., z^0_n)$. We can reformulate this theorem in the following form:

THEOREM 1.12 (IMPLICIT FUNCTIONS).– *Let*

$$\Omega \subset \mathbb{C}^n \times \mathbb{C}^p = \{(z, w) : z = (z_1, ..., z_n) \in \mathbb{C}^n, w(w_1, ..., w_p) \in \mathbb{C}^p\},$$

be an open set and $g : \Omega \longrightarrow \mathbb{C}^p$ *be a holomorphic function. Let* $(a, b) \in \Omega$*. Suppose that:*

$$\det\left(\frac{\partial g}{\partial w}(a, b)\right) \neq 0, \qquad g(a, b) = 0.$$

Then, there exists a neighborhood $U(a)$ *of* a *in* $\mathbb{C}^n$ *and a neighborhood* $V(b)$ *of* b *in* $\mathbb{C}^p$*, with* $U(a) \times V(b) \subset \Omega$*, such that there exists a holomorphic function* $f : U(a) \longrightarrow V(b)$*, with* $b = f(a)$ *and for all* $z \in U(a)$*,* $g(z, f(z)) = 0$*.*

PROOF.– Simply apply the theorem for real implicit functions to the application $g : \Omega \longrightarrow \mathbb{R}^{2n}$, $\Omega \subset \mathbb{R}^{2n} \times \mathbb{R}^{2p}$. Let $(z, w) \in \Omega$ and set

$$\Omega_z = \{w \in \mathbb{C}^p : (z, w) \in \Omega\}, \qquad \Omega_w = \{z \in \mathbb{C}^n (z, w) \in \Omega\}.$$

Hence, the partial maps:

$$g(z, .) : \Omega_z \longrightarrow \mathbb{R}^{2p}, \quad w \longmapsto g(z, w), \quad (z \text{ fixed}),$$

and

$$g(., w) : \Omega_w \longrightarrow \mathbb{R}^{2p}, \quad z \longmapsto g(z, w), \quad (w \text{ fixed}),$$

are $\mathbb{C}$-differentiable. By decomposing g and w into real and imaginary parts: $g = u + iv$, $w = x + iy$, we obtain (as in the proof of the previous theorem),

$$\det \frac{\partial(u, v)}{\partial(x, y)}(a, b) = \left|\det \frac{\partial g}{\partial w}\right|^2 \neq 0,$$

because by hypothesis $\det \dfrac{\partial g}{\partial w} \neq 0$. By theorem 1.11, there exists a neighborhood $U(a)$ of a in $\mathbb{C}^n$ and a neighborhood $V(b)$ of b in $\mathbb{C}^p$, $(a, b) \in U(a) \times V(b) \subset \Omega$ such that there exists a unique function $f : U(a) \longrightarrow V(b) \subset \mathbb{R}^{2p}$ of class C^1 such that for $z \in U(a)$, we have $g(z, f(z)) = 0$. It suffices to show that f is holomorphic in $U(a)$. Let $z = \alpha + i\beta$. Using the formula for differentiating a composite function, we obtain for all $z \in U(a)$,

$$\frac{dg}{dz} = -\left(\frac{\partial(u, v)}{\partial(x, y)}(z, f(z))\right)^{-1} o \left(\frac{\partial(u, v)}{\partial(\alpha, \beta)}(z, f(z))\right).$$

Since the partial maps $f(z, .)$ and $g(., w)$ are C-differentiable, then they satisfy the Cauchy–Riemann equations. From the above formula, we deduce that the map $f : U(a) \longrightarrow V(b) \subset \mathbb{R}^{2p}$ also satisfies the Cauchy–Riemann equations. Therefore, f is holomorphic. ■

Let $n, p \in \mathbb{N}^*$, Ω be an open set of $\mathbb{C}^n$ and $f : \Omega \longrightarrow \mathbb{C}^p$ be a holomorphic map. Recall that the rank of f at the point $a \in \Omega$ is the rank of the linear application $df(a)$. In other words, it is the rank of the Jacobian matrix of f at a, i.e.

$$rank_a f = rank \begin{pmatrix} \frac{\partial f_1}{\partial z_1}(a) & \dots & \frac{\partial f_1}{\partial z_n}(a) \\ \vdots & & \vdots \\ \frac{\partial f_p}{\partial z_1}(a) & \dots & \frac{\partial f_p}{\partial z_n}(a) \end{pmatrix}.$$

DEFINITION 1.11.– *We say that $a \in \Omega$ is a regular point of f if the rank of f in a is the maximum possible, i.e.* $\min(n, p)$. *Otherwise, we say that it is a critical point or singular point of f.*

PROPOSITION 1.13.– *Let $f : \mathbb{C}^n \longrightarrow \mathbb{C}^p$, $n \leq p$, be a holomorphic map and suppose that* 0 *is a regular point of f. If*

$$\det \begin{pmatrix} \frac{\partial f_{k+1}}{\partial w_1}(0) & \dots & \frac{\partial f_{k+1}}{\partial w_p}(0) \\ \vdots & & \vdots \\ \frac{\partial f_p}{\partial w_1}(0) & \dots & \frac{\partial f_p}{\partial w_n}(0) \end{pmatrix} \neq 0, \qquad k \equiv p - n,$$

then there exists a biholomorphic function $g : \mathbb{C}^p \longrightarrow \mathbb{C}^p$ such that:

$$(gof)(w_1, ..., w_n) = (0, ..., 0, w_1, ..., w_n).$$

PROOF – We identify the space $\mathbb{C}^n$ with $\{0\} \times \mathbb{C}^n \subset \mathbb{C}^p$ by identifying the coordinates $(w_1, ..., w_n)$ on $\mathbb{C}^n$ with $(0, z_{p-n+1}, ..., z_p)$, , $z_{p-n+j} = w_j$. Let $\varphi : \mathbb{C}^p \longrightarrow \mathbb{C}^p$ be the map defined by

$$\varphi(z) = f(z_{p-n+1}, ..., z_p) + (z_1, ..., z_{p-n}, 0).$$

Since φ is biholomorphic, it suffices to choose $g = \varphi^{-1}$. ■

PROPOSITION 1.14.– *Let $f : \mathbb{C}^n \longrightarrow \mathbb{C}^k$, $n \geq k$, be a holomorphic map and suppose that* 0 *is a regular point of f. If*

$$\det \begin{pmatrix} \frac{\partial f_1}{\partial z_1}(0) & \dots & \frac{\partial f_1}{\partial z_k}(0) \\ \vdots & & \vdots \\ \frac{\partial f_k}{\partial z_1}(0) & \dots & \frac{\partial f_k}{\partial z_k}(0) \end{pmatrix} \neq 0,$$

then there exists a biholomorphic function $g : \mathbb{C}^n \longrightarrow \mathbb{C}^n$ *such that:*

$$(fog)(z_1, ..., z_k, ..., z_n) = (z_1, ..., z_k).$$

PROOF.– Let us first consider the case where $n > k$ and let φ be the map defined by $\varphi : \mathbb{C}^n \longrightarrow \mathbb{C}^n$, $z \longmapsto (f(z), z_{k+1}, ..., z_n)$. (We can consider $(f_1, .., f_k, z_{k+1}, .., z_n)$ as a coordinate system on a neighborhood of 0 in $\mathbb{C}^n$.) The point 0 is a regular point of the function φ, and the latter is biholomorphic by virtue of the local inversion theorem. It is therefore sufficient to choose $g = \varphi^{-1}$ and write $\varphi^{-1} = (u, v)$ where $u : \mathbb{C}^n \longrightarrow \mathbb{C}^k$ and $v : \mathbb{C}^n \longrightarrow \mathbb{C}^{n-k}$. So, $f(u(z), v(z)) = (z_1, ..., z_k)$ and $b(z) = (z_{k+1}, ..., z_n)$. Thus, g is the desired map. Finally, if $n = k$ then the function f is obviously biholomorphic and it suffices to set $g = f^{-1}$. ■

See also the constant rank theorem stated and proven in the context of complex manifolds (section 2.4).

1.6. Exercises

EXERCISE 1.1.– *Let* $pr_k : \mathbb{C}^n \longrightarrow \mathbb{C}$, $(z_1, ..., z_n) \longmapsto z_k$ *be the projection. Show that* pr_k *is holomorphic and determine* $dpr_k(a)$ *for all* $a \in \mathbb{C}^n$.

Solution: Let e_k be the kth canonical basis vector. For $\varepsilon > 0$, $a \in \mathbb{C}^n$, there exists $\delta > 0$ such that:

$$\|z - a\| \leq \delta \Longrightarrow 0 = |pr_k(z) - pr_k(a) - (z - a|_k)| \leq \varepsilon\|z - a\| \leq \varepsilon\delta.$$

Hence, the function pr_k is holomorphic and $dpr_k(a) = e_k$.

EXERCISE 1.2.– *Show that the geometric series in several variables:* $\sum_{\alpha\in\mathbb{N}^n} z^\alpha$ *converges uniformly and absolutely on all compact subsets of the unit polycylinder* D.

Solution: We repeat the reasoning used in example 1.1. For $z \in D$, we have

$$\sum_{\alpha\in\mathbb{N}^n} z^\alpha = \sum_{k=0}^{\infty}\left(\sum_{|\alpha|=k} z^\alpha\right) = \sum_{k_1=0}^{\infty}\cdots\sum_{k_n=0}^{\infty} z_1^{\alpha_1}\cdots z_n^{\alpha_n},$$

hence,

$$\sum_{\alpha\in\mathbb{N}^n} z^\alpha = \left(\sum_{k_1=0}^{\infty} z_1^{\alpha_1}\right)\cdots\left(\sum_{k_n=0}^{\infty} z_n^{\alpha_n}\right) = \prod_{j=1}^{n}\frac{1}{1 - z_j} = \frac{1}{1 - z},$$

and the above holds for every z with $|z_j| < 1$. Any compact set on D is contained in a closed polydisk centered at 0 of radius $1 - \varepsilon$ (where $\varepsilon > 0$). Note that the same fact for each factor is true in one dimension. Hence, the series in question converges uniformly and absolutely on this closed polydisk.

EXERCISE 1.3.– *Let $f : \Omega \longrightarrow \mathbb{C}$ be a holomorphic function on an open set $\Omega \subset \mathbb{C}^n$. Show that:* $\dfrac{\partial f}{\partial z_j} = \dfrac{\partial f}{\partial x_j} = -i\dfrac{\partial f}{\partial y_j}$, *where* $z_j = x_j + iy_j$, $j = 1, ..., n$.

Solution: The partial (complex) derivative of f at $a = (a_1, ..., a_n) \in \mathbb{C}$ with respect to z_j, $j = 1, ..., n$, is given by

$$\frac{\partial f}{\partial z_j}(a) = \lim_{h \to 0} \frac{f(a_1, ..., a_j + h, ..., a_n) - f(a_1, ..., a_n)}{h},$$

where the limit is taken for all h in a perforated disk centered at the origin in $\mathbb{C}$. We obtain the same limit if we make $h = h_1 + ih_2$, $(h_1, h_2 \in \mathbb{R})$, tend towards 0 through real values only or imaginary values only. Hence, we have

$$\frac{\partial f}{\partial z_j}(a) = \lim_{h_1 \to 0} \frac{f(a_1, ..., a_j + h_1, ..., a_n) - f(a_1, ..., a_n)}{h_1} = \frac{\partial f}{\partial x_j}(a),$$

and

$$\frac{\partial f}{\partial z_j}(a) = \lim_{h_2 \to 0} \frac{f(a_1, ..., a_j + ih_2, ..., a_n) - f(a_1, ..., a_n)}{ih_2} = -i\frac{\partial f}{\partial y_j}(a).$$

EXERCISE 1.4.– *Let* $\sum_{\alpha \in \mathbb{N}^n} c_\alpha (z - a)^\alpha$, $c_\alpha \in \mathbb{C}$, *be a power series that converges for* $z = \zeta$. *We assume that* $\{|c_\alpha \zeta^\beta|\}$ *is bounded. Let* $D(a, r)$ *be an open polydisk of* $\mathbb{C}^n$ *with center* a *and radius* r *with* $r_j < |w_j - a_j|$. *Show that this series converges normally on* $\overline{D}(a, r)$. *What can we deduce from this for the case where* $n = 1$?

Solution: All we need to do is to repeat the proof of proposition 1.4. To have the same result for the case $n = 1$, it is sufficient that the series in question converges in $z = \zeta$.

EXERCISE 1.5.– *Show that the inversion of matrices:*

$$GL_n(\mathbb{C}) \longrightarrow GL_n(\mathbb{C}), \quad z \longmapsto z^{-1},$$

is a holomorphic mapping.

Hint: Use the well-known Cramer's rule.

EXERCISE 1.6.– *Let Ω_1 and Ω_2 be two domains of $\mathbb{C}^n$ such that: $\emptyset \neq \Omega_1 \subset \Omega_2$. Show that if a holomorphic function on Ω_1 admits an extension on Ω_2, then this extension is necessarily unique.*

Solution: This can be seen as follows. Suppose there are two holomorphic extensions g_1 and g_2 on Ω_2. Taking the identity principle (theorem 1.5) into account, we immediately deduce that $g_1 = g_2$ on Ω_2.

EXERCISE 1.7.– *Let $\Omega \subset \mathbb{C}^n$ be open and $f : \Omega \longrightarrow \mathbb{C}$ be a locally constant function. Show that f is holomorphic and determine $df(a)$ for all $a \in \Omega$.*

Solution: By hypothesis, the function f is locally constant. So for all $\varepsilon > 0$, there exists $\delta > 0$ such that for all $z \in \Omega$ and $a \in \Omega$, $f(z) = f(a)$, $\|z - a\| \leq \delta$, implies $0 = |f(z) - f(a)| \leq \varepsilon\|z - a\| \leq \varepsilon\delta$. Hence, the function f is holomorphic and $df(a) = 0$.

EXERCISE 1.8.– *Let $\Omega \subset \mathbb{C}^n$ be a connected open set and $U \subset \Omega$ be a non-empty open set. Show that if the holomorphic function f on Ω vanishes at a point of U as well as all its partial derivatives, then this function is identically zero on Ω. Deduce that if f and g are two holomorphic functions on Ω and if $f = g$ on U, then $f = g$ on Ω.*

Solution: Let V be the set of points where all partial derivatives of f are zero. By hypothesis, the function f is holomorphic, so (see proposition 1.6) it is developable in a convergent power series at $(z - a)$ in a polydisk $D(a, r) \subset \Omega$,

$$\forall z \in D(a,r), \quad f(z) = \sum_{\alpha \in \mathbb{N}^\alpha} c_\alpha (z - a)^\alpha,$$

where the coefficients are determined by $c_\alpha = \dfrac{\partial^\alpha f(a)}{\alpha!}$ (the point $a \in V$ being arbitrary). Since all partial derivatives of f vanish (by hypothesis) at a, then $c_\alpha = 0$ which implies that the power series above is zero. Consequently, the function f is identically zero in $D(a, r) \subset \Omega$ and hence V is open. Note that V is also closed in Ω because all partial derivatives of f are continuous. Therefore, V is an open and a closed set of Ω. Since Ω is connected, then $V = \Omega$ or $V = \emptyset$. Now, $U \subset V$; therefore, $V \neq \emptyset$ and the only valid case is $V = \Omega$. Regarding the second question, note that this is another form of the uniqueness theorem. This result follows immediately from the previous question. (As f and g are holomorphic, and therefore developable into convergent power series, we can reason using the coefficients of these series.)

EXERCISE 1.9.– *Let $\sum c_\alpha z^\alpha$, $\sum d_\alpha z^\alpha$ be two convergent power series in $\Omega \subset \mathbb{C}^n$. Show that if these series coincide in a domain of the neighborhood of $0 \in \Omega$, then $c_\alpha = d_\alpha$ for all α. (This result is often cited as the identity theorem for power series.)*

Solution: Let $\Delta = \{z \in \Omega : \sum c_\alpha z^\alpha = \sum d_\alpha z^\alpha$ on a neighborhood of $0 \in \Omega\}$. Note that Δ is an open subset of Ω and it is nonempty. By hypothesis, the series $\sum c_\alpha z^\alpha$ and $\sum d_\alpha z^\alpha$ converge in Ω to $f(z)$ and $g(z)$ respectively, and by proposition 1.6, f and g are holomorphic in Ω as well as on Δ. The coefficients c_α and d_α are determined explicitly in theorem 1.2 and taking into account corollary 1.1, the partial derivatives of f and g of any order are holomorphic on Ω. Still according to the same corollary, we deduce that we have for all α,

$$c_\alpha = \frac{\partial^\alpha f(0)}{\alpha!} = \frac{\partial^\alpha g(0)}{\alpha!} = d_\alpha.$$

EXERCISE 1.10.– *Show (Montel theorem) that if $\mathcal{F}$ is a bounded family of holomorphic functions on an open $\Omega \subset \mathbb{C}^n$, then every sequence of functions in $\mathcal{F}$ has a sub-sequence that converges uniformly on compact subsets.*

Hint: Recall that a sequence of functions (f_j) is bounded on Ω if there exists a number $M > 0$ such that $|f_j| \leq M$, for all j. The Montel theorem can be proved using the Arzela–Ascoli theorem, well known in topology: if K is a compact metric space and (f_j) is a sequence of complex-valued functions which is pointwise bounded and equicontinuous on K, then (f_j) is uniformly bounded on K and has a uniformly convergent sub-sequence.

EXERCISE 1.11.– *Let $\Omega \subset \mathbb{C}^n$ be open and $f : \Omega \longrightarrow \mathbb{C}^p$ be a biholomorphic function. Show that $n = p$. What can we deduce from this concerning the rank of $df(a)$ and* $\det df(a)$*?*

Solution: The derivatives $d(gof)$ and $d(fog)$ of the composites gof and fog are (see proposition 1.11):

$$d(gof)(z) = dg(f(z))of(z), \qquad d(fog)(z) = df(g(z))og(z).$$

Let $a \in \Omega$ and $b = f(a)$, we have $id_{\mathbb{C}^n} = d(gof)(a)$ and $id_{\mathbb{C}^p} = d(fog)(b)$. Let us reason by the absurd and assume that $n \neq p$, which implies that neither df nor dg would be of maximum rank. Without restricting generality, we assume that the rank of $df(a)$ is strictly less than n (this can obviously be done by taking into account the symmetry of the above equations), and there would therefore exist a non-zero vector $u \in \mathbb{C}^n$ with $df(a)u = 0$ which contradicts the fact that $id_{\mathbb{C}^n} = d(gof)(a)$. We deduce that $n = p$ and $\det df(a) \neq 0$ since the rank of $df(a)$ is maximal.

2

Complex Manifolds and Analytic Varieties

2.1. Preliminaries

Recall that a smooth manifold (also known as a differential manifold) is a space in which some neighborhood of every point is homeomorphic to an open subset of $\mathbb{R}^n$, such that the transitions between those open sets are given by smooth functions. Similarly, a complex manifold is a space in which some neighborhood of every point is homeomorphic to an open subset of $\mathbb{C}^n$, such that the transitions between those open sets are given by holomorphic functions. In other words, it is described as a space obtained by patching open sets in $\mathbb{C}^n$ together by biholomorphic maps. It is also described as a space that is locally the set of common zeros of holomorphic functions without singularities. Any complex (analytic) manifold of dimension n has a canonical structure of a real orientable differential manifold of dimension equal to $2n$. In rough terms, a manifold is a space that looks locally like a euclidean space. An algebraic variety can be defined similarly as a space that looks locally like the zero set of a collection of polynomials. An analytic variety in a complex manifold is a subset of the manifold that is the common zeros of holomorphic functions, possibly with singularities.

DEFINITION 2.1.– *Let M be a Hausdorff topological space with a countable basis. We say that M is a complex manifold of (complex) dimension n if there exists an open cover $\{U_\alpha\}$, indexed by a set I and for each $\alpha \in I$ a homeomorphism $\varphi_\alpha : U_\alpha \longrightarrow \mathbb{C}^n$, such that for all α, β in I with $U_\alpha \cap U_\beta \neq \emptyset$, the mapping*

$$\varphi_{\beta\alpha} \equiv \varphi_\beta o \varphi_\alpha^{-1} : \varphi_\alpha (U_\alpha \cap U_\beta) \longrightarrow \varphi_\beta (U_\alpha \cap U_\beta),$$

is a biholomorphism (i.e. $\varphi_{\beta\alpha}$ and its inverse $\varphi_{\beta\alpha}^{-1}$ are holomorphic). The pair $(U_\alpha, \varphi_\alpha)$ is called a chart on M.

In other words, a complex manifold M is a differentiable manifold admitting an open covering $\{U_\alpha\}$ and coordinate maps $\varphi_\alpha : U_\alpha \longrightarrow \mathbb{C}^n$ such that $\varphi_\beta o \varphi_\alpha^{-1}$ is holomorphic on $\varphi_\alpha (U_\alpha \cap U_\beta)$ for any pair α, β.

The sets $\varphi_\alpha (U_\alpha \cap U_\beta)$ and $\varphi_\beta (U_\alpha \cap U_\beta)$ are open in $\mathbb{C}^n$. We call $(U_\alpha, \varphi_\alpha)$ a (holomorphic) local coordinate system on M and for a point p in U_α, we call U_α a coordinate neighborhood of p. Note that $\varphi_\alpha(p)$ is a point of $\mathbb{C}^n$. Let us denote the ith coordinate of $\varphi(p)$ by $z_\alpha(p)$. Hence, we have

$$\varphi_\alpha(p) = \left(z_\alpha^1(p), \ldots, z_\alpha^n(p)\right).$$

This gives n functions (called local coordinates of p, with respect to φ_α):

$$z_\alpha^1 : U_\alpha \longrightarrow \mathbb{C}, p \longmapsto z_\alpha^1(p), \qquad z_\alpha^n : U_\alpha \longrightarrow \mathbb{C}, p \longmapsto z_\alpha^n(p).$$

Let $(U_\alpha, \varphi_\alpha) = \left(U_\alpha, z_\alpha^1, \ldots, z_\alpha^n\right)$ and $(U_\beta, \varphi_\beta) = \left(U_\beta, z_\beta^1, \ldots, z_\beta^n\right)$ be two maps on a manifold M of dimension n and suppose that: $U_\alpha \cap U_\beta \neq \emptyset$. Since the inverse map φ_α^{-1} of $\varphi_\alpha(U_\alpha \cap U_\beta)$ on $U_\alpha \cap U_\beta$ and the map φ_β of $U_\alpha \cap U_\beta$ on $\varphi_\beta(U_\alpha \cap U_\beta)$ are homeomorphisms, then the map

$$\begin{aligned} &\varphi_{\beta\alpha} \equiv \varphi_\beta o \varphi_\alpha^{-1} : \mathbb{C}^n \supset \varphi_\alpha (U_\alpha \cap U_\beta) \longrightarrow \varphi_\beta (U_\alpha \cap U_\beta) \subset \mathbb{C}^n, \\ &u \longmapsto \varphi_{\beta\alpha}(u) = \varphi_\beta o \varphi_\alpha^{-1}(u), \end{aligned}$$

is a homeomorphism (as composed of two homeomorphisms φ_β and φ_α^{-1}). We pass from one map to the other by means of this application. Let

$$\begin{aligned} &u = \left(u^1, ..., u^n\right), \quad \varphi_{\beta\alpha}(u) = \left(\varphi_{\beta\alpha}^1 \left(u^1, ..., u^n\right), ..., \varphi_{\beta\alpha}^n \left(u^1, ..., u^n\right)\right), \\ &p \in U_\alpha \cap U_\beta, \quad \varphi_\alpha(p) = \left(z_\alpha^1(p), \ldots, z_\alpha^n(p)\right), \quad \varphi_\beta(p) = \left(z_\beta^1(p), \ldots, z_\beta^n(p)\right). \end{aligned}$$

Note that: $\varphi_\alpha(p) \in \varphi_\alpha (U_\alpha \cap U_\beta)$, so we can set $u = \varphi_\alpha(p)$ in the expression above. We obtain

$$\varphi_{\beta\alpha} (\varphi_\alpha(p)) = \varphi_\beta(p),$$

i.e.

$$\left(\varphi_{\beta\alpha}^1 \left(z_\alpha^1(p), \ldots, z_\alpha^n(p)\right), ..., \varphi_{\beta\alpha}^n \left(z_\alpha^1(p), \ldots, z_\alpha^n(p)\right)\right) = \left(z_\beta^1(p), \ldots, z_\beta^n(p)\right).$$

Hence,

$$z_\beta^j(p) = \varphi_{\beta\alpha}^j \left(z_\alpha^1(p), \ldots, z_\alpha^n(p)\right), \quad 1 \leq j \leq n, \quad p \in U_\alpha \cap U_\beta,$$

i.e.

$$z_\beta^j = \varphi_{\beta\alpha}^j \left(z_\alpha^1, \ldots, z_\alpha^n\right), 1 \leq j \leq n.$$

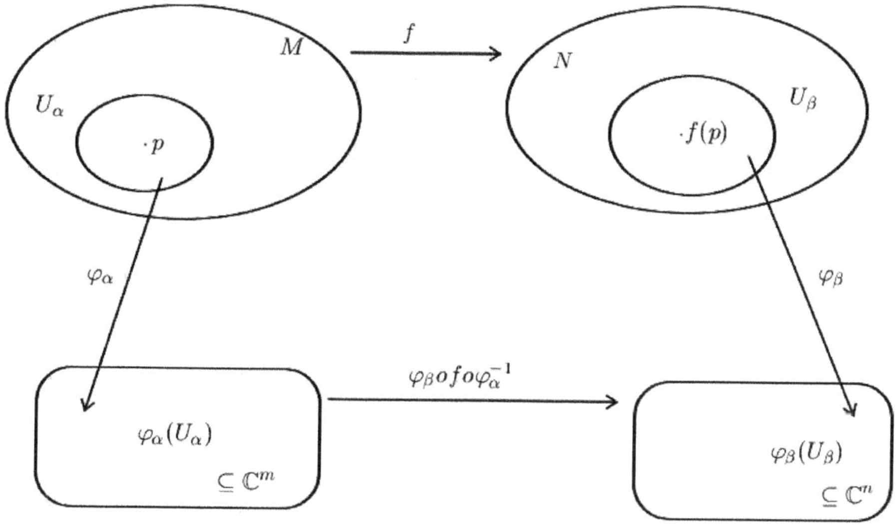

Figure 2.1. *Holomorphic map*

The map $\varphi_{\beta\alpha}$ is called a transition function or coordinate change. It is also said that the functions $\varphi_{\beta\alpha}^j$, $1 \leq j \leq n$, are called a change of charts (on $U_\alpha \cap U_\beta$) from coordinates $z_\alpha^1, \ldots, z_\alpha^n$ to coordinates $z_\beta^1, \ldots, z_\beta^n$. The above formulas are called a change of chart formulas or transition formulas.

DEFINITION 2.2.– *Let M, N be two complex manifolds of dimensions m and n respectively and $f : M \longrightarrow N$ be a continuous map. We say that f is holomorphic if for every chart $(U_\alpha, \varphi_\alpha)$ of M and every chart (U_β, φ_β) of N such that $f(U_\alpha) \subset U_\beta$, the map*

$$\varphi_\beta o \left(f\right)|_{U_\alpha} o \varphi_\alpha^{-1} : \varphi_\alpha\left(U_\alpha\right) \longrightarrow \varphi_\beta\left(U_\beta\right),$$

is holomorphic. A biholomorphic map is a bijective holomorphic map f such that f^{-1} is also holomorphic.

DEFINITION 2.3.– *Two charts $(U_\alpha, \varphi_\alpha)$ and (U_β, φ_β) are said to be compatible if either they are disjoint (i.e. $U_\alpha \cap U_\beta = \varnothing$), or the followings maps (defined if $U_\alpha \cap U_\beta \neq \varnothing$),*

$$\varphi_{\beta\alpha} : \varphi_\alpha\left(U_\alpha \cap U_\beta\right) \rightarrow \varphi_\beta\left(U_\alpha \cap U_\beta\right),\ u \mapsto \varphi_{\beta\alpha}(u) = \varphi_\beta o \varphi_\alpha^{-1}(u),$$

$$\varphi_{\alpha\beta} : \varphi_\beta\left(U_\alpha \cap U_\beta\right) \rightarrow \varphi_\alpha\left(U_\alpha \cap U_\beta\right),\ u \mapsto \varphi_{\alpha\beta}(u) = \varphi_\alpha o \varphi_\beta^{-1}(u),$$

are biholomorphic. An atlas of the manifold M is a set of charts $(U_\alpha, \varphi_\alpha)$ of M which are pairwise compatible and whose domains U_α cover the whole manifold M.

The compatibility relation introduced above is an equivalence relation for atlases. Two atlases on M are equivalent if their union is an atlas. A complex (analytic) manifold is a topological manifold together with a complex analytic structure, i.e. with any class of equivalent atlases. The manifold M is called connected or compact if the underlying topological space is connected or compact.

On a connected compact complex manifold, holomorphic functions are constant according to the maximum principle. A complex manifold is an even dimensional orientable differential manifold. However, for a given even dimensional oriented manifold, it is not always clear whether or not we can make it a complex manifold.

Let $(U_\alpha, \varphi_\alpha) = \left(U_\alpha, z^1_\alpha, \ldots, z^m_\alpha\right)$, $(U_\beta, \varphi_\beta) = \left(U_\beta, z^1_\beta, \ldots, z^n_\beta\right)$. The map $\varphi_\beta o\,(f)|_{U_\alpha} o\varphi_\alpha^{-1}$ is defined by n functions:

$$z^j_\beta = f^j\left(z^1_\alpha, \ldots, z^m_\alpha\right), \quad 1 \leq j \leq n.$$

This expression is said to be the local expression of f in the charts $(U_\alpha, \varphi_\alpha)$ and (U_β, φ_β).

PROPOSITION 2.1.– *Let M and N be two complex manifolds of dimensions m and n, respectively. Then, the product $M \times N = \{(p, q) : p \in M, q \in N\}$ is a complex manifold of dimension $m + n$.*

PROOF.– Let $(U_\alpha, \varphi_\alpha)$ be an atlas of M and (V_β, ψ_β) be an atlas of N. We know how to equip the product $M \times N$ of the topological spaces M and N with a topology and that $M \times N$ is a separated space with a countable basis. Let $(U_\alpha \times V_\beta)$ be an open covering of $M \times N$. The map

$$\varphi_\alpha \times \psi_\beta : U_\alpha \times V_\beta \longrightarrow \mathbb{C}^{m+n}, \quad (p, q) \longmapsto (\varphi_\alpha(p), \psi_\beta(q)).$$

is a homeomorphism from $U_\alpha \times V_\beta$ to an open set $\varphi_\alpha(U_\alpha) \times \psi_\beta(V_\beta)$ of $\mathbb{C}^{m+n}$ (note that $\varphi_\alpha(U_\alpha) \times \psi_\beta(V_\beta) = (\varphi_\alpha \times \psi_\beta)(U_\alpha \times V_\beta)$ is an open set because $\varphi_\alpha(U_\alpha)$ and $\psi_\beta(V_\beta)$ are too). So, $M \times N$ is a (topological) manifold of dimension $m + n$. We can easily show that $M \times N$ is a complex manifold, and we call it the product of the manifolds M and N. Let $(U_\alpha, \varphi_\alpha)$ and (U_β, φ_β) be compatible charts on M, i.e. $U_\alpha \cap U_\beta = \emptyset$ or else the maps (defined if $U_\alpha \cap U_\beta \neq \emptyset$),

$$\varphi_{\beta\alpha} : \varphi_\alpha(U_\alpha \cap U_\beta) \longrightarrow \varphi_\beta(U_\alpha \cap U_\beta), \quad u \longmapsto \varphi_{\beta\alpha}(u) = \varphi_\beta o \varphi_\alpha^{-1}(u),$$

$$\varphi_{\alpha\beta} : \varphi_\beta(U_\alpha \cap U_\beta) \longrightarrow \varphi_\alpha(U_\alpha \cap U_\beta), \quad u \longmapsto \varphi_{\alpha\beta}(u) = \varphi_\alpha o \varphi_\beta^{-1}(u),$$

are diffeomorphisms. Similarly, let (V_α, ψ_α) and (V_β, ψ_β) be compatible charts on N, i.e. $V_\alpha \cap V_\beta = \emptyset$ or the maps (defined if $V_\alpha \cap V_\beta \neq \emptyset$),

$$\psi_{\beta\alpha} : \psi_\alpha(V_\alpha \cap V_\beta) \longrightarrow \psi_\beta(V_\alpha \cap V_\beta), \quad u \longmapsto \psi_{\beta\alpha}(u) = \psi_\beta o \psi_\alpha^{-1}(u),$$

$$\psi_{\alpha\beta} : \psi_\beta(V_\alpha \cap V_\beta) \longrightarrow \psi_\alpha(V_\alpha \cap V_\beta), \quad u \longmapsto \psi_{\alpha\beta}(u) = \psi_\alpha o \psi_\beta^{-1}(u),$$

are diffeomorphisms. Let us see the case where $U_\alpha \cap U_\beta = \emptyset$ and/or $V_\alpha \cap V_\beta = \emptyset$. If $U_\alpha \cap U_\beta = V_\alpha \cap V_\beta = \emptyset$, then

$$(U_\alpha \times V_\alpha) \cap (U_\beta \times V_\beta) = (U_\alpha \cap U_\beta) \times (V_\alpha \cap V_\beta) = \emptyset.$$

For other cases, i.e. $U_\alpha \cap U_\beta = \emptyset$ or $V_\alpha \cap V_\beta = \emptyset$, we agree that $(U_\alpha \cap U_\beta) \times (V_\alpha \cap V_\beta) = \emptyset$. Now suppose that $U_\alpha \cap U_\beta \neq \emptyset$ and $V_\alpha \cap V_\beta \neq \emptyset$. The sets $\varphi_\alpha \times \psi_\alpha[(U_\alpha \times V_\alpha) \cap (U_\beta \times V_\beta)]$ and $\varphi_\beta \times \psi_\beta[(U_\alpha \times V_\alpha) \cap (U_\beta \times V_\beta)]$ are open sets of $\mathbb{C}^{m+n}$ (indeed, we have

$$\begin{aligned}\varphi_\alpha \times \psi_\alpha[(U_\alpha \times V_\alpha) \cap (U_\beta \times V_\beta)] &= \varphi_\alpha \times \psi_\alpha[(U_\alpha \cap U_\beta) \times \psi_\alpha(V_\alpha \cap V_\beta)], \\ &= \varphi_\alpha(U_\alpha \cap U_\beta) \times \psi_\alpha(V_\alpha \cap V_\beta),\end{aligned}$$

and since $\varphi_\alpha(U_\alpha \cap U_\beta)$ is an open set of $\mathbb{C}^m$ and $\psi_\alpha(V_\alpha \cap V_\beta)$ is an open set of $\mathbb{C}^n$, the result follows. The same thing for $\varphi_\beta \times \psi_\beta[(U_\alpha \times V_\alpha) \cap (U_\beta \times V_\beta)]$). The map

$$(\varphi_\beta \times \psi_\beta) o (\varphi_\alpha \times \psi_\alpha)^{-1} : \varphi_\alpha \times \psi_\alpha[(U_\alpha \times V_\alpha) \cap (U_\beta \times V_\beta)]$$

$$\longrightarrow \varphi_\beta \times \psi_\beta[(U_\alpha \times V_\alpha) \cap (U_\beta \times V_\beta)]$$

is a diffeomorphism. Indeed, we have

$$(\varphi_\beta \times \psi_\beta) o (\varphi_\alpha \times \psi_\alpha)^{-1} = (\varphi_\beta o \varphi_\alpha^{-1}) \times (\psi_\beta o \psi_\alpha^{-1}) = \varphi_{\beta\alpha} \times \psi_{\beta\alpha},$$

and the result follows from the fact that $\varphi_{\beta\alpha}$ and $\psi_{\beta\alpha}$ are diffeomorphisms. The same argument is made for the map $(\varphi_\alpha \times \psi_\alpha) o (\varphi_\beta \times \psi_\beta)^{-1}$. In conclusion, the charts $(U_\alpha \times V_\alpha, \varphi_\alpha \times \psi_\alpha)$ and $(U_\beta \times V_\beta, \varphi_\beta \times \psi_\beta)$ on $M \times N$ are also compatible and $M \times N$ is a complex manifold of dimension $m + n$. ■

DEFINITION 2.4.– *Let $f : M \longrightarrow N$ be a holomorphic map, and let m and n be dimensions of M and N, respectively. (i) The map f is an immersion if $m \leq n$ and every point of M is a regular point of f. (ii) The map f is an embedding if, furthermore, it is a homeomorphism onto $f(M)$ with the topology induced from that of N. (iii) The map f is a submersion if $m \geq n$ and every point of M is a regular point of f. (iv) The map f is a covering map if it is surjective and if each point of N has a neighborhood U such that each connected component of $f^{-1}(U)$ is mapped biholomorphically onto U by f. (In this case, $m = n$ and f is an immersion and a submersion.)*

For a holomorphic map $f : M \longrightarrow N$, we may define regular and singular points of f (see also definition 1.11), choosing local coordinate systems, and the definition does not depend on the choice of the coordinate systems. Any map $f : M \longrightarrow \{p\}$ of a complex manifold M is holomorphic, and every point in M is a regular point of f (see section 2.2.1).

An analytic variety in a complex manifold is a subset of the manifold that consists of the common zeros of holomorphic functions possibly with singularities. More precisely, an analytic variety in a complex manifold M is a subset V of M such that every point a in M admits a neighborhood U such that there exists a holomorphic map $f : U \longrightarrow \mathbb{C}^n$ with $V \cap U = f^{-1}(0)$. In other words, we have the following definition:

DEFINITION 2.5.– *A subset V of an open set $U \subset \mathbb{C}^n$ is called an analytic variety in U if, for any $p \in U$, there exists a neighborhood Ω of p in U such that $V \cap \Omega$ is the common zero locus of a finite collection of holomorphic functions $f_1, ..., f_n$ on Ω. In particular, V is called an analytic hypersurface if V is locally the zero locus of a single nonzero holomorphic function f.*

A complex analytic variety is a generalization of a complex manifold and more general than analytic manifolds in allowing singularities. Sometimes, instead of analytic variety, we will say (if there is no confusion) simply a variety. Note that a variety in M is a closed set in a complex manifold M. If every point a in a subset V of M admits a neighborhood U such that there exists a holomorphic map $f : U \longrightarrow \mathbb{C}^n$ with $V \cap U = f^{-1}(0)$, then V is a variety in some open set in M, and in particular, if V is a closed set in M, it is a variety in M. Furthermore, the set of regular points of a variety V is an open set in V (this set has a natural structure of complex submanifold of M, defined in section 2.5), and the set of singular points of V is a closed set in V and in M. The set of singular points of V is an analytic variety, and when this set is empty, we will say that the variety V is non-singular. It should be noted that the word variety (in French) and manifold (in English) refer to geometric spaces, but they are not exactly equivalent. In French (and other languages), the term variety can refer to a differentiable variety or an algebraic variety, whereas in English, manifold generally refers to a differentiable variety.

2.2. Examples of complex manifolds

2.2.1. $\mathbb{C}^n$ *and open subsets*

The space $\mathbb{C}^n$ is a complex manifold of dimension n. Indeed, $(\mathbb{C}^n, id.)$ is a chart on $\mathbb{C}^n$ where $id. : \mathbb{C}^n \longrightarrow \mathbb{C}^n$, and $\{(\mathbb{C}^n, id.)\}$ is an atlas on $\mathbb{C}^n$ (which is contained in a unique maximal atlas). In particular, the set $\{p\}$ consisting of a point p is a connected complex manifold of dimension 0, with a global chart homeomorphic to $\mathbb{C}^0 = \{0\}$. Similarly, any domain of $\mathbb{C}^n$ is a complex manifold of dimension n.

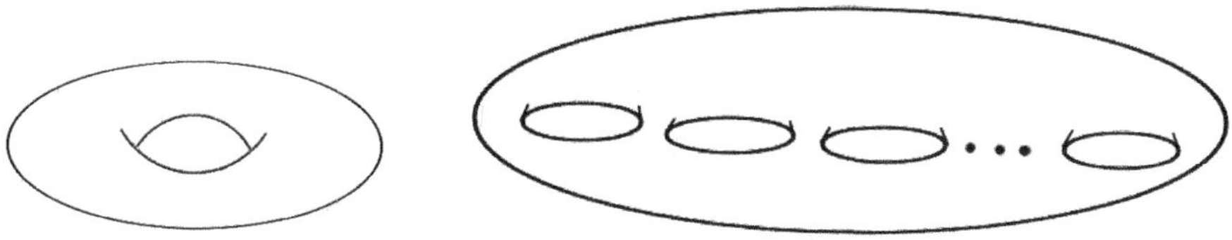

Figure 2.2. *Elliptic and hyperelliptic curve*

2.2.2. *Complex algebraic curves or compact Riemann surfaces*

Recall that complex algebraic curves C or compact Riemann surfaces which we also note X are one-dimensional complex manifolds equipped with atlases whose coordinate changes are holomorphic maps. The locus of zeros in $\mathbb{C}\times\mathbb{C}$ of a polynomial equation $P(w,z) = 0$ is a Riemann surface. We consider the equation:

$$P(w,z) \equiv p_0(z)w^n + p_1(z)w^{n-1} + \cdots + p_n(z) = 0,$$

where $P(w,z) \in \mathbb{C}[w,z]$ is a polynomial with two complex variables w and z, of degree n in w and irreducible, and $p_0(z) \neq 0$, $p_1(z)$,...,$p_n(z)$ are polynomials in z. The Riemann surface $X = \{(w,z) : P(w,z) = 0\}$ is homeomorphic to a torus with g holes. The number g is called the genus of X. An important particular case is represented by the elliptic curves ($g = 1$) and hyperelliptic curves of genus g whose affine equation is $P(w,z) = w^2 - p_n(z) = 0$, where $p_n(z)$ is a polynomial without multiple roots, of degree $n = 2g+1$ or $2g+2$. If $gradP \equiv \left(\frac{\partial P}{\partial w}, \frac{\partial P}{\partial z}\right) \neq 0$, then the complex algebraic curve C is non-singular. For a detailed study of compact Riemann surfaces or complex algebraic curves, see, for example, Lesfari (2023).

2.2.3. *Spheres*

Let

$$S^2 = \{(\xi,\zeta,\eta) \in \mathbb{C}^3 : \xi^2 + \zeta^2 + \eta^2 = 1\},$$

be the Riemann sphere. Study S^2 using the notion of stereographic projection and deduce that it is a complex manifold. Let us look imaginatively at the special case of the Riemann sphere S^2 (unit sphere of the affine space $\mathbb{R}^3$), using the notion of stereographic projection. Let $p_0 = (0,0,1)$ be a point (north pole) of the sphere

$$S^2 = \{(\xi,\zeta,\eta) \in \mathbb{R}^3 : \xi^2 + \zeta^2 + \eta^2 = 1\},$$

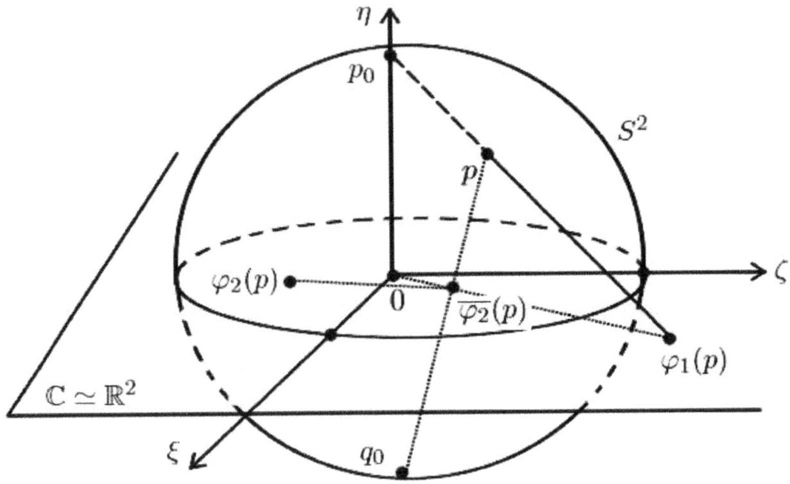

Figure 2.3. *Stereographic projection*

and $q_0 = (0, 0, -1)$ be a point of the sphere S^2 distinct from the points p_0 and q_0. Consider the charts (U_1, φ_1) and (U_2, φ_2) where

$$\varphi_1 : \mathcal{U}_1 = S^2 \backslash \{p_0\} \longrightarrow \mathbb{C}, \quad (\xi, \zeta, \eta) \longmapsto \frac{\xi + i\zeta}{1 - \eta}, \quad \eta < 1,$$

$$\varphi_2 : \mathcal{U}_2 = S^2 \backslash \{q_0\} \longrightarrow \mathbb{C}, \quad (\xi, \zeta, \eta) \longmapsto \frac{\xi - i\zeta}{1 + \eta}, \quad \eta > -1.$$

The function φ_1 (respectively φ_2) is the stereographic projection relative to the north pole p_0 (respectively south pole q_0). (Note that φ_2 is the composite of the stereographic projection relative to the south pole q_0 followed by the symmetry $(\xi, \zeta, \eta) \longmapsto (\xi, -\zeta, \eta)$ which, in the figure above, is the conjugation in the equatorial plane identified with $\mathbb{C} \simeq \mathbb{R}^2$.).The maps φ_1 and φ_2 are homeomorphisms. Note that $\mathcal{U}_1, \mathcal{U}_2$ are connected and that $\mathcal{U}_1 \cap \mathcal{U}_2 = S^2 \backslash \{p_0, q_0\} \neq \emptyset$, so, S^2 is also connected and $\varphi_1(\mathcal{U}_1 \cap \mathcal{U}_2) = \varphi_2(\mathcal{U}_1 \cap \mathcal{U}_2) = \mathbb{C} \backslash \{0\}$. These sets are open, and we show that

$$\varphi_{21} = \varphi_2 o \varphi_1^{-1} : \mathbb{C} \backslash \{0\} \longrightarrow \mathbb{C} \backslash \{0\}, \quad z \longmapsto \frac{1}{z},$$

$$\varphi_{12} = \varphi_1 o \varphi_2^{-1} : \mathbb{C} \backslash \{0\} \longrightarrow \mathbb{C} \backslash \{0\}, \quad z \longmapsto \frac{1}{z}.$$

Indeed, to determine the change of charts, for example, φ_{21}, we proceed as follows: let $z = x + iy \in \mathbb{C}$ and determine $\varphi_1^{-1}(z)$.

$$z = x + iy = \frac{\xi + i\zeta}{1 - \eta} \Longleftrightarrow \xi = x(1 - \eta) \text{ and } \zeta = y(1 - \eta).$$

Or

$$1 = \xi^2 + \zeta^2 + \eta^2 = (1 - \eta)^2(x^2 + y^2) + \eta^2 = (1 - \eta)^2|z|^2 + \eta^2,$$

i.e.

$$(1 + |z|^2)\eta^2 - 2|z|^2\eta + |z|^2 - 1 = 0,$$

hence

$$\eta = \frac{|z|^2 \pm 1}{1 + |z|^2} = \begin{cases} 1 \text{ to be rejected because this solution corresponds} \\ \text{to the excluded point } p_0. \\ \dfrac{|z|^2 - 1}{1 + |z|^2} \text{ acceptable.} \end{cases}$$

So, the only valid case is $\eta = \dfrac{|z|^2 - 1}{1 + |z|^2}$. Therefore,

$$\xi = x\left(1 - \frac{|z|^2 - 1}{1 + |z|^2}\right) = \frac{2x}{1 + |z|^2}, \qquad \zeta = y\left(1 - \frac{|z|^2 - 1}{1 + |z|^2}\right) = \frac{2y}{1 + |z|^2},$$

and so

$$\varphi_1^{-1}(z) = \begin{pmatrix} \xi \\ \zeta \\ \eta \end{pmatrix} = \begin{pmatrix} \frac{2x}{1+|z|^2} \\ \frac{2y}{1+|z|^2} \\ \frac{|z|^2 - 1}{1+|z|^2} \end{pmatrix}.$$

Consequently, we have

$$\varphi_{21}(z) = \varphi_2 o \varphi_1^{-1}(z) = \frac{\xi - i\zeta}{1 + \eta}.$$

Replacing ξ and ζ with their expressions obtained above and noting that

$$1 + \eta = 1 + \frac{|z|^2 - 1}{1 + |z|^2} = \frac{2|z|^2}{1 + |z|^2},$$

we obtain

$$\varphi_{21}(z) = \frac{x - iy}{|z|^2} = \frac{\overline{z}}{z\overline{z}} = \frac{1}{z}.$$

The same reasoning applies to

$$\varphi_{12}(z) = \varphi_1 o \varphi_2^{-1}(z) = \frac{1}{z}.$$

Obviously, the maps φ_{21} and φ_{12} are biholomorphic. The maps $(\mathcal{U}_1, \varphi_1)$, $(\mathcal{U}_2, \varphi_2)$ are thus compatible and since $\mathcal{U}_1 \cup \mathcal{U}_2 = S^2$, they form an atlas. Therefore, S^2 is a complex manifold.

Study the sphere

$$S^n = \{(z_1, ..., z_{n+1}) \in \mathbb{C}^{n+1} : z_1^2 + \cdots + z_{n+1}^2 = 1\},$$

of dimension n. The atlas is made up of two maps (U_1, φ_1), (U_2, φ_2) with $U_1 = S^n \backslash \{p_0\}$, $U_2 = S^n \backslash \{q_0\}$, where $p_0 = (0, ..., 0, 1)$, $q_0 = (0, ..., 0, -1)$, while φ_1, φ_2 are the maps $\varphi_1 : U_1 \longrightarrow \mathbb{C}^n$, $\varphi_2 : U_2 \longrightarrow \mathbb{C}^n$ defined by the stereographic projection from the points p_0 and q_0,

$$\varphi_1(z_1, ..., z_{n+1}) = \left(\frac{z_1}{1 - z_{n+1}}, ..., \frac{z_n}{1 - z_{n+1}}\right),$$

$$\varphi_2(z_1, ..., z_{n+1}) = \left(\frac{z_1}{1 + z_{n+1}}, ..., \frac{z_n}{1 + z_{n+1}}\right),$$

respectively. Does the sphere S^n have a complex manifold structure?

Note: Recall that an n-dimensional complex manifold, which is also called a holomorphic manifold, is a topological manifold of dimension $2n$ equipped with an atlas whose transition functions are all holomorphic. A complex manifold of dimension 1 is called a complex curve, and one of dimension 2 is called a complex surface. A complex manifold of dimension 3 or higher is sometimes called a threefold, fourfold, etc., complex. Complex manifolds are defined in the same way as differentiable manifolds, with the difference that the local coordinate charts must take their values in $\mathbb{C}^n$ and overlap holomorphically. This may seem like a minor change to the definition of differentiable manifolds, but in reality, the holomorphicity requirement changes everything. For example, on a connected compact complex manifold, the only global holomorphic functions are constants, and the space of holomorphic sections of a holomorphic fibered vector is always finite dimensional. While any differentiable manifold can be smoothly immersed in a euclidean space, only certain complex manifolds can be holomorphically immersed in $\mathbb{C}^n$ or in a complex projective space $\mathbb{P}^n(\mathbb{C})$. While some classes of manifolds are really just slight variations on the theme of differentiable manifolds, it turns out that nearly everything changes when we move into the holomorphic category. Since any complex (analytic) manifold of dimension n has a canonical structure of a real orientable differential manifold of dimension equal to $2n$, when it is necessary to distinguish between the dimension of an n-dimensional complex manifold and the dimension of its underlying topological $2n$-manifold, we can possibly specify that n is the complex dimension and $2n$ is the real dimension. It should be noted that the same topological space can be equipped with several different differentiable structures. For example, for the canonical topology, the sphere S^7 of dimension 7 can be equipped with 28 different differentiable structures (these 28 differentiable manifolds are homeomorphic to S^7 but not diffeomorphic to each other. They are called Milnor spheres), S^8 with two differentiable structures and S^{31} with more than 16 millions. On the other hand, $\mathbb{R}^4$, S^1, S^2, S^4 and S^6 have a unique differentiable structure. For S^3, the problem is related to the Poincaré conjecture (any compact, simply connected three-dimensional manifold is homeomorphic to a three-dimensional hypersurface), which was solved in 2003 by Perelman.

2.2.4. *Projective spaces*

DEFINITION 2.6.– *A line is a complex linear subspace of $\mathbb{C}^{n+1}$ of complex dimension one. The complex projective n-space,*

$$\mathbb{P}^n(\mathbb{C}) = \left\{[Z_0 : ... : Z_n] : (Z_0, ..., Z_n) \in \mathbb{C}^{n+1}\backslash\{0\}\right\},$$

is defined to be the set of complex vector lines passing through the origin with coordinates in $\mathbb{C}^{n+1}$, where $[Z_0 : ... : Z_n]$ denotes the line generated by the vector.

We will see that the complex projective space $\mathbb{P}^n(\mathbb{C})$ is a complex analytic manifold. The topology on the space $\mathbb{P}^n(\mathbb{C})$ is the quotient topology inherited from the natural projection

$$\mathbb{C}^{n+1}\backslash\{0\} \longrightarrow \mathbb{P}^n(\mathbb{C}), \quad (Z_0, ..., Z_n) \longmapsto [Z_0 : ... : Z_n].$$

The topological space $\mathbb{P}^n(\mathbb{C})$ is separable and compact. Furthermore, it is covered by $n+1$ open $\mathcal{U}_0, ..., \mathcal{U}_n$ where

$$\mathcal{U}_j = \{[Z_0 : ... : Z_n] \in \mathbb{P}^n(\mathbb{C}) : Z_j \neq 0\},$$

is the space of lines for which $Z_j \neq 0$ (i.e. the set of all lines not contained in the complex hyperplane $\{Z_j = 0\}$). Consider the map (local coordinate in $\mathcal{U}_j$, $j = 0, ..., n$),

$$\varphi_j : \mathcal{U}_j \longrightarrow \mathbb{C}^n, [Z_0 : ... : Z_n] \longmapsto \left(\frac{Z_0}{Z_j}, ..., \frac{Z_{j-1}}{Z_j}, \frac{Z_{j+1}}{Z_j}, ..., \frac{Z_n}{Z_j}\right) \equiv (z_1, ..., z_n),$$

with

$$z_k = \begin{cases} \frac{Z_{k-1}}{Z_j} \text{ if } k \leq j \\ \frac{Z_k}{Z_j} \text{ if } k > j \end{cases}$$

where $1 \leq k \leq n$. It is obvious that $\varphi_j(\mathcal{U}_j) = \mathbb{C}^n$. The map φ_j is a homeomorphism whose inverse is given by

$$(z_1, ..., z_n) \longmapsto [z_1, ...z_j, 1, z_{j+1}, ..., z_n].$$

So, the space $\mathbb{P}^n(\mathbb{C})$ is a topological manifold of dimension n and the pair $(\mathcal{U}_j, \varphi_j)$, $0 \leq j \leq n$, is a chart on $\mathbb{P}^n(\mathbb{C})$. As

$$\mathcal{U}_j \cap \mathcal{U}_k = \{[Z_0 : ... : Z_n] \in \mathbb{P}^n(\mathbb{C}) : Z_j \neq 0 \text{ and } Z_k \neq 0\}, \quad j \neq k$$

then (for $k > j$),

$$\varphi_j(\mathcal{U}_j \cap \mathcal{U}_k) = \{(z_1, ..., z_n) \in \mathbb{C}^n : z_k \neq 0\},$$
$$\varphi_k(\mathcal{U}_j \cap \mathcal{U}_k) = \{(z_1, ..., z_n) \in \mathbb{C}^n : z_{j+1} \neq 0\}.$$

The transition maps are defined by $\varphi_{kj} \equiv \varphi_k o \varphi_j^{-1}$ with

$$\varphi_{kj} : \varphi_j(\mathcal{U}_j \cap \mathcal{U}_k) \longrightarrow \varphi_k(\mathcal{U}_j \cap \mathcal{U}_k), (z_1, ..., z_n) \longmapsto \left(\frac{z_1}{z_k}, ..., \widehat{\frac{z_k}{z_k}}, ..., \frac{1}{z_k}, ..., \frac{z_n}{z_k}\right),$$

where $z_k \neq 0$ and the sign $\widehat{}$ means to omit. For $\varphi_{jk} \equiv \varphi_j o \varphi_k^{-1}$, we just need to swap the indices. We can see that the maps φ_{kj} and φ_{jk} are analytic. The charts $(\mathcal{U}_j, \varphi_j)$, $(\mathcal{U}_k, \varphi_k)$ are therefore pairwise compatible and since $\bigcup_{j=0}^{n} \mathcal{U}_j = \mathbb{P}^n(\mathbb{C})$, they form an atlas. Thus, $\mathbb{P}^n(\mathbb{C})$ becomes a complex analytic manifold.

For another description of the complex projective space $\mathbb{P}^n(\mathbb{C})$, we have the following definition:

DEFINITION 2.7.– *The complex projective space is defined as the quotient set of* $\mathbb{C}^{n+1} \backslash \{0\}$ *by the equivalence relation*

$$[Z_0 : ... : Z_n] \sim [\lambda Z_0 : ... : \lambda Z_n], \quad \lambda \in \mathbb{C}^*.$$

Namely, two points of $\mathbb{C}^{n+1} \backslash \{0\}$ *are equivalent if and only if they are complex linearly dependent, i.e. lie on the same line. The space* $\mathbb{P}^n(\mathbb{C})$ *is also written*

$$\mathbb{P}^n(\mathbb{C}) = \frac{\{[Z] \neq 0 \in \mathbb{C}^{n+1}\}}{[Z] \sim [\lambda Z]}.$$

We denote by $[Z_0 : ... : Z_n]$ *the equivalence class of* $(Z_0, ..., Z_n)$ *and we say that* $Z_0, ..., Z_n$ *are the homogeneous coordinates of the corresponding point of the space* $\mathbb{P}^n(\mathbb{C})$.

Instead of considering the complex projective space $\mathbb{P}^n(\mathbb{C})$ as the quotient set of $\mathbb{C}^{n+1}\backslash\{0\}$ by the above equivalence relation, we may also view it as a compactification of $\mathbb{C}^n$. Let

$$H_j = \{[Z_0 : \dots : Z_n] \in \mathbb{P}^n(\mathbb{C}) : Z_j = 0\}, \quad 0 \leq j \leq n,$$

be the hyperplane of $\mathbb{P}^n(\mathbb{C})$. It is a projective subspace of dimension $n-1$. The map

$$\varphi_j : \mathbb{P}^n(\mathbb{C})\backslash H_j \longrightarrow \mathbb{C}^n, \quad [Z_0 : \dots : Z_n] \longmapsto \left(\frac{Z_0}{Z_j}, \dots, \frac{Z_{j-1}}{Z_j}, \frac{Z_{j+1}}{Z_j}, \dots, \frac{Z_n}{Z_j}\right),$$

shows that $\mathbb{P}^n(\mathbb{C})\backslash H_j$ is naturally isomorphic to $\mathbb{C}^n$. Let

$$\mathcal{U}_j = \mathbb{P}^n(\mathbb{C})\backslash H_j = \{[Z_0 : \dots : Z_n] \in \mathbb{P}^n(\mathbb{C}) : Z_j \neq 0\}.$$

Since H_j is closed, then its complement $\mathcal{U}_j$ is open. We equip $\mathbb{P}^n(\mathbb{C})$ with the quotient topology of that of $\mathbb{C}^{n+1}\backslash\{0\}$, i.e. a subset of $\mathbb{P}^n(\mathbb{C})$ is closed or open if and only if its inverse image by the projection $\mathbb{C}^{n+1}\backslash\{0\} \longrightarrow \mathbb{P}^n(\mathbb{C})$ is. Furthermore, the map φ_j is a homeomorphism. Hence, an explicit set of charts of $\mathbb{P}^n(\mathbb{C})$ is provided by $(\mathcal{U}_j, \varphi_j)$. The coordinates z_k above are the affine coordinates on $\mathcal{U}_j = \mathbb{P}^n(\mathbb{C})\backslash H_j$. Each H_j is isomorphic to $\mathbb{P}^{n-1}(\mathbb{C})$ (just omit the j-th coordinate). From which, we consider set-theoretically $\mathbb{P}^n(\mathbb{C})$ as the disjoint union,

$$\mathbb{P}^n(\mathbb{C}) = \mathcal{U}_j \cup H_j = (\mathbb{P}^n(\mathbb{C})\backslash H_j) \cup H_j \simeq \mathbb{C}^n \cup H_j.$$

Note that $\bigcap_{j=0}^{n} H_j = \emptyset$, because no point $[Z_0 : \dots : Z_n]$ has all of its homogeneous coordinates Z_j equal to zero. Hence,

$$\bigcup_{j=0}^{n} \mathcal{U}_j = \bigcup_{j=0}^{n} (\mathbb{P}^n(\mathbb{C})\backslash H_j) = \mathbb{P}^n(\mathbb{C}).$$

(Indeed, let $\xi \in \mathbb{P}^n(\mathbb{C})$. If $Z_0, \dots, Z_n$ is a homogeneous coordinate system of ξ, then there exists j such that: $Z_j \neq 0$ and $\xi \in \mathcal{U}_j$.) We have an explicit set of charts defining $\mathbb{P}^n(\mathbb{C})$ as the union of $(n+1)$ copies of $\mathbb{C}^n$ (i.e. $\mathbb{P}^n(\mathbb{C}) = \mathbb{C}^n \cup \mathbb{P}^{n-1}(\mathbb{C}) = \mathbb{C}^n \cup \mathbb{C}^{n-1} \cup \dots \cup \mathbb{C}^0$). As before, we deduce that the space $\mathbb{P}^n(\mathbb{C})$ is a complex analytic manifold.

Yet, another description of complex projective space $\mathbb{P}^n(\mathbb{C})$ is obtained from the unit sphere S^{2n+1} of the space $\mathbb{C}^{n+1}$, by identifying the points $z \sim e^{i\theta}z$ (each line in $\mathbb{C}^{n+1}$ intersects S^{2n+1} in a circle S^1). Topologically, $\mathbb{P}^n(\mathbb{C})$ can be obtained equivalently as the quotient space S^{2n+1}/S^1, i.e.

$$\mathbb{P}^n(\mathbb{C}) \cong S^{2n+1}/S^1,$$

where S^1 acts as a multiplication by complex numbers of unit length, and it follows that $\mathbb{P}^n(\mathbb{C})$ is compact.

Note: In section 2.2.3, we have imaginatively analyzed the special case of the Riemann sphere S^2, i.e. the complex projective line $\mathbb{P}^1(\mathbb{C})$, using the notion of stereographic projection. From the reasoning that has been done, we see that $\mathbb{P}^1(\mathbb{C})$ is diffeomorphic to S^2.

2.2.5. *Grassmannians*

Let $G_{n,k}$, $0 \leq k \leq n$, be a complex grassmannian, i.e. the set of all k-dimensional complex linear subspaces of $\mathbb{C}^n$. It is a compact complex manifold of dimension $k(n-k)$. For $k = 1$, we obtain the complex projective space $\mathbb{P}^{n-1}(\mathbb{C})$ (studied in section 2.2.4). We will show that we can equip $G_{n,k}$ with a complex manifold structure. (Thus, grassmannian manifolds generalize projective spaces.) Indeed, let $v_1, ..., v_k$ be linearly independent vectors in a subspace Δ of $G_{n.k}$ and consider the matrix

$$A = A(\Delta) = (v_{ij})_{\substack{1\leq i\leq k\\ 1\leq j\leq n}},$$

of rank k whose rows are the coordinates of these vectors in the canonical basis of $\mathbb{C}^n$. This allows us to embed $G_{n,k}$ into the matrix space. Let A_I ($I = \{i_1, ..., i_k\}$ be a subset of $\{1, 2, ..., n\}$), the subset of $G_{n,k}$ or the submatrix (we keep the same notation A_I) of order k formed by the columns $i_1, ..., i_k$ of A. Since $v_1, ..., v_k$ are linearly independent vectors, then there exists a multi-index such that: $\det A_I \neq 0$. Two matrices A and B define a subspace of $G_{n,k}$ if and only if $B = CA$ where $C \in GL(n, \mathbb{R})$ is an invertible matrix of order k (we write $A \sim B$). This is indeed an equivalence relation and therefore, $G_{n,k}$ can be interpreted (modulo the relation $A \sim B$) as a set of equivalence classes $[A]$ of matrices A of order $n \times k$ of rank k. If $v'_1, ..., v'_k$ are other linearly independent vectors in Δ, then $B = CA$ where $C \in GL(n, \mathbb{R})$ and we have $A'_I = BA_I$. Hence,

$$A_I^{-1}B = A_I^{-1}C^{-1}CA = A_I^{-1}A,$$

which means that the matrix $(A_I)^{-1}A$ does not depend on the subspace Δ and the multi-index I. For any multi-index I, we consider the set

$$U_I = \{[A] \in G_{n,k} : \det A_I \neq 0\} \subset G_{n,k},$$

of these subspaces $\Delta \in G_{n,k}$ for which $\det A_I \neq 0$. These sets form an open covering of $G_{n,k}$ (if $\det A_I \neq 0$, then $\det B_I \neq 0$ for any matrix B whose elements are sufficiently close to those of A). We obtain a homeomorphism

$$\varphi_I : A_I \longrightarrow \mathbb{R}^{k(n-k)},$$

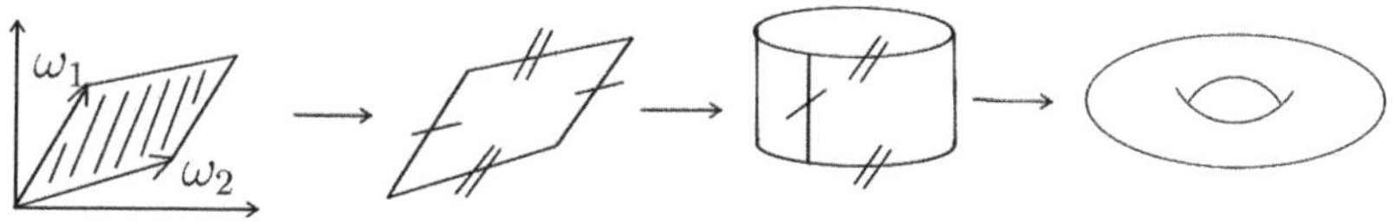

Figure 2.4. *Torus*

where $\varphi_I([A]) = (A_I^{-1}A)_{I^\top}$ is the submatrix of $A_I^{-1}A$ of order $k(n-k)$ obtained by removing from $A_I^{-1}A$ the columns that correspond to the indices of I (the set of these columns forms the unit matrix of order k). We show that the family $\{U_I, \varphi_I\}$ is an atlas of charts of $G_{n,k}$. For the transition functions, if I and J are two multi-indices, then the map

$$\varphi_J o \varphi_I^{-1} : \varphi_I(U_I \cap U_J) \longrightarrow \varphi_J(U_I \cap U_J), \quad \varphi_J o \varphi_I^{-1}(A_{I^\top}) = (A_J^{-1}A)_{J^\top},$$

(with A: matrix of order $k \times n$ and A_I: unit matrix). Thus, the grassmannian $G_{n,k}$ is a complex manifold of dimension $(n-k)k$. Furthermore, $G_{n,k}$ and $G_{n,n-k}$ are diffeomorphic. (See also exercise 2.9 for another construction of grassmannians with a slight change of notation, the complex grassmannians which are defined in a similar manner and admit several interesting properties.)

2.2.6. *Tori*

Let $\mathbb{T}^n$ be a complex torus of dimension n, i.e. the direct product of n circles, and from section 2.2.3 and proposition 2.1, we deduce that $\mathbb{T}^n$ is a complex manifold. The torus $\mathbb{T}^n$ is the quotient $\mathbb{C}^n/\Lambda$ of $\mathbb{C}^n$ by a discrete commutative free subgroup Λ of maximal rank $2n$ of $\mathbb{C}^n$, i.e. $\Lambda \simeq \mathbb{Z}^{2n}$. We write

$$\Lambda = \{m_1\omega_1 + \cdots + m_{2n}\omega_{2n} : m_j \in \mathbb{Z}, j = 1, ..., 2n\},$$

where $\omega_1, ..., \omega_{2n} \in \mathbb{C}^n \simeq \mathbb{R}^{2n}$ are $2n$ linearly independent $\mathbb{R}$-complex vectors. We can equip T^n with an analytic manifold structure (it is compact). Thus, we have

$$\mathbb{T}^n = (\mathbb{R}/\mathbb{Z})^{2n} = \mathbb{C}^n/\mathbb{Z}^{2n} = \left(S^1\right)^{2n},$$

this tells us about the topology of $\mathbb{T}^n$. For a complex torus of dimension 1, i.e. the quotient of $\mathbb{C}$ by the equivalence relation determined by Λ,

$$z_1, z_2 \in \mathbb{C}, \quad z_1 \sim z_2 \text{mod.}\Lambda \Longleftrightarrow z_1 - z_2 \in \Lambda,$$

is a torus denoted by $\mathbb{C}/\Lambda$, homeomorphic to $S^1 \times S^1$, visualized by gluing the sides of a square or parallelogram together (see Figure 2.4).

2.3. Tangent spaces and tangent bundles

Let us first recall that if f is a differentiable function of the variables x and y, then we have $df = \frac{\partial f}{\partial x}dx + \frac{\partial f}{\partial y}dy$. Since $z = x + iy, \overline{z} = x - iy$, hence

$$dx = \frac{1}{2}(dz + d\overline{z}), \qquad dy = \frac{1}{2i}(dz - d\overline{z}),$$

and consequently,

$$df = \frac{1}{2}\left(\frac{\partial f}{\partial x} - i\frac{\partial f}{\partial y}\right)dz + \frac{1}{2}\left(\frac{\partial f}{\partial x} + i\frac{\partial f}{\partial y}\right)d\overline{z}.$$

Let us introduce the two operators

$$\frac{\partial}{\partial z} = \frac{1}{2}\left(\frac{\partial}{\partial x} - i\frac{\partial}{\partial y}\right), \qquad \frac{\partial}{\partial \overline{z}} = \frac{1}{2}\left(\frac{\partial}{\partial x} + i\frac{\partial}{\partial y}\right).$$

So,

$$df = \frac{\partial f}{\partial z}dz + \frac{\partial f}{\partial \overline{z}}d\overline{z}.$$

Let us now return to the manifold M and instead of the vectors $\frac{\partial}{\partial x_i}, \frac{\partial}{\partial y_i}$, we can introduce the operators

$$\frac{\partial}{\partial z_j} = \frac{1}{2}\left(\frac{\partial}{\partial x_j} - i\frac{\partial}{\partial y_j}\right), \qquad \frac{\partial}{\partial \overline{z_i}} = \frac{1}{2}\left(\frac{\partial}{\partial x_j} + i\frac{\partial}{\partial y_j}\right).$$

In this case, any vector v of the tangent space T_pM to M at a point p can be written in the form

$$v = \sum_{j=1}^{n}\left(\alpha_j\left(\frac{\partial}{\partial z_j}\right)_p + \overline{\alpha}_j\left(\frac{\partial}{\partial \overline{z}_j}\right)_p\right),$$

and the space T_pM is of real dimension $2n$.

Consider the germs of holomorphic functions f, then $\left(\frac{\partial}{\partial z_1}, ..., \frac{\partial}{\partial z_n}\right)$ is a basis of the tangent space T_pM, while $(dz_1, ..., dz_n)$ is a basis of the dual space of $\mathbb{C}$-linear functions. The complex dimension of T_pM and the dual space is equal to n. The union of the tangent spaces to M at p is the tangent bundle, which has dimension $2n$. The differential of a holomorphic function being $\mathbb{C}$-linear, to define the tangent bundle it suffices to reason in a way analogous to the real case. Let us briefly recall that if $\mathcal{O}_p$ is the set of germs of functions f in the neighborhood of a point $p \in M$ and if

$\gamma : [0,1] \longrightarrow M$ is a differentiable path with $\gamma(0) = p$, then we have the equivalence relation:

$$\gamma_1 \sim \gamma_2 \Longleftrightarrow \frac{d}{dt} f o \gamma_1(t)\Big|_{t=0} = \frac{d}{dt} f o \gamma_2(t)\Big|_{t=0}, \quad \forall f \in \mathcal{O}_p$$

The equivalence classes modulo this equivalence relation are called jets of order 1. A tangent vector to M at p is a functional $v : \mathcal{O}_p \longrightarrow \mathbb{R}$ such that:

$$v(f) = \frac{d}{dt} f o \gamma(t)\Big|_{t=0},$$

where $f \in \mathcal{O}_p$ and γ belongs to an equivalence class modulo the equivalence relation above. The set of all tangent vectors to M at p is the tangent space denoted by T_pM.

Another definition is to consider the germs of differentiable functions f using the above definition of tangent vector. Instead of the tangent space T_pM of dimension n, we will have a space on $\mathbb{C}$ of (complex) dimension $2n$ and which we denote by $\mathcal{T}_pM$. Any vector v in this tangent space to M at p is written in the form

$$v = \sum_{j=1}^{n} \left(\alpha_j \left(\frac{\partial}{\partial z_j} \right)_p + \beta_j \left(\frac{\partial}{\partial \overline{z}_j} \right)_p \right), \quad (\alpha_j, \beta_j \in \mathbb{C}$$

The dual space of $\mathcal{T}_pM$ is such that:

$$df = \sum_{j=1}^{n} \left(\frac{\partial f}{\partial z_j} dz_j + \frac{\partial f}{\partial \overline{z}_j} d\overline{z}_j \right),$$

and its basis is $(dz_1, ..., dz_n, d\overline{z}_1, ..., d\overline{z}_n)$. The latter is none other than the dual basis of $\left(\frac{\partial}{\partial z_1}, ..., \frac{\partial}{\partial z_n}, \frac{\partial}{\partial \overline{z}_1}, ..., \frac{\partial}{\partial \overline{z}_n} \right)$.

2.4. Constant rank theorem

THEOREM 2.1 (CONSTANT RANK THEOREM).– *Let M and N be two complex manifolds of dimensions m and n, respectively, and $f : M \longrightarrow N$ be a holomorphic map with constant rank r. Then, for each $p \in M$, there exist charts $(U, \varphi) = (U, z_1, ..., z_m)$ for M centered at p and $(V, \psi) = (V, w_1, ..., w_n)$ for N centered at $f(p)$ such that $f(U) \subseteq V$, in which the map f is expressed in the local coordinates $z_1, ..., z_m$ and $w_1, ..., w_n$ by*

$$w_j = \begin{cases} z_j \text{ for } 1 \leq j \leq r \\ 0 \text{ for } r+1 \leq j \leq n \end{cases}$$

and f has a coordinate representation of the form:

$$(z_1, ..., z_m) \longmapsto \psi o f o \varphi^{-1}(z_1, ..., z_m) = (z_1, ..., z_r, 0, ..., 0).$$

PROOF.– Suppose that the map f is expressed in the charts (U, φ), (V, ψ) by $w_1 = f_1(z_1, ..., z_m), ..., w_n = f_n(z_1, ..., z_m)$. By hypothesis, we have

$$rg_p f = rg \begin{pmatrix} \frac{\partial f_1}{\partial z_1}(p) & \cdots & \frac{\partial f_1}{\partial z_m}(p) \\ \vdots & & \vdots \\ \frac{\partial f_n}{\partial z_1}(p) & \cdots & \frac{\partial f_n}{\partial z_m}(p) \end{pmatrix} = r.$$

So, there exists an invertible square matrix of order r extracted from the Jacobian matrix above. Without restricting the generality, we can choose $\det\left(\frac{\partial f_i}{\partial z_j}(p)\right)_{1\leq i,j\leq r} \neq 0$ and therefore the matrix $\left(\frac{\partial f_i}{\partial z_j}(p)\right)_{1\leq i,j\leq r}$ is invertible (otherwise, we will use other notations). Let $u_1, ..., u_m$ be functions defined in a neighborhood U of p by

$$u_1 = f_1(z_1, ..., z_m), ..., u_r = f_r(z_1, ..., z_m), u_{r+1} = z_{r+1}, ..., u_m = z_m.$$

The determinant of the Jacobian matrix of the map $z \longmapsto u$ is (in the neighborhood of a point $p \in U$)

$$\det\begin{pmatrix} \left(\frac{\partial f_i}{\partial z_j}(p)\right)_{1\leq i,j\leq r} & * \\ 0 & I_{m-k} \end{pmatrix} = \det\left(\frac{\partial f_i}{\partial z_j}(p)\right)_{1\leq i,j\leq r} \neq 0,$$

where I_{m-k} is the unit matrix of order $m-k$. Since this Jacobian matrix is invertible, we deduce from the local inversion theorem that there exists a neighborhood U_p of p on which the map $x \longmapsto u$ is a diffeomorphism and therefore a change of local coordinates in U_p. By replacing the neighborhood U_p by U and the coordinates $u_1, ..., u_m$ by $z_1, ..., z_m$, we obtain maps on which the equality $w = f(z)$ is equivalent to (in other words, the map f is expressed by): $w_1 = z_1, ..., w_r = z_r, w_{r+1} = f_1(z_1, ..., z_m), ..., w_m = f_m(z_1, ..., z_m)$. In these charts, the Jacobian matrix of the map $z \longmapsto w$ is written as:

$$\begin{pmatrix} I_r & 0 \\ * & \left(\frac{\partial f_i}{\partial z_j}(p)\right) \end{pmatrix}, \quad r+1 \leq n, r+1 \leq j \leq m.$$

We will now use the assumption of a constant rank equal to r. Since the composite by a diffeomorphism leaves the rank unchanged, we deduce that the matrix above has rank r for any neighborhood of p (which can be assimilated to U) which implies that the derivatives $\frac{\partial f_i}{\partial z_j}$, $r+1 \leq n, r+1 \leq j \leq m$, must be identically zero (because otherwise the rank would be $\geq r+1$). Therefore, in U (or more precisely in $\varphi(U)$), the f_i, $r+1 \leq n$, depend only on $z_1, ..., z_r$ (because we can admit that the neighborhood p is connected) and therefore the map f is expressed in coordinates z and w by

$$w_1 = z_1, ..., w_r = z_r, w_{r+1} = f_1(z_1, ..., z_r), ..., w_m = f_m(z_1, ..., z_r).$$

Let

$$v_1 = w_1, ..., v_r = w_r, v_{r+1} = w_{r+1} - f_{r+1}(w_1, ..., w_r), ..., v_n = w_n - f_n(w_1, ..., w_r).$$

The functions $f_{r+1}(w_1, ..., w_r), ..., f_n(w_1, ..., w_r)$ are defined on a neighborhood V_q of the point $q = f(p)$. Note that

$$det \left(\frac{\partial v_i}{\partial w_j}(p)\right)_{1 \leq i,j \leq n} = \det \begin{pmatrix} 1 \cdots 0 \; 0 \cdots 0 \\ \vdots \; \ddots \; \vdots \; \vdots \; \ddots \; \vdots \\ 0 \cdots 1 \; 0 \cdots 0 \\ * \cdots * \; 1 \cdots 0 \\ \vdots \; \ddots \; \vdots \; \vdots \; \ddots \; \vdots \\ * \cdots * \; 0 \cdots 1 \end{pmatrix} = 1 \neq 0.$$

Hence, the functions $v_1, ..., v_n$ defined above are local coordinates on V_q. By replacing the neighborhood V_q by V and the coordinates $v_1, ..., v_n$ by $w_1, ..., w_n$, we obtain maps on which the equality $w = f(z)$ is written (in other words, the map f is expressed by) $w_1 = z_1, ..., w_r = z_r, w_{r+1} = 0, ..., w_n = 0$ and the proof ends. ■

2.5. Submanifolds, subvarieties and examples

DEFINITION 2.8.– *A subset N of an m-dimensional complex manifold M is called a complex submanifold of dimension n, if for every point $p \in N$, there exists a local chart (U, φ) on M (U being a neighborhood of p and $\varphi : U \longrightarrow \varphi(U) \subset \mathbb{C}^m$ a homeomorphism, with $\varphi(p) = 0$) such that:*

$$\varphi(U \cap N) = \{(z_1, ..., z_n) \in \varphi(U) : z_{n+1}, ..., z_m = 0\}.$$

The codimension of N is $m - n$. A complex hypersurface is a complex submanifold of codimension 1.

A complex submanifold is a complex manifold of dimension n. Indeed, to obtain a complex chart of N, it suffices (taking into account the chart (U, φ) above), to consider the chart $(U \cap N, \varphi|_{U \cap N})$ where $\varphi|_{U \cap N} : U \cap N \longrightarrow \mathbb{C}^n$. The compatibility of these charts follows from the compatibility of the charts on the manifold M. A generalization of submanifold is the following:

DEFINITION 2.9.– *An analytic subvariety V of a complex manifold M is a subset given locally as the common zeroes of finitely many holomorphic functions, i.e. for any $p \in V$, there is an open set $U \subset M$ and holomorphic functions $f_1, ..., f_n$ on U such that:*

$$V \cap U = \{z \in U : f_1(z) = \cdots = f_n(z) = 0\}.$$

An analytic subvariety V is called a hypersurface if it is locally the zero locus of a holomorphic function.

A complex submanifold is an analytic subvariety; we just choose U to be the domain of the adapted chart and f_j to be $z_{n+1}, ..., z_m$.

DEFINITION 2.10.– *Let V be an analytic subvariety of an m-dimensional complex manifold M. A point $p \in V$ is called a regular point (see also definition 1.11) if we can find open $U \subset M$ and holomorphic functions $f_1, ..., f_n$ on U such that:*

$$V \cap U = \{z \in U : f_1(z) = \cdots = f_n(z) = 0\},$$

and

$$rank \frac{\partial(f_1, ..., f_n)}{\partial(z_1, ..., z_m)}(p) = n.$$

In this case, V is locally near p a complex submanifold of dimension $m - n$. (Indeed, without loss of generality, assume $rank \frac{\partial(f_1,...,f_n)}{\partial(z_1,...,z_n)}(p) \neq 0$, then we can choose a new compatible coordinate system $(f_1, ..., f_k, z_{k+1}, ..., z_n)$. This is an adapted chart for V near p.) The locus of regular points of N is denoted by N_{reg}. Its complement in N is called the singular locus, and its elements are called singular points of N.

Using the implicit function theorem and the constant rank theorem, we can give two ways to obtain local characterization of complex submanifolds.

PROPOSITION 2.2.– *Let M, N both be complex manifolds of dimension m and n, respectively.*

a) If $m > n$ and $f : M \longrightarrow N$ is a holomorphic map with constant rank n (the rank of f is maximal on all $f^{-1}(q)$, $q \in f(M) \subseteq N$) and each $p \in f^{-1}(q)$, then $f^{-1}(q)$ is a complex submanifold of M of dimension $m - n$.

b) If $m \leq n$ and $f : N \longrightarrow M$ is an injective holomorphic map with constant rank n on all N (the rank of f is maximal) and N is compact, then $f(N)$ is a submanifold of M and $f : N \longrightarrow f(N)$ is biholomorphic (in other words, $f(N)$ is isomorphic to N and f is an embedding).

PROOF.– *a)* Note that the rank of the map f is maximal on a neighborhood of $f^{-1}(q)$ on M, since by hypothesis its rank is maximal on all $f^{-1}(q)$. The hypotheses of the constant rank theorem being satisfied, we deduce that for each $p \in M$ there exist charts $(U, \varphi) = (U, z_1, ..., z_m)$ for M centered at p and $(V, \psi) = (V, w_1, ..., w_n)$ for N centered at $f(p)$ such that $f(U) \subseteq V$, in which the map f has a coordinate representation of the form:

$$(z_1, ..., z_m) \longmapsto \psi o f o \varphi^{-1}(z_1, ..., z_m) = (z_1, ..., z_n).$$

Hence,

$$
\begin{aligned}
\varphi(U \cap f^{-1}(q)) &= \left\{z \in \varphi(U) : \varphi^{-1}(z) \in f^{-1}(q)\right\}, \\
&= \left\{z \in \varphi(U) : f(\varphi^{-1}(z)) = q\right\}, \\
&= \left\{z \in \varphi(U) : \psi(f()\varphi^{-1}(z)) = \psi(q) = 0\right\}, \\
&= \left\{z = (z_1, ..., z_m) \in \varphi(U) : z_1 = \cdots = z_n = 0\right\}.
\end{aligned}
$$

We deduce that if $z \in \varphi(U \cap f^{-1}(q))$, then $z = (0, ..., 0, z_{n+1}, ..., z_m)$, and taking into account φ we obtain an appropriate chart for $f^{-1}(q)$ in p (possibly by permuting the coordinates in C^m), and $f^{-1}(q)$ is a submanifold.

b) By hypothesis, N is compact, hence $N\backslash\Omega$ is also compact where Ω is an open set of N (note that $N\backslash\Omega$ is closed in N). Since f is continuous (by hypothesis it is holomorphic), then $f(N\backslash\Omega)$ is closed in $f(N)$ because $f(N\backslash\Omega)$ is compact in $f(N) \subseteq M$ and M is a Hausdorff topological space (every pair of points can be separated by two non-intersecting neighborhoods). Note that $f(\Omega)$ is open in $f(N)$ (since its complement is closed). We deduce that $f : N \longrightarrow f(N)$ is a homeomorphism. According to the constant rank theorem, we deduce (taking into account that the rank of f is n on all N) that for $p \in N$ there exist charts $(U, \varphi) = (U, z_1, ..., z_m)$ for M and $(V, \psi) = (V, w_1, ..., w_n)$ for N centered at p such that $f(V) \subseteq U$, in which the map f has a coordinate representation of the form:

$$(z_1, ..., z_n) \longmapsto \varphi o f o \psi^{-1}(z_1, ..., z_n) = (z_1, ..., z_n, 0, ..., 0).$$

Since $f(V)$ is open in $f(N)$ (because f is a homeomorphism and V is open in N), so there is an open set W of M such that $f(V) = f(N) \cap W = f(N) \cap (U \cap W)$ because $f(V) \subseteq U$. Hence,

$$\varphi(f(N) \cap (U \cap W)) = \varphi(f(V)) = \{z \in \varphi(U \cap W) : z = (z_1, ..., z_n, 0, ..., 0)\},$$

and we deduce that $f(N)$ has an appropriate chart and it is a submanifold. ■

EXAMPLE 2.1 (LOCAL CHARACTERIZATION OF SUBMANIFOLDS).– *Suppose M is a complex manifold. A subset $N \subseteq M$ is an embedded complex manifold if and only if each point of N has a neighborhood U in M such that $U \cap N$ is an embedded complex submanifold of U.*

EXAMPLE 2.2 (GRAPHS OF HOLOMORPHIC MAPS).– *Suppose $f : M \longrightarrow N$ is a holomorphic map. The graph of f is the set $\Gamma(f) = \{(p, f(p)) \subseteq M \times N : p \in M\}$ This is a complex submanifold of $M \times N$ biholomorphic to M, because the map from M to $M \times N$ given by $p \longmapsto (p, f(p))$ is a holomorphic embedding.*

EXAMPLE 2.3 (REGULAR LEVEL SETS).– *Suppose M and N are complex manifolds, $f : M \longrightarrow N$ is a holomorphic map, and $c \in N$ is a regular value of f (meaning* *that $df(p)$ is surjective for each $p \in f^{-1}(c)$. Then, the level set $f^{-1}(c)$ is a complex submanifold of M whose codimension is equal to the dimension of N.*

EXAMPLE 2.4.– *Let $X = \{(z, w) \in \mathbb{C}^2 : w^2 = P(z)\}$, where $P(z)$ is a complex polynomial of degree > 1, without multiple roots. The set X is a complex submanifold of $\mathbb{C}^2$.*

2.6. Exercises

EXERCISE 2.1.– *Show that the space of matrices $M_n(\mathbb{C})$ is a complex manifold.*

Solution: Note that $M_n(\mathbb{C}) \cong \mathbb{C}^{n^2}$ and, from section 2.2.1, $M_n(\mathbb{C})$ with the coordinate chart $\varphi : (a_{ij})_{ij} \longmapsto (a_{11}, a_{12}, ..., a_{nn})$ is a complex manifold.

EXERCISE 2.2.– *Let M be a complex manifold. Show that if $\emptyset \neq N \subset M$ is a connected open set, then N is also a complex manifold and* $\dim N = \dim M$.

Solution: Let $(U_\alpha, \varphi_\alpha)$ be an atlas of M, i.e. a set of charts of M which are pairwise compatible and where domains U_α cover the whole manifold M. Let $\emptyset \neq N \subset M$ be a connected open set and covered by a family of domains U_α. The set N is also a complex manifold with atlas (V_α, ψ_α) where $V_\alpha = N \cap U_\alpha$ and $\psi_\alpha = \varphi_\alpha|_{N\cap U_\alpha}$. The transition functions:

$$\psi_{\beta\alpha} : \psi_\alpha(V_\alpha \cap V_\beta) \longrightarrow \psi_\beta(V_\alpha \cap V_\beta),$$

$$u \longmapsto \psi_{\beta\alpha} = \psi_\beta o \psi_\alpha^{-1}(u) = \varphi_\beta|_{N\cap U_\beta} o \varphi_\alpha^{-1}|_{N\cap U_\alpha}(u),$$

are biholomorphic since they are just restrictions of the $\varphi_\beta o \varphi_\alpha^{-1}$.

EXERCISE 2.3.– *Show that any holomorphic function on a compact connected complex manifold should be a constant, and deduce that there are no compact complex submanifolds of $\mathbb{C}^m$ of positive dimension.*

EXERCISE 2.4.– *Let $SL_n(\mathbb{C}) = \{A$ matrix of order n :* $\det A = 1\}$ *be the unimodular (or special linear) group. Show that $SL_n(\mathbb{C})$ is a (closed) complex submanifold (of dimension $n^2 - 1$) of the space $GL(n, \mathbb{C})$ of complex square matrices.*

Solution: Note that the group $SL_n(\mathbb{C})$ is the inverse image of $\{1\}$ by the (continuous) application $\det : M_n(\mathbb{C}) \longrightarrow \mathbb{C}$, so $SL_n(\mathbb{C})$ is closed in $GL(n, \mathbb{C})$. This map being a submersion above 1, we deduce that $SL_n(\mathbb{C})$ is a complex submanifold of dimension $n^2 - 1$.

EXERCISE 2.5.– *Show that a subset V of a complex manifold M is a complex submanifold of M if and only if it is a non-singular variety in some open set of M.*

EXERCISE 2.6.– *A subset M in the projective space*

$$\mathbb{P}^n(\mathbb{C}) = \left\{[Z_0 : \cdots : Z_n](Z_0, ..., Z_n) \in \mathbb{C}^{n+1}\backslash\{0\}\right\},$$

is an algebraic variety if there exist homogeneous polynomials[1] $f_1, ..., f_n$ *to* $n+1$ *variables such that:*

$$M = \{[Z_0 : \cdots : Z_n] : f_1(Z_0, ..., Z_n) = \cdots = f_k(Z_0, ..., Z_n) = 0\}.$$

Show that an algebraic variety M in $\mathbb{P}^n(\mathbb{C})$ is closed and that it is an analytic variety.

Solution: Let $f_1, ..., f_n$ be homogeneous polynomials and M be a subset defined by these polynomials. We know that (see section 2.2.4) the space $\mathbb{P}^n(\mathbb{C})$ is covered by $n+1$ open spaces $U_j = \{[Z_0 : \cdots : Z_n] \in \mathbb{P}^n(\mathbb{C}) : Z_j \neq 0\}\}$ (set of lines for which $Z_j \neq 0$). Without restricting the generality, we assume that $j = 0$, i.e. that $Z_0 \neq 0$. Therefore, $f_k(Z_0, ..., Z_n) = 0$ if and only if $f_k\left(1, \frac{Z_1}{Z_0}, ..., \frac{Z_n}{Z_0}\right) = 0$. Let

$$\varphi_k : \mathbb{C}^n \longrightarrow \mathbb{C}, \quad (z_1, ..., z_n) \longmapsto f_k(1, z_1, ..., z_n), \quad \left(z_1 = \frac{Z_1}{Z_0}, ..., z_n = \frac{Z_n}{Z_0}\right).$$

Since φ_k is a polynomial, then φ_k is holomorphic. The function defined by $g_k = \varphi_k o \psi_0$ (where $\psi_0 : U_0 \longrightarrow \mathbb{C}^n$, $[Z_0 : \cdots : Z_n] \longmapsto \frac{1}{Z_0}(Z_1, ..., Z_n) = (z_1, ..., z_n)$) is a homeomorphism) is holomorphic on the open set U_0 and, hence, we have $M \cap U_0 = \{Z \in \mathbb{P}^n(\mathbb{C}) : g_k(Z) = 0\}$. Therefore, M is closed and is an analytic variety.

EXERCISE 2.7.– *(Non-singular surfaces defined by equations). Let Σ be a non-singular surface of dimension k defined in $\mathbb{C}^n$ by*

$$\Sigma - \{(z_1, ..., z_n) \in \mathbb{C}^n : f_1(z_1, ..., z_n) = 0, ..., f_{n-k}(z_1, ..., z_n) = 0\},$$

where $f_1, ..., f_{n-k}$ are holomorphic functions. Let $p = (z_1^0, ..., z_n^0)$ be a non-singular point on the surface Σ (i.e. the rank of the Jacobian matrix $\left(\frac{\partial f_i}{\partial z_j}\right)$, $1 \leq i \leq n-k$, $1 \leq j \leq n$, is exactly equal to $n-k$). Show that the surface Σ is a complex manifold.

Solution: Without restricting the generality, because if necessary, we can change the numbering of the coordinates, we assume that the determinant of the matrix of order $n-k$, $\left(\frac{\partial f_i}{\partial z_j}\right)_p$, $k+1 \leq j \leq n$, is non-zero. Hence, there exists a neighborhood of p included in Σ such that the missing variables, i.e. $z_1, ..., z_n$ are local coordinates on Σ. In the neighborhood of p, we have (by the implicit function theorem),

1 If f is a homogeneous polynomial, then $f(Z_0, ..., Z_n) = 0$ implies $f(\lambda Z_0, ..., \lambda Z_n) = 0$ for all $\lambda \in \mathbb{C}$.

$z_{k+1} = z_{k+1}(\xi_1, ..., \xi_k), ..., z_n = z_n(\xi_1, ..., \xi_k)$, where $\xi_1 = z_1, ..., \xi_k = z_k$. We thus obtain a parametric representation (provided by the variables $\xi_1, ..., \xi_k$) of Σ in the neighborhood of p. In other words, in the neighborhood of the point $p \in \Sigma$, we can introduce local coordinates $\xi_1, ..., \xi_k$. In order not to complicate the notations, we consider two domains of the form $U_j \subset \Sigma$, $U_l \subset \Sigma$, neighborhoods of p. The domain U_j is equipped with the coordinates $\xi_1, ..., \xi_k$ and we have $z_j = \varphi_i(\xi_1, ..., \xi_k)$, $1 \leq i \leq n-k$, where φ_i are holomorphic functions. Similarly, the domain U_l is equipped with the coordinates $z_1, ..., z_k$ and we have $z_l = \psi_i(\eta_1, ..., \eta_k)$, $1 \leq i \leq n-k$, where ψ_i are holomorphic functions. On the intersection $U_j \cap U_l$, the transition functions are (we assume that $1 < j < l$):

$$\xi_1 = \eta_1 = z_1, ..., \xi_{j-1} = \eta_{j-1} = z_{j-1},$$

$$\varphi_1(\xi_1, ..., \xi_k) = \eta_j = z_j,$$

$$\xi_j = \eta_{j+1} = z_{j+1}, ..., \xi_{l-2} = \eta_{l-1} = z_{l-1},$$

$$\xi_{l-1} = \psi_1(\eta_1, ..., \eta_k) = z_l,$$

$$\xi_l = \eta_l = z_{l+1}, ..., \xi_k = \eta_k = z_n.$$

The Jacobian of a differentiable application whose reciprocal is differentiable is non-zero and the result follows from it.

EXERCISE 2.8.– *Show that the unitary special group,*

$$SU(2) = \{A \in GL(2, \mathbb{C}) : A^*A = I, \det A = 1\},$$

is an analytic variety of dimension 3.

Solution: Let $A = \begin{pmatrix} a & b \\ c & d \end{pmatrix} \in SU(2)$, where $a, b, c, d \in \mathbb{C}$. Note that the equality $A^*A = I$ gives $|a|^2 + |c|^2 = 1$, $|b|^2 + |d|^2 = 1$, $\overline{a}b + \overline{c}d = 0$. In addition, we have $\det A = ad - bc = 1$ from which,

$$\overline{a}ad - \overline{a}bc = \overline{a} \Longrightarrow \overline{a}ad - \overline{c}dc = \overline{a} \Longrightarrow (|a|^2 + |c|^2)d = \overline{a} \Longrightarrow d = \overline{a}.$$

Similarly, we have

$$\overline{b}ad - \overline{b}bc = \overline{b} \Longrightarrow -c\overline{d}d - \overline{b}bc = \overline{b} \Longrightarrow -c(|d|^2 + |b|^2) = \overline{b} \Longrightarrow c = -\overline{b}.$$

Hence,

$$SU(2) = \left\{ \begin{pmatrix} a & b \\ -\overline{b} & \overline{a} \end{pmatrix} \in GL(2, \mathbb{C}) : |a|^2 + |b|^2 = 1 \right\}.$$

Let $a = x_1 + ix_2, b = x_3 + ix_4, (x_1, x_2, x_3, x_4 \in \mathbb{R})$ from which,

$$|a|^2 + |b|^2 = 1 \Longleftrightarrow x_1^2 + x_2^2 + x_3^2 + x_4^2 = 1.$$

Therefore, $SU(2)$ is homeomorphic to the sphere $S^3 \subset \mathbb{R}^4$ by the application

$$S^3 \longrightarrow SU(2), \quad (x_1, x_2, x_3, x_4) \longmapsto \begin{pmatrix} x_1 + ix_2 & x_3 + ix_4 \\ -x_3 + ix_4 & x_1 - ix_2 \end{pmatrix}.$$

Since S^3 is a complex manifold (section 2.2.3), we deduce that the same is true for $SU(2)$ of dimension 3.

EXERCISE 2.9.– *Let (see also section 2.2.5) $G_{n,k}$, $0 \leq k \leq n$, be a (complex) grassmannian, i.e. the set of k-dimensional subspaces of $\mathbb{C}^n$ passing through 0. $G_{n,k}$ can be considered as the space of spheres with center 0 and dimension $k-1$ contained in the sphere S^{n-1}; these spheres correspond one-to-one to the k-dimensional vector subspaces of $\mathbb{C}^n$.*

a) What do $G_{n,0}$, $G_{n,n}$ and $G_{n,1}$ represent?

b) Show that $G_{n,k}$ can be interpreted as a set of matrices of order $n \times k$ of rank k, modulo the relation $A \sim B$ if there exists a regular matrix C of order k such that: $B = CA$. Deduce that such a given interpretation allows us to introduce coordinates in $G_{n,k}$. Let π be an element of $G_{n,k}$ and $A = (a_{ij})$, $1 \leq i \leq n, 1 \leq j \leq k$, the corresponding matrix, $I = \{i_1, ..., i_k\} \subset \{1, ..., n\}$. Let

$$P_I = \det \begin{pmatrix} a_{i_1 1} & \cdots & a_{i_1 k} \\ \vdots & \ddots & \vdots \\ a_{i_k 1} & \cdots & a_{i_k k} \end{pmatrix}. \qquad [2.1]$$

Show that the numbers P_I are not all zero and that they are defined up to a non-zero complex multiplicative factor. Show that the data of the minors P_I of the matrix A allows us to define the corresponding π plane. What can we say about the coordinates [2.1] and what is their number?

c) Let

$$U_I = \{\pi \in G_{n,k} : P_I(\pi) \neq 0\}.$$

Show that the points of $U_{\{1,...,n\}}$ can be represented by matrices

$$\begin{pmatrix} 1 & 0 & 0 & \cdots & 0 & * & \cdots & * \\ 0 & 1 & & & \vdots & * & & \vdots \\ \vdots & & \ddots & & \vdots & \vdots & \ddots & \vdots \\ \vdots & & & \ddots & 0 & \vdots & & \ddots \vdots \\ 0 & \cdots & \cdots & \cdots & 1 & * & \cdots & * \end{pmatrix},$$

i.e. of the form (E, Z) *where* E *is the unit matrix of order* k *and* $Z = (z_{ij})$ *for any* $(n-k) \times k$ *matrix. Show that the domains* U_I *constitute a covering of* $G_{n,k}$ *and that the elements of the matrix* Z *(see question d)) determine a bijective map* $\varphi_I : U_I \longrightarrow \mathbb{C}^{(n-k)k}$.

d) Deduce that the family $\{(U_I, \varphi_I)\}$ *is an atlas of maps of* $G_{n,k}$ *and that, consequently,* $G_{n,k}$ *is a compact and connected complex manifold of dimension* $(n-k)k$.

Response: *a)* $G_{n,1}$ and $G_{n,n}$ are points, while $G_{n,1}$ is identified as a topological space, with the complex projective space $\mathbb{P}^{n-1}(\mathbb{C})$ studied in section 2.2.4. For questions *b)* to *d)*, it is sufficient to adopt here what was done in section 2.2.5 (taking into account some notational changes).

EXERCISE 2.10.– *a) Show that any complex* n*-manifold* M *is orientable as a real differentiable manifold* V.

b) What can we say about the reciprocal of this result?

Solution: *a)* The transition functions

$$\varphi_{\beta\alpha} \equiv \varphi_\beta o \varphi_\alpha^{-1} : \mathbb{C}^n \supset \varphi_\alpha (U_\alpha \cap U_\beta) \longrightarrow \varphi_\beta (U_\alpha \cap U_\beta) \subset \mathbb{C}^n,$$

$$u \longmapsto \varphi_{\beta\alpha}(u) = \varphi_\beta o \varphi_\alpha^{-1}(u),$$

on M are holomorphic as well as the inverse. By identifying $\mathbb{C}^n$ with $\mathbb{R}^{2n}$, $(z_1, ..., z_n) \longmapsto (x_1, y_1, ..., x_n, y_n)$, where $z_k = x_k + iy_k$, we get a real manifold structure on M. By calculating the Jacobian matrices of the transition functions for the two structures, we show (taking into account the Cauchy–Riemann equations) that $\det(J_{\mathbb{R}}\varphi_{\beta\alpha}) = |\det(J_{\mathbb{C}}\varphi_{\beta\alpha})|^2 > 0$ (since $\det(J_{\mathbb{C}}\varphi_{\beta\alpha}) \neq 0$; $\varphi_{\beta\alpha}$ has an inverse). Note that $J_{\mathbb{R}}\varphi_{\beta\alpha}$ is the Jacobian matrix of the transition function which we will note $\widetilde{\varphi_{\beta\alpha}}$ on V. Each transition function of V has a positive Jacobian determinant, so V has a positive atlas and is positively oriented. Another way to show that a complex manifold is orientable is to introduce a metric. More precisely, we know that any complex manifold admits an hermitian metric. So, there exists a nowhere vanishing $2n$ (n is the complex dimension of M) form on any complex manifold of real dimension $2n$.

b) The reciprocal of the result obtained in the previous question is not true in general. In other words, not all orientable manifolds are complex manifolds. To see this, consider the example of the sphere S^n which is orientable, but S^3 cannot be a complex manifold because $\dim S^3 = 3$ is odd. It should be noted, however, that only S^2 is a complex manifold (see section 2.2.3). More generally, it can be shown that any compact orientable differentiable manifold of dimension 2 admits a complex structure.

We have just shown in *a)* that a complex manifold is an orientable differential manifold of even dimensions. However, for an orientable manifold of even dimensions, it is not obvious that it is a complex manifold. Topological obstacles to an almost complex structure may exclude all even-dimensional spheres except S^2 and S^6. We know that S^2 is a complex manifold, but the case of S^6 remains open.

3

Sheaves and Cohomology

3.1. Sheaves

A sheaf $\mathcal{F}$ on a topological space M is the data for any open $\mathcal{U}$ of M of an Abelian group $\mathcal{F}(\mathcal{U})$, whose elements are called (global) sections of $\mathcal{F}$ on $\mathcal{U}$, and for any inclusion $\mathcal{V} \subset \mathcal{U}$ of a restriction homomorphism $r_{\mathcal{U}\mathcal{V}} : \mathcal{F}(\mathcal{U}) \longrightarrow \mathcal{F}(\mathcal{V})$, such that:

(i) $\mathcal{F}(\emptyset) = 0$, $r_{\mathcal{U}\mathcal{U}} = Id_{\mathcal{F}(\mathcal{U})}$ (identity) and for all inclusions $\mathcal{W} \subset \mathcal{V} \subset \mathcal{U}$, we have

$$r_{\mathcal{U}\mathcal{W}} = r_{\mathcal{V}\mathcal{W}} \circ r_{\mathcal{U}\mathcal{V}}.$$

(ii) if $\mathcal{U}$ is the union of open sets $\mathcal{U}_\alpha$ and if s_α is an element of $\mathcal{F}(\mathcal{U}_\alpha)$ satisfying the relation:

$$r_{\mathcal{U}_\alpha \cap \mathcal{U}_\beta, \mathcal{U}_\alpha}(s_\alpha) = r_{\mathcal{U}_\alpha \cap \mathcal{U}_\beta, \mathcal{U}_\beta}(s_\beta),$$

then there exists a unique element s of $\mathcal{F}(\mathcal{U})$ satisfying for all α, $r_{\mathcal{U}_\alpha \mathcal{U}}(s) = s_\alpha$ (rebonding condition).

The definition of a sheaf is valid in the case of Abelian groups, rings, modules and in a much more general context such as sheaves of sets. We will denote this with either $r_{\mathcal{U}\mathcal{V}}(s)$ or $s|_{\mathcal{V}}$. For $\mathcal{U} = \emptyset$, we have $\mathcal{F}(\emptyset) = 0$. The notation $\Gamma(\mathcal{U}, \mathcal{F})$ is sometimes used instead of $\mathcal{F}(\mathcal{U})$. An element of $\Gamma(\mathcal{U}, \mathcal{F})$ is called a section of $\mathcal{F}$ over $\mathcal{U}$. The elements of $\Gamma(M, \mathcal{F})$ are called the global sections of $\mathcal{F}$; these elements are defined on the open sets that coincide on their intersection.

EXAMPLE 3.1.– *The space of continuous functions on the open sets of a topological space is a sheaf. The set $C^\infty(\mathcal{U})$ of functions of class C^∞ on an open set $\mathcal{U}$ of a differentiable manifold is a sheaf. Let us also cite as examples the sheaf $\mathcal{O}_M(E)$ of holomorphic sections on a holomorphic vector bundle E on a complex manifold M,* *the sheaf Ω^p_M of holomorphic p-forms on a complex manifold M and the sheaf $\mathcal{A}^{p,q}_M$ of differential forms of type (p,q) on a complex manifold M.*

EXAMPLE 3.2.– *Let K be an Abelian group and $p \in K$. The skyscraper sheaf, denoted by K_p, is defined by $K_p(\mathcal{U}) = K$ if $p \in \mathcal{U}$ and $= 0$ otherwise, where $\mathcal{U}$ is an open set of M. If $p \in \mathcal{U} \subset \mathcal{V}$, $r_{\mathcal{U}\mathcal{V}}(p) = p$, and if $\mathcal{U} \subset \mathcal{V}$ with $p \notin \mathcal{U}$, $r_{\mathcal{U}\mathcal{V}}(p) = 0$. The Abelian group structure is addition in K and 0 is the trivial group.*

EXAMPLE 3.3.– *The set of locally constant functions with values in an Abelian group G is a sheaf (see exercise 3.2).*

EXAMPLE 3.4.– *The set of holomorphic functions (respectively holomorphic everywhere non-zero) on a complex manifold M is a sheaf denoted by $\mathcal{O}_M$ (respectively $\mathcal{O}^*_M$) (see exercise 3.3).*

Let $\varphi : M \longrightarrow N$ be a continuous map between two topological spaces and let $\mathcal{F}$ be a sheaf on M. We define the direct image $\varphi_*\mathcal{F}$ on N of the sheaf $\mathcal{F}$ by the map φ as being the sheaf defined by

$$\varphi_*\mathcal{F}(\mathcal{U}) = \mathcal{F}(\varphi^{-1}(\mathcal{U})),$$

for any open set $\mathcal{U}$ of N.

If $\mathcal{F}$ and $\mathcal{G}$ are sheaves of Abelian groups, a sheaf morphism $\varphi : \mathcal{F} \longrightarrow \mathcal{G}$ is a collection of morphisms $\varphi_\mathcal{U} : \mathcal{F}(\mathcal{U}) \longrightarrow \mathcal{G}(\mathcal{U})$ for any open set $\mathcal{U}$ of M, commuting to the restriction. In other words, $\mathcal{U} \subset \mathcal{V}$, then the diagram:

$$\begin{array}{ccc} \mathcal{F}(\mathcal{V}) & \xrightarrow{\varphi_\mathcal{V}} & \mathcal{G}(\mathcal{V}) \\ {\scriptstyle r_{\mathcal{U}\mathcal{V}}}\downarrow & & \downarrow{\scriptstyle r_{\mathcal{U}\mathcal{V}}} \\ \mathcal{F}(\mathcal{U}) & \xrightarrow{\varphi_\mathcal{U}} & \mathcal{G}(\mathcal{U}) \end{array}$$

is commutative.

The kernel of a morphism $\varphi : \mathcal{F} \longrightarrow \mathcal{G}$ of sheaves of Abelian groups is a subsheaf of $\mathcal{F}$ defined on each open set $\mathcal{U}$ as the kernel of the morphism on $\mathcal{U}$,

$$\mathcal{U} \longmapsto \ker(\varphi)(\mathcal{U}) = \ker\left(\varphi|_\mathcal{U} : \mathcal{F}(\mathcal{U}) \longrightarrow \mathcal{G}(\mathcal{U})\right).$$

This is indeed a sheaf, and it is zero if and only if φ is injective. Concerning the image and the cokernel, it is more complicated because since the gluing condition is not generally satisfied, then in general they are not sheaves. However, we define Im φ as being the set of sections of $\mathcal{G}$ which come locally from sections of $\mathcal{F}$. In other words, an element g of $\mathcal{G}(\mathcal{U})$ is in $(\mathrm{Im}\,\varphi)(\mathcal{U})$ if there exists an open cover $(\mathcal{U}_\alpha)$ of $\mathcal{U}$ and elements s_α of $\mathcal{F}(\mathcal{U}_\alpha)$ such that: $(s_\alpha) = g|_{\mathcal{U}_\alpha}$ for all α. Im φ is the sheaf associated with

$$\mathcal{U} \longmapsto (\mathrm{Im}\,\varphi)(\mathcal{U}) = \mathrm{Im}\,(\varphi|_{\mathcal{U}} : \mathcal{F}(\mathcal{U}) \longrightarrow \mathcal{G}(\mathcal{U}))\,.$$

We reason similarly for coker(φ). Consider an open covering $(\mathcal{U}_\alpha)$ of $\mathcal{U}$ and elements s_α of $\mathcal{F}(\mathcal{U}_\alpha)$. Let us assume that the family $(\mathcal{U}_\alpha, s_\alpha)$ is admissible, i.e. for all α, β, the elements $s_\alpha|_{\mathcal{U}_\alpha\cap\mathcal{U}_\beta} - s_\beta|_{\mathcal{U}_\alpha\cap\mathcal{U}_\beta}$ are in $\varphi(\mathcal{G}(\mathcal{U}_\alpha \cap \mathcal{U}_\beta)$. When the union of such families is admissible, we say that they are equivalent. Then, an element of the cokernel $(\mathrm{coker}(\varphi))(\mathcal{U})$ is an equivalence class of admissible families. Coker(φ) is the sheaf associated with

$$\mathcal{U} \longmapsto (\mathrm{coker})(\mathcal{U}) = \mathrm{coker}\,(\varphi|_{\mathcal{U}} : \mathcal{F}(\mathcal{U}) \longrightarrow \mathcal{G}(\mathcal{U}))\,.$$

The quotient sheaf of the sheaf $\mathcal{G}$ by the subsheaf $\mathcal{F} \subset \mathcal{G}$ is the sheaf whose section s over an open set $\mathcal{U}$ is given by an open covering $(\mathcal{U}_\alpha)$ of $\mathcal{U}$ and a family of sections $s_\alpha \in \mathcal{G}(\mathcal{U}_\alpha)$ coinciding on the intersections:

$$s_\alpha|_{\mathcal{U}_\alpha\cap\mathcal{U}_\beta} = s_\beta|_{\mathcal{U}_\alpha\cap\mathcal{U}_\beta}\,.$$

Let $\mathcal{F}$, $\mathcal{G}$, $\mathcal{H}$ be three sheaves of Abelian groups on M. We say that the sequence

$$\mathcal{F} \xrightarrow{\varphi} \mathcal{G} \xrightarrow{\psi} \mathcal{H}$$

of morphisms of the sheaves is exact in $\mathcal{G}$ if $\mathrm{Im}\,\varphi = \ker\psi$. The set of morphisms of the form

$$0 \longrightarrow \mathcal{F} \xrightarrow{\varphi} \mathcal{G} \xrightarrow{\psi} \mathcal{H} \longrightarrow 0$$

exact in $\mathcal{F}$, $\mathcal{G}$, $\mathcal{H}$ is said to be an exact set of sheaves.

EXAMPLE 3.5.– *The sequence*

$$0 \longrightarrow \mathcal{F} \longrightarrow \mathcal{G} \longrightarrow \mathcal{G}/\mathcal{F} \longrightarrow 0$$

is exact where $\mathcal{F}$ is a subsheaf of $\mathcal{G}$. Similarly, if $N \subset M$ is a submanifold of a complex manifold M and $\mathcal{I}_N$ is the sheaf of ideals of $\mathcal{O}_M$ consisting of holomorphic functions on M that vanish on N, then the sequence of sheaves

$$0 \longrightarrow \mathcal{I}_N \longrightarrow \mathcal{O}_M \longrightarrow \mathcal{O}_N \longrightarrow 0$$

is exact.

It should be noted that if

$$0 \longrightarrow \mathcal{F} \longrightarrow \mathcal{G} \longrightarrow \mathcal{H} \longrightarrow 0$$

is exact, the last arrow is only locally surjective and it is therefore false to say that the sequence

$$0 \longrightarrow \mathcal{F}(\mathcal{U}) \longrightarrow \mathcal{G}(\mathcal{U}) \longrightarrow \mathcal{H}(\mathcal{U}) \longrightarrow 0$$

is also exact for an open set $\mathcal{U}$.

EXAMPLE 3.6.– *The following exponential sequence is exact:*

$$0 \longrightarrow \mathbb{Z} \xrightarrow{i} \mathcal{O}_M \xrightarrow{\exp} \mathcal{O}_M^* \longrightarrow 0,$$

where $\mathcal{O}_M$ is the sheaf of holomorphic functions on a complex manifold M, $\mathcal{O}_M^*$ is the sheaf of non-vanishing holomorphic functions on M, i is the trivial inclusion, exp *is the exponential map* $\exp f = e^{2\pi\sqrt{-1}f}$ *and $\mathbb{Z}$ (also noted $\mathbb{Z}_M$) is the sheaf of locally constant functions on M with integral values. This sequence does not give an exact sequence for $\mathbb{C}^*$. Indeed, we cannot define the logarithm on all $\mathbb{C}^*$.*

Moreover, for any exact sequence

$$0 \longrightarrow \mathcal{F} \xrightarrow{\varphi} \mathcal{G} \xrightarrow{\psi} \mathcal{H} \longrightarrow 0$$

of sheaves of Abelian groups, there exists a morphism

$$\delta^* : H^k(M, \mathcal{H}) \longrightarrow H^{k+1}(M, \mathcal{H}),$$

such that the long sequence in cohomology

$$\begin{array}{l} 0 \longrightarrow H^0(M,\mathcal{F}) \xrightarrow{\varphi^*} H^0(M,\mathcal{G}) \xrightarrow{\psi^*} H^0(M,\mathcal{H}) \\ \xrightarrow{\delta^*} H^1(M,\mathcal{F}) \xrightarrow{\varphi^*} H^1(M,\mathcal{G}) \xrightarrow{\psi^*} H^1(M,\mathcal{H}) \\ \qquad\qquad\qquad\qquad \vdots \\ \xrightarrow{\delta^*} H^k(M,\mathcal{F}) \xrightarrow{\varphi^*} H^k(M,\mathcal{G}) \xrightarrow{\psi^*} H^k(M,\mathcal{H}) \xrightarrow{\delta^*} \cdots \end{array}$$

is exact. Moreover, for morphisms $\lambda_1, \lambda_2, \mu_1, \mu_2, \nu_1, \nu_2$ of exact sequences

$$\begin{array}{ccccccccc} 0 \longrightarrow & \mathcal{F}_1 & \xrightarrow{\varphi_1} & \mathcal{G}_1 & \xrightarrow{\psi_1} & \mathcal{H}_1 & \longrightarrow 0 \\ & \downarrow\lambda_1 & & \downarrow\mu_1 & & \downarrow\nu_1 & \\ 0 \longrightarrow & \mathcal{F}_2 & \xrightarrow{\varphi_2} & \mathcal{G}_2 & \xrightarrow{\psi_2} & \mathcal{H}_2 & \longrightarrow 0 \\ & \downarrow\lambda_2 & & \downarrow\mu_2 & & \downarrow\nu_2 & \\ 0 \longrightarrow & \mathcal{F}_3 & \xrightarrow{\varphi_3} & \mathcal{G}_3 & \xrightarrow{\psi_3} & \mathcal{H}_3 & \longrightarrow 0 \end{array}$$

the following diagrams:

$$\begin{array}{ccccccccc} 0\longrightarrow & H^0(M,\mathcal{F}_1) & \xrightarrow{\varphi_1^*} & H^0(M,\mathcal{G}_1) & \xrightarrow{\psi_1^*} & H^0(M,\mathcal{H}_1) & \xrightarrow{\delta_1} & H^1(M,\mathcal{F}_1) & \xrightarrow{\varphi_1^*} \cdots \\ & \downarrow & & \downarrow & & \downarrow & & \downarrow & \\ 0\longrightarrow & H^0(M,\mathcal{F}_2) & \xrightarrow{\varphi_2^*} & H^0(M,\mathcal{G}_2) & \xrightarrow{\psi_2^*} & H^0(M,\mathcal{H}_2) & \xrightarrow{\delta_2} & H^1(M,\mathcal{F}_2) & \xrightarrow{\varphi_2^*} \cdots \\ & \downarrow & & \downarrow & & \downarrow & & \downarrow & \\ 0\longrightarrow & H^0(M,\mathcal{F}_3) & \xrightarrow{\varphi_3^*} & H^0(M,\mathcal{G}_3) & \xrightarrow{\psi_3^*} & H^0(M,\mathcal{H}_3) & \xrightarrow{\delta_3} & H^1(M,\mathcal{F}_3) & \xrightarrow{\varphi_3^*} \cdots \end{array}$$

are commutative and we have

$$\lambda_1^* \circ \lambda_2^* = (\lambda_1 \circ \lambda_2)^*, \quad \mu_1^* \circ \mu_2^* = (\mu_1 \circ \mu_2)^*, \quad \nu_1^* \circ \nu_2^* = (\nu_1 \circ \nu_2)^*.$$

A coherent sheaf $\mathcal{F}$ on an algebraic variety M is a sheaf that satisfies the conditions: (i) $\mathcal{F}$ is a sheaf of $\mathcal{G}_M$-modules; and (ii) for any point of M, there exists an exact sequence on a neighborhood $\mathcal{U}$,

$$\mathcal{O}_{\mathcal{U}}^{\oplus k} \longrightarrow \mathcal{O}_{\mathcal{U}}^{\oplus l} \longrightarrow \mathcal{F} \longrightarrow 0,$$

of $\mathcal{G}_{\mathcal{U}}$-modules (this condition means that the sheaf is locally of finite presentation).

EXAMPLE 3.7.– *The sheaf of holomorphic sections is a coherent sheaf. The skyscraper sheaf (see example 3.2) is a coherent sheaf.*

EXAMPLE 3.8.– *The sheaf $\mathcal{O}_M^*$ is not a coherent sheaf. Indeed, condition (i) above is not satisfied.*

EXAMPLE 3.9.– *The sheaf of rational fractions on M is a sheaf of $\mathcal{G}_M$-modules but is not a coherent sheaf. Indeed, condition (ii) above is not satisfied.*

We denote by $\mathcal{L}$ the sheaf of sections of a line bundle on a manifold M. This sheaf is isomorphic to $\mathcal{O}_{\mathcal{U}}$ where $\mathcal{U}$ is an open covering of M, in other words $\mathcal{L}$ is locally free. The converse is true. Such a sheaf is said to be invertible.

3.2. Cech cohomology

Let $\mathcal{F}$ be a sheaf on a topological space M. Consider the set $C^k(\mathcal{U},\mathcal{F})$ of k-cochains of degree k with values in $\mathcal{F}$, i.e. the set of maps that associates with each k-tuple of open sets of the covering $\mathcal{U}$ a section on their intersection. In other words, we have

$$C^k(\mathcal{U},\mathcal{F}) = \prod_{\alpha_0,\ldots,\alpha_k} \mathcal{F}(\mathcal{U}_{\alpha_0} \cap \ldots \cap \mathcal{U}_{\alpha_k}).$$

We have $C^0(\mathcal{U},\mathcal{F}) = \prod_\alpha \mathcal{F}(\mathcal{U}_\alpha)$ and $C^1(\mathcal{U},\mathcal{F}) = \prod_{\alpha,\beta} \mathcal{F}(\mathcal{U}_\alpha \cap \mathcal{U}_\beta)$. The (coboundary) operator is the homomorphism $\delta : C^k(\mathcal{U},\mathcal{F}) \longrightarrow C^{k+1}(\mathcal{U},\mathcal{F})$, for $s \in C^k(\mathcal{U},\mathcal{F})$, defined by

$$\delta s(\mathcal{U}_{\alpha_0},\ldots,\mathcal{U}_{\alpha_k}) = \sum_{j-0}^{k|1}(-1)^j s\left(\mathcal{U}_{\alpha_0},\ldots,\widehat{\mathcal{U}_{\alpha_j}},\ldots,\mathcal{U}_{\alpha_{k+1}}\right)\Big|_{\mathcal{U}_{\alpha_0}\cap\ldots\cap\mathcal{U}_{\alpha_{k+1}}}.$$

We have $\delta^2 = 0$ and the complex (of Cech)

$$C^0(\mathcal{U},\mathcal{F}) \xrightarrow{\delta} C^1(\mathcal{U},\mathcal{F}) \xrightarrow{\delta} C^2(\mathcal{U},\mathcal{F}) \xrightarrow{\delta} \cdots$$

with

$$\delta : C^0(\mathcal{U},\mathcal{F}) \to C^1(\mathcal{U},\mathcal{F}), s_{\alpha\beta} \mapsto s_\beta|_{\mathcal{U}_\alpha\cap\mathcal{U}_\beta} - s_\alpha|_{\mathcal{U}_\alpha\cap\mathcal{U}_\beta},$$

$$\delta : C^1(\mathcal{U},\mathcal{F}) \to C^2(\mathcal{U},\mathcal{F}), s_{\alpha\beta\gamma} \mapsto s_{\beta\gamma}|_{\mathcal{U}_\alpha\cap\mathcal{U}_\beta\cap\mathcal{U}_\gamma} - s_{\alpha\gamma}|_{\mathcal{U}_\alpha\cap\mathcal{U}_\beta\cap\mathcal{U}_\gamma} + s_{\alpha\beta}|_{\mathcal{U}_\alpha\cap\mathcal{U}_\beta\cap\mathcal{U}_\gamma},$$

$$\vdots$$

$$\delta : C^k(\mathcal{U},\mathcal{F}) \to C^{k+1}(\mathcal{U},\mathcal{F}), s_{\alpha_0\ldots\alpha_{k+1}} \mapsto \sum_{j=0}^{k+1}(-1)^j\, s_{\alpha_0\ldots\widehat{\alpha_j}\ldots\alpha_{k+1}}\big|_{\mathcal{U}_{\alpha_0}\cap\ldots\cap\mathcal{U}_{\alpha_{k+1}}}.$$

For any sheaf $\mathcal{F}$ of Abelian groups on M and for any open covering $\mathcal{U}$ of M, we can find a map $\mathcal{F} \longrightarrow C^0(\mathcal{U},\mathcal{F})$, so that the sequence

$$0 \longrightarrow \mathcal{F} \longrightarrow C^0(\mathcal{U},\mathcal{F}) \xrightarrow{\delta} C^1(\mathcal{U},\mathcal{F}) \xrightarrow{\delta} \cdots$$

is exact.

The Cech cohomology of the sheaf $\mathcal{F}$ for the covering $\mathcal{U}$ is the cohomology of the above complex,

$$H^k(\mathcal{U},\mathcal{F}) = \frac{\ker\left[\delta : C^k(\mathcal{U},\mathcal{F}) \longrightarrow C^{k+1}\mathcal{U},\mathcal{F})\right]}{\operatorname{Im}\left[\delta : C^{k-1}\mathcal{U},\mathcal{F}) \longrightarrow C^k\mathcal{U},\mathcal{F})\right]}.$$

These groups depend on the covering $\mathcal{U}$.

We define the Cech cohomology of the space M as the inductive limit on all coverings of the above cohomology,

$$H^k(M,\mathcal{F}) = \lim_{\overrightarrow{\mathcal{U}}} H^k(\mathcal{U},\mathcal{F}).$$

The passage to the limit in this definition is delicate in practice, although its advantage is that there is no longer any dependence on the covering. However, Leray theorem states that if the open sets $\mathcal{U}_\alpha$ of a covering $\mathcal{U}$ of M are such that: $H^k(A,\mathcal{F}) = 0$ for all $k > 0$ and any finite intersection $A \equiv \mathcal{U}_{\alpha_1} \cap \ldots \cap \mathcal{U}_{\alpha_k}$ of open sets $\mathcal{U}_\alpha$, then the groups $H^k(\mathcal{U},\mathcal{F})$ and $H^k(M,\mathcal{F})$ are isomorphic for all k.

EXAMPLE 3.10.– *The Mittag–Leffler problem, where Cech's cohomology groups appear mutually, consists of finding on a complex manifold a meromorphic function having prescribed poles that are holomorphic elsewhere. Precisely, for any sequence of complex numbers p_n such that: $\lim_{n\to\infty} p_n = \infty$ and any sequence of functions f_n of the form:*

$$f_n(z) = \sum_{k=1}^{k_n} \frac{a_k}{(z-p_n)^k},$$

there exists a meromorphic function whose points p_n are the only poles and whose principal part (of the Laurent expansion) coincides with f_n at each pole f_n. Note that we have an exact sequence:

$$0 \longrightarrow \mathcal{O}_{\mathbb{C}} \longrightarrow \mathcal{M}_{\mathbb{C}} \longrightarrow \mathcal{M}_{\mathbb{C}}/\mathcal{O}_{\mathbb{C}} \longrightarrow 0$$

where $\mathcal{M}_{\mathbb{C}}$ is the sheaf of meromorphic functions on $\mathbb{C}$. Similarly, the sequence:

$$0 \longrightarrow H^0(\mathbb{C}, \mathcal{O}_{\mathbb{C}}) \longrightarrow H^0(\mathbb{C}, \mathcal{M}_{\mathbb{C}}) \longrightarrow H^0(\mathbb{C}, \mathcal{M}_{\mathbb{C}}/\mathcal{O}_{\mathbb{C}}) \longrightarrow 0$$

is exact since $H^1(\mathbb{C}, \mathcal{O}_{\mathbb{C}}) = 0$. The finite set of f_n having the above form determines on $\mathbb{C}$ a section of $\mathcal{M}_{\mathbb{C}}/\mathcal{O}_{\mathbb{C}}$. Furthermore, this can be raised as a global meromorphic function on $\mathbb{C}$, which confirms the Mittag–Leffler problem.

EXAMPLE 3.11.– *The group $H^0(M, \mathcal{F})$ is canonically isomorphic to $\Gamma(M, \mathcal{F})$ (see exercise 3.4).*

EXAMPLE 3.12.– *For any sheaf of functions of class C^∞ on a manifold M, the cohomology groups of degree $k \geq 1$ are zero (see exercise 3.5).*

EXAMPLE 3.13.– *The set of line bundles, modulo isomorphism, on M is equipped with a commutative group structure for the tensor product called the Picard group and denoted by Pic(M). The group operation in $H^1(M, \mathcal{O}_M^*)$ corresponds to the tensor product of line bundles, and these latter have a one-to-one correspondence with the elements of the Cech cohomology group $H^1(M, \mathcal{O}_M^*)$. In other words, Pic(M) is isomorphic to $H^1(M, \mathcal{O}_M^*)$ (see proposition 4.8). To determine this group, a standard tool is the exact exponential sequence of sheaves mentioned in example 3.6.*

3.3. De Rham cohomology

The de Rham cohomology $H^p_{DR}(M)$ is by definition the quotient of the closed p-forms by the differentials of the $(p-1)$-forms. Let Ω^p be the space of complex-valued p-forms on a complex manifold M and $\mathcal{Z}^p$ be the subspace of closed complex-valued p-forms on M, and $d : \Omega^p \longrightarrow \Omega^{p+1}$. The quotient group:

$$H^p_{DR}(M) = \frac{\mathcal{Z}^p}{d\Omega^{p-1}},$$

is called the de Rham cohomology group of M.

A p-form is a global section of the sheaf of p-forms, so it is an element of $H^0(M, \Omega^p)$ because this last cohomology group is identified with the global sections. Therefore,

$$H^p_{DR}(M) = \frac{H^0(M, \mathcal{Z}^p)}{d_* H^0(M, \Omega^{p-1})},$$

where d_* is the cohomology map associated with the map $d : \Omega^p \longrightarrow \mathcal{Z}^{p+1}$. By virtue of the Poincaré lemma, any closed form is locally exact. Therefore, the sequence of sheaves

$$0 \longrightarrow \mathcal{Z}^p \longrightarrow \Omega^p \xrightarrow{d} \mathcal{Z}^{p+1} \longrightarrow 0,$$

is exact where $\mathcal{Z}^0 \equiv \mathbb{R}$ is the sheaf of locally constant functions and Ω^0 is the sheaf of $\mathcal{C}^\infty$ functions. The long exact sequence in cohomology associated with the above sequence is

$$H^q(M, \Omega^p) \xrightarrow{d^*} H^q(M, \mathcal{Z}^{p+1}) \xrightarrow{\partial} H^{q+1}(M, \mathcal{Z}^p) \longrightarrow H^{q+1}(M, \Omega^k),$$

where d^* denotes the exterior differential and ∂ is the edge operator of the long exact sequence. The sheaf Ω^p admits partitions of unity, so for $q \geq 1$, $H^q(M, \Omega^p) = 0$ and hence $H^q(M, \mathcal{Z}^{p+1}) = H^{q+1}(M, \mathcal{Z}^p)$. For the special case $q = 0$, we therefore have

$$H^0(M, \Omega^p) \xrightarrow{d^*} H^0(M, \mathcal{Z}^{p+1}) \xrightarrow{\partial} H^1(M, \mathcal{Z}^p) \longrightarrow 0,$$

and

$$H^1(M, \Omega^p) = \frac{H^0(M, \mathcal{Z}^{p+1})}{d_* H^0(M, \Omega^p)}.$$

Therefore,

$$H^p_{DR}(M) = H^1(M, \mathcal{Z}^{p-1}) = H^{p-1}(M, \mathcal{Z}^1) = H^p(M, \mathcal{Z}^0) = H^p(M, \mathbb{R}).$$

The de Rham cohomology of a manifold is isomorphic to its Cech cohomology with coefficients in $\mathbb{R}$.

3.4. Dolbeault cohomology

At each point of a complex manifold M, we define local coordinates z_j and $\overline{z}_j$ as well as the tangent bundle $\mathbb{C}\left[\frac{\partial}{\partial z_j}, \frac{\partial}{\partial \overline{z}_j}\right]$ which we denote by TM. This admits the decomposition $TM = T'M \oplus T''M$ where $T'M$ (respectively $T''M$) is the holomorphic (respectively antiholomorphic) part of TM generated by $\frac{\partial}{\partial z_j}$ (respectively $\frac{\partial}{\partial \overline{z}_j}$). Similarly, the cotangent bundle T^*M (dual of TM) admits the decomposition $T^*M = T'^*M \oplus T''^*M$ into holomorphic and antiholomorphic parts. The points of this bundle are the linear forms on the fibers of $TM \longrightarrow M$. A p-differential form is a section of the bundle $\Lambda^p T^*M$ (the exterior powers of T^*M). The points of the latter are the alternating p-multilinear forms on T^*M. If $z_1, ..., z_n$ are holomorphic local coordinates on a complex manifold M of dimension n, then the $dz_1, ..., dz_n, d\overline{z}_1, ..., d\overline{z}_n$ form a local basis of the space of differential forms. A differential form on M of type $(p, q))$ is given locally by

$$\omega = \sum_{\substack{1\leq j_1<...<j_p\leq n\\ 1\leq k_1<...<k_q\leq n}} f_{j_1...j_p k_1...k_q} dz_{j_1} \wedge ... \wedge dz_{j_p} \wedge d\overline{z}_{k_1} \wedge ... \wedge d\overline{z}_{k_q},$$

where the $f_{j_1...j_p k_1...k_q}$ are complex-valued functions of class C^∞. This definition does not depend on the choice of holomorphic local coordinates. Obviously, a form of type $(0, 0)$ is a function. The set of differential forms of type (p, q) on M is a complex vector bundle denoted by $A_M^{p,q}$. It should be noted that only $A_M^{p,0}$ is holomorphic. The differential of ω is

$$\begin{aligned} d\omega &= \sum_{\substack{1\leq j_1<...<j_p\leq n\\ 1\leq k_1<...<k_q\leq n}} df_{j_1...j_p k_1...k_q} \wedge dz_{j_1} \wedge ... \wedge dz_{j_p} \wedge d\overline{z}_{k_1} \wedge ... \wedge d\overline{z}_{k_q}, \\ &= \sum_{\substack{1\leq j_1<...<j_p\leq n\\ 1\leq k_1<...<k_q\leq n}} \partial f_{j_1...j_p k_1...k_q} \wedge dz_{j_1} \wedge ... \wedge dz_{j_p} \wedge d\overline{z}_{k_1} \wedge ... \wedge d\overline{z}_{k_q} \\ &\quad + \sum_{\substack{1\leq j_1<...<j_p\leq n\\ 1\leq k_1<...<k_q\leq n}} \overline{\partial} f_{j_1...j_p k_1...k_q} \wedge dz_{j_1} \wedge ... \wedge dz_{j_p} \wedge d\overline{z}_{k_1} \wedge ... \wedge d\overline{z}_{k_q}, \\ &= \partial\omega + \overline{\partial}\omega, \end{aligned}$$

where $\partial \equiv \frac{\partial}{\partial z} dz$ and $\overline{\partial} = \frac{\partial}{\partial \overline{z}} d\overline{z}$. Since $\partial f_{j_1...j_p k_1...k_q}$ is of type $(1, 0)$ and $\overline{\partial} f_{j_1...j_p k_1...k_q}$ is of type $(0, 1)$, then the component $\partial\omega$ of $d\omega$ is of type $(p + 1, q)$ and $\overline{\partial}\omega$ is of type $(p, q + 1)$. We therefore have the following decomposition: $d = \partial + \overline{\partial}$ where the operators $\partial : A_M^{p,q} \longrightarrow A_M^{p+1,q}$ and $\overline{\partial} : A_M^{p,q} \longrightarrow A_M^{p,q+1}$ are independent of the choice of holomorphic local coordinates. The relation $d^2 = 0$ means here that: $\partial^2 = \overline{\partial}^2 = \partial\overline{\partial} + \overline{\partial}\partial = 0$. Indeed, from the relations $d^2 = 0$ and $d = \partial + \overline{\partial}$, we derive $0 = d^2 = \partial^2 + \overline{\partial}^2 + \partial\overline{\partial} + \overline{\partial}\partial$. Note that if ω is a form of type (p, q), then $\partial^2\omega$ is of type $(p + 2, q)$, $\overline{\partial}^2\omega$ is of type $(p, q + 2)$ and $(\partial\overline{\partial} + \overline{\partial}\partial)$ is of type $(p + 1, q + 1)$. The result follows from the fact that the sum of these three forms is zero.

The operator $\overline{\partial}$ satisfies Leibniz's rule,

$$\overline{\partial}(\omega \wedge \lambda) = \overline{\partial}\omega \wedge \lambda + (-1)^{deg\omega} \omega \wedge \overline{\partial}\lambda.$$

Noting that $\partial\omega = \overline{\overline{\partial}\overline{\omega}}$, we also obtain:

$$\partial(\omega \wedge \lambda) = \partial\omega \wedge \lambda + (-1)^{deg\omega} \omega \wedge \partial\lambda.$$

Moreover, the differential form ω is said to be $\overline{\partial}$-closed if $\overline{\partial}\omega = 0$ and $\overline{\partial}$-exact if there exists a form λ such that: $\omega = \overline{\partial}\lambda$. Every $\overline{\partial}$-exact form is $\overline{\partial}$-closed. The analogue of the Poincaré lemma for the $\overline{\partial}$ operator can be stated as follows: if ω is a form of type (p,q) of class $\mathcal{C}^1$, $q > 0$ with $\overline{\partial}\omega = 0$, then we can find locally on M a form λ of class $\mathcal{C}^1$ of type $(p, q-1)$ such that: $\omega = \overline{\partial}\lambda$. Recall that the Poincaré lemma implies the nullity of cohomology groups in the case of star-shaped open sets.

The Dolbeault cohomology $H^{p,q}_{\overline{\partial}}(M)$ is by definition the quotient of the (p,q)-closed $\overline{\partial}$ forms on M by the (p,q)-exact $\overline{\partial}$-forms on M. Let $\mathcal{Z}^{p,q}_{\overline{\partial}}(M)$ be a space of $\overline{\partial}$-closed forms of type (p,q). With these notations, we define the Dolbeault cohomology group by

$$H^{p,q}_{\overline{\partial}}(M) = \frac{\mathcal{Z}^{p,q}_{\overline{\partial}}(M)}{A^{p,q-1}_{M}}$$

for each (p,q).

We have an isomorphism (Dolbeault theorem):

$$H^{p,q}_{\overline{\partial}}(M) \simeq H^q(M, \Omega^p)$$

for all integers p, q, where Ω^p is a sheaf of holomorphic p-forms on a complex manifold M.

3.5. Connections and curvature

Let M be a complex manifold. The set of alternating multilinear maps of degree p on the tangent space T_xM, $x \in M$, forms a vector space denoted by $\Lambda^p T^*_x M$. The set of these spaces forms a vector bundle over M, denoted by ΛT^*M. A differential p-form can be redefining itself as a global section of the bundle ΛT^*M. Let $\mathcal{A}^p(\mathcal{L})$ be the set of differential p-forms over M with values in the bundle $\mathcal{L}$, i.e. the sections of the bundle $(\Lambda T^*M) \otimes \mathcal{L}$. A connection on a bundle $\mathcal{L}$ is an operator $D : \mathcal{A}^0(\mathcal{L}) \longrightarrow \mathcal{A}^1(\mathcal{L})$ satisfying the Leibniz rule

$$D(f.s) = df \otimes s + f.Ds,$$

for any function f of class $\mathcal{C}^\infty$ on M and for any section $s \in \mathcal{A}^0(\mathcal{L})$. One of the interests of the connection is that it provides a way to derive sections of the bundle with respect to a basis vector and also a tool to realize parallel transport. For a section s and a tangent vector $v \in T_x(M)$, $x \in M$, and then $(Ds)(v)$ can be seen as the derivative of s in the direction v. In a local trivialization $\mathcal{U} \times \mathbb{C}^n \longrightarrow \pi^{-1}(\mathcal{U})$ of the bundle $\pi : \mathcal{L} \longrightarrow M$, the connection D is written in the form $De_j = \sum \theta_{jl} \otimes e_l$, where $e \equiv (e_j)$ is a basis of $\mathcal{L}$ with respect to the open set $\mathcal{U} \subset M$ and $(\theta_{jl}) \equiv \theta_e$ is a matrix (called the connection matrix) whose coefficients are 1-forms. Since the sections of $\mathcal{L}$ are identified with $s = \sum s_j e_j$, then by Leibniz's rule, the connection D is written in the form:

$$Ds = \sum ds_j \otimes e_j + \sum \alpha_j De_j = \sum_l \left(ds_l + \sum_j s_j \theta_{jl} \right) \otimes e_l.$$

If $e' \equiv (e'_j)$ is another basis such that: $e' = f(e)$, then

$$\theta_{e'} = df.f^{-1} + f\theta_e f^{-1}.$$

Using the decomposition of 1-forms

$$\mathcal{A}^1(\mathcal{L}) = \mathcal{A}^{1,0}(\mathcal{L}) \oplus \mathcal{A}^{0,1}(\mathcal{L}),$$

into holomorphic subsets D' and antiholomorphic subsets D'' of D, we can define these subsets so that D splits into the sum: $D = D' + D''$, of a connection of type $(1,0)$ and a connection of type $(0,1)$.

The curvature of the connection D on the bundle $\mathcal{L}$ is the operator defined by $D^2 : \mathcal{L} \longrightarrow \mathcal{A}^2(\mathcal{L})$. Let $f \in \mathcal{C}^\infty(M)$ and $s \in \mathcal{A}^0(\mathcal{L})$ be given. We have

$$\begin{aligned} D^2(f.s) &= D(df \otimes s + f.Ds), \\ &= (d^2 \otimes s - df \wedge Ds) + (df \wedge Ds + f.D^2 s), \\ &= f.D^2 s, \end{aligned}$$

which shows that D^2 is $\mathcal{C}^\infty$-linear. The value of $D^2 s$ at each point depends only on the section s at that point and not on neighboring values. Therefore, if two sections s_1 and s_2 have the same value at a point, then $D^2 s_1$ and $D^2 s_2$ have the same value. Therefore, D^2 defines a global section Θ of the bundle $\Lambda^2 T^* M \otimes \mathcal{L} \otimes \mathcal{L}^*$. The form Θ is called the curvature form of the bundle. If $e \equiv (e_j)$ is a basis of $\mathcal{L}$ with respect to $\mathcal{U}$, then

$$D^2 e_j = \sum \Theta_{jl} \otimes e_l \in \mathcal{A}^1(\mathcal{L}),$$

where $(\Theta_{jl}) \equiv \Theta_e$ is a matrix of 2-forms on M, called the curvature matrix of D with respect to the basis (e_j). If $e' \equiv (e'_j)$ is another basis such that: $e' = f(e)$, then the curvature matrix $\Theta_{e'}$ with respect to the basis (e'_j) is related to Θ_e by the relation: $\Theta_{e'} = f\Theta_e f^{-1}$. We have

$$D^2 e_j = D(De_j) = D\left(\sum \Theta_{jl} \otimes e_l\right) = \sum_l ((d\Theta_{jl}) \otimes e_l - \Theta_{lj} \wedge De_l)$$

so,

$$D^2 e_j = \sum_l (d\Theta_{jl}) \otimes e_l - \sum_{l,k} (\Theta_{jl} \wedge \Theta_{kl}) \otimes e_k = \sum_l ((d\Theta_{kj}) + (\Theta \wedge \Theta_{kj})) \otimes e_k.$$

Therefore, locally, D^2 is given by the curvature matrix and we thus obtain the Cartan structure equation $\Theta_e = d\theta_e - \theta_e \wedge \theta_e$.

3.6. Curvature form, first Chern class of line bundles

Let $\mathcal{L}$ be a line bundle on a complex manifold M, equipped with any connection, and let Θ be the curvature form of this connection. We previously showed that $\Theta = d\theta + \theta \wedge \theta$. Since $\mathcal{L}$ is a line bundle, then $\theta \wedge \theta = 0$ and $\Theta = d\theta$ is a closed 2-form on M since it is locally exact. It is therefore quite natural to consider its class $[\Theta]$ in the de Rham cohomology. Moreover, the exact sequence (see example 3.6):

$$0 \longrightarrow \mathbb{Z} \xrightarrow{i} \mathcal{O}_M \xrightarrow{\exp} \mathcal{O}_M^* \longrightarrow 0,$$

gives a long exact sequence in cohomology and provides the definition of the Chern class in cohomology

$$H^1(M,\mathbb{Z}) \longrightarrow H^1(M,\mathcal{O}_M) \longrightarrow H^1(M,\mathcal{O}_M^*) \overset{c_1=\partial}{\longrightarrow} H^2(M,\mathbb{Z}) \longrightarrow H^2(M,\mathcal{O}_M)$$

For $\mathcal{L} \in H^1(M,\mathcal{O}_M^*)$ (or $Pic(M)$ since these groups are isomorphic), the class $c_1(\mathcal{L}) = \partial\mathcal{L} \in H^2(M,\mathbb{Z})$ is the first Chern class of $\mathcal{L}$. This is a topological invariant that depends only on $\mathcal{L}$ as a topological complex line bundle. In fact, $c_1(\mathcal{L})$ can be seen as an element of $H^2(M,\mathbb{R})$ and the latter is identified according to the above with $H^2_{DR}(M)$. To calculate $c_1(\mathcal{L})$, we consider an open covering $\{\mathcal{U}_\alpha\}$ of M, giving a trivialization of $\mathcal{L}$. We choose this covering to be fine enough so that the intersections $\mathcal{U}_\alpha \cap \mathcal{U}_\beta$ are simply connected. Let $g_{\alpha\beta}$ be the transition functions between the open sets of this covering. The boundary of the cocycle $(g_{\alpha\beta})$ being zero, then by setting $h_{\alpha\beta} = \frac{1}{2\pi i}\log g_{\alpha\beta}$, we can search for the boundary of $(g_{\alpha\beta})$ in $\mathcal{C}^2(M,\mathbb{Z})$. It is given by the cocycle

$$z_{\alpha\beta\gamma} = \frac{1}{2\pi i}(\log g_{\alpha\beta} + \log g_{\beta\gamma} - \log g_{\alpha\gamma}).$$

Since the complex logarithm admits an infinite number of determinations, we can find a determination on each of the simply connected open sets mentioned above.

Let e_α be the trivialization corresponding to section $1 \in \mathbb{C}$ in the open $\mathcal{U}_\alpha$. By definition, we have $e_\alpha = g_{\alpha\beta}e_\beta$. Choosing any connection D on $\mathcal{L}$, then locally we express it by a 1-form $\lambda_\alpha \in \mathcal{A}^1(M)$ with $De_\alpha = \lambda_\alpha \otimes e\alpha$. On $\mathcal{U}_\alpha \cap \mathcal{U}_\beta$, we have

$$\lambda_\alpha \otimes e\alpha = De_\alpha = D(g_{\alpha\beta}e_\beta) = dg_{\alpha\beta} \otimes e_\beta + g_{\alpha\beta}De_\beta = \frac{dg_{\alpha\beta}}{g_{\alpha\beta}} \otimes e_\alpha + g_{\alpha\beta}\lambda_\beta \otimes e_\beta$$

so,

$$\lambda_\alpha \otimes e\alpha = d\log g_{\alpha\beta} \otimes e_\alpha + \lambda_\beta \otimes e_\alpha.$$

Hence, $\lambda_\alpha - \lambda_\beta = d\log g_{\alpha\beta}$, is the boundary of the 0-chain (λ_α) and hence the curvature form is $\Theta = d\lambda_\alpha = \lambda_\beta$. The curvature Θ is given as a 2-closed form $\Theta = d\lambda_\alpha$ with $\lambda_\alpha - \lambda_\beta = d\log g_{\alpha\beta}$ and the Chern class $c_1(\mathcal{L})$ is given by the Cech cocycle $(z_{\alpha\beta\gamma})$. To explain the de Rham isomorphism, it suffices to use the exact sequences:

$$0 \longrightarrow \mathbb{R} \longrightarrow \Omega^0 \xrightarrow{d} \mathcal{Z}^1 \longrightarrow 0,$$

where Ω^0 is the bundle of the 0-forms or the functions of $\mathcal{C}^\infty$ and

$$0 \longrightarrow \mathcal{Z}^1 \longrightarrow \Omega^1 \xrightarrow{d} \mathcal{Z}^2 \longrightarrow 0,$$

by virtue of Poincaré lemma. These sequences respectively provide the boundary operators

$$H^1(\mathcal{Z}^1) \xrightarrow{\partial} H^2(\mathbb{R}), \qquad \frac{H^0(\mathcal{Z}^2)}{dH^0(\Omega^1)} \xrightarrow{\partial} H^1(\mathcal{Z}^1).$$

For the first sequence, the boundary operator is calculated by writing $\Theta = d\lambda_\alpha$ locally and therefore $\{\lambda_\alpha - \lambda_\beta\} \in H^1(\mathcal{Z}^1)$. Now, $\lambda_\alpha - \lambda_\beta = d\log g_{\alpha\beta}$ so

$$(-\log g_{\alpha\beta} - \log g_{\beta\gamma} + \log g_{\alpha\gamma}) = -2\pi i c_1(\mathcal{L}).$$

For the second sequence, the boundary operator associates an element of $H^1(\mathcal{Z}^1)$ with a closed 2-form. We can therefore choose as a closed 2-form Θ and write it locally (as before) on each open $\mathcal{U}_\alpha$ of M, as the differential of a 1-form λ_α. We have $\Theta = d\lambda_\alpha$ on $\mathcal{U}_\alpha$ and the boundary of this 0-chain of 1-forms is the 1-chain $(\lambda_\alpha - \lambda_\beta) = d\log g_{\alpha\beta} \in H^1(\mathcal{Z}^1)$. Consequently, we have just shown that if $\mathcal{L}$ is a line bundle over M with any connection and if Θ is the curvature form of the connection, then we have

$$c_1(\mathcal{L}) = \frac{i}{2\pi}\Theta \in H^2_{DR}(M).$$

Note that this cohomology class is independent of the connection.

3.7. The Poincaré dual

Let $V \subset M$ be a hypersurface on the manifold M of dimension n. The homology class $(V) \in H_{2n-2}(M,\mathbb{Z})$ is defined by the following linear functional: $\varphi \longmapsto \int_V \varphi$, on $H^{2n-2}(M,\mathbb{Z})$ and we denote by $\eta_V \in H^2(M,\mathbb{C})$ the Poincaré dual. The class of this is called the fundamental class of the hypersurface V. In the special case of a divisor $\mathcal{D} = \sum k_j V_j$ over M, its fundamental class is given by the Poincaré dual $\eta_{\mathcal{D}} = \sum k_j.\eta_{V_j}$. Let $\mathcal{L}$ be a line bundle and $\mathcal{D}$ be a divisor over M. Using the Stokes–Cartan theorem, we can show that if $\mathcal{L} = [\mathcal{D}]$, then $c_1(\mathcal{L}) = \eta_{\mathcal{D}} \in H^2_{DR}(M)$, where $\eta_{\mathcal{D}}$ is the Poincaré dual of $\mathcal{D}$.

3.8. Exercises

EXERCISE 3.1.– *Show that the set of locally constant functions with values in an Abelian group G is a sheaf.*

Solution: Let $\mathcal{U}$ be a non-empty open set of M, $G(\mathcal{U}) = G$ and $G(\emptyset) = 0$. For $\mathcal{U} \subset \mathcal{V}$, we have $r_{\mathcal{U}\mathcal{V}} =$ identity. On the other hand, $U \longrightarrow \mathcal{F}(\mathcal{U}) = \mathbb{C}$ is not a sheaf.

EXERCISE 3.2.– *Show that the set of holomorphic functions (respectively holomorphic everywhere non-zero) on a complex manifold M is a sheaf denoted by $\mathcal{O}_M$ (respectively $\mathcal{O}_M^*$).*

Solution: Let

$$\mathcal{O}_M(\mathcal{U}) = \{f : \mathcal{U} \longrightarrow \mathbb{C} : f \text{ holomorphic}\},$$

and

$$\mathcal{O}_M^*(\mathcal{U}) = \{f : \mathcal{U} \longrightarrow \mathbb{C} : f \text{ holomorphic}, f(z) \neq 0, \forall z \in \mathcal{U}\},$$

where $\mathcal{U}$ is an open of M. If $\mathcal{U} \subset \mathcal{V}$, then $r_{\mathcal{U}\mathcal{V}}(f) = f|_{\mathcal{U}}$. The Abelian group structure on $\mathcal{O}_M(\mathcal{U})$ is addition (respectively multiplication).

EXERCISE 3.3.– *Show that the group $H^0(M, \mathcal{F})$ is canonically isomorphic to $\Gamma(M, \mathcal{F})$ (the elements of the latter are the global sections of $\mathcal{F}$; see section 3.1 for notations if necessary).*

Solution: Indeed, we have $H^0(M, \mathcal{F}) = \ker \delta$ with

$$\delta : C^0(\mathcal{U}, \mathcal{F}) \longrightarrow C^1(\mathcal{U}, \mathcal{F}), \quad s_{\alpha\beta} \longmapsto s_\beta|_{\mathcal{U}_\alpha \cap \mathcal{U}_\beta} - s_\alpha|_{\mathcal{U}_\alpha \cap \mathcal{U}_\beta},$$

and $\delta s_{\alpha\beta} = 0$ if and only if $s_\beta = s_\alpha$ on $\mathcal{U}_\alpha \cap \mathcal{U}_\beta$ for all α, β. Moreover, $\mathcal{F}$ being a sheaf, then the map

$$\Gamma(M, \mathcal{F}) \longrightarrow \ker \delta, \quad s \longmapsto s|_{\mathcal{U}_\alpha},$$

is bijective, hence the result.

EXERCISE 3.4.– *Show that for any sheaf of functions of class C^∞ on a manifold M, the cohomology groups of degree $k \geq 1$ are zero.*

Solution: Indeed, consider a cocycle $\sigma \in C^k(M, \mathcal{U})$ where $\mathcal{U} = (\mathcal{U}_\alpha)$ is an open covering of M and $\delta\sigma = 0$. Therefore, there exists a partition of unity ρ_α subordinate to the covering. We set

$$s_\alpha(\mathcal{U}_{\alpha_0}, ..., \mathcal{U}_{\alpha_{k-1}}) = \sum_\beta \rho_\beta \sigma(\mathcal{U}_\beta, \mathcal{U}_{\alpha_0}, ..., \mathcal{U}_{\alpha_{k-1}}),$$

and we directly verify that $\delta s_\alpha = \sigma$.

4

Divisors and Line Bundles

4.1. Divisors

DEFINITION 4.1.– *A (Weil) divisor on a complex manifold M is a formal linear combination*

$$\mathcal{D} = \sum_j n_j \mathcal{D}_j, \quad n_j \in \mathbb{Z},$$

where the $\mathcal{D}_j$ are irreducible subvarieties (or hypersurfaces) of codimension 1 in M.

This is an element of the free Abelian group generated by irreducible subvarieties of codimension 1 in M. It is a locally finite combination, i.e. each point of M has a neighborhood that meets only a finite number of $\mathcal{D}_j$ appearing in $\mathcal{D}$. Recall that a hypersurface of M is an analytic subvariety of codimension 1; i.e. of dimension equal to $\dim M - 1$, and it is called irreducible when it is not a union of distinct subvarieties of this hypersurface.

For example, for the special case of divisors on a Riemann surface (or on a complex algebraic curve) $\mathcal{C}$ which is a complex 1-dimensional manifold, a divisor on $\mathcal{C}$ is a formal combination of the type

$$\mathcal{D} = \sum_j n_j p_j, \quad n_j \in \mathbb{Z},$$

with (p_j) a locally finite family of points of $\mathcal{C}$. This notation used in the definition of the divisor is useful and denotes a finite (unordered) collection of points, and integers related to each of these points. Such a notation is an element of the free additive group having as its basis the elements of $\mathcal{C}$, which means in particular that two divisors are equal if, and only if, the points which appear there are the same except for the order, and that the coefficients of the same points are equal.

In this book, we often deal with compact varieties, which means that the above sum is finite for divisors. The set of divisors on M forms a group denoted $\mathrm{Div}(M)$. We say that a divisor $\mathcal{D}$ is positive or effective, and we denote $\mathcal{D} \geq 0$ if all integers n_j are positive.

Let p be a point on an irreducible hypersurface $\mathcal{U} \subset M$ and choose τ_p as the local function at p. For any holomorphic function f defined in the neighborhood of p, we define the order of f along $\mathcal{U}$ at p as the unique integer n such that: $f = \tau_p^n g$, where g is a holomorphic function at p and does not vanish along $\mathcal{U}$. The integer n is independent of the choice of τ_p. The order, which we denote by $\mathrm{ord}_{\mathcal{U}}(f)$, is well defined: $\mathrm{ord}_{\mathcal{U}}(f) = n$. We easily verify that

$$\mathrm{ord}_{\mathcal{U}}(f_1 f_2) = \mathrm{ord}_{\mathcal{U}}(f_1) + \mathrm{ord}_{\mathcal{U}}(f_2).$$

Let f be a meromorphic function on $\mathcal{U}$. We assume that this function can be written locally in the form $f = \dfrac{g}{h}$ where g and h are the two holomorphic functions. We have

$$\mathrm{ord}_{\mathcal{U}}(f) = \mathrm{ord}_{\mathcal{U}}(g) - \mathrm{ord}_{\mathcal{U}}(h).$$

When $\mathrm{ord}_{\mathcal{U}}(f) = n > 0$, we will say that the function f has a zero of order n on $\mathcal{U}$. Similarly, if $\mathrm{ord}_{\mathcal{U}}(f) = -k < 0$, we will say that f has a pole of order k on $\mathcal{U}$. To the function f, we make a divisor denoted by (f) correspond by setting

$$(f) = \sum_{\mathcal{U}} \mathrm{ord}_{\mathcal{U}}(f).\mathcal{U}.$$

Any divisor of a meromorphic function is said to be a principal divisor.

DEFINITION 4.2.– *A (Cartier) divisor $\mathcal{D}$ on a complex manifold M is the data of an open covering O_α of M and a family of meromorphic functions f_α such that on each intersection $O_\alpha \cap O_\beta$, the function $\dfrac{f_\alpha}{f_\beta}$ is holomorphic invertible.*

Let $\mathcal{M}_M$ be the sheaf of meromorphic functions on a complex manifold M and $\mathcal{M}_M^*$ be the (multiplicative) sheaf of invertible meromorphic functions on M. Let $\{f_\alpha\}$ be a global section of $\mathcal{M}_M^*/\mathcal{O}_M^*$, i.e. a collection of meromorphic functions $f_\alpha \neq 0$ defined on open sets O_α with $\cup O_\alpha = M$ and $\dfrac{f_\alpha}{f_\beta} \in H^0\left(O_\alpha \cap O_\beta, \mathcal{O}_M^*|_{O_\alpha \cap O_\beta}\right)$. With this terminology, we can translate the definition of a (Cartier) divisor on M by saying that it is a global section of the quotient sheaf $\mathcal{M}_M^*/\mathcal{O}_M^*$. Therefore, for any hypersurface $\mathcal{U} \subset M$, the order of f_α on $\mathcal{U}$ coincides with that of f_β on $\mathcal{U}$, and consequently we can associate with $\{f_\alpha\}$ the divisor

$$\mathcal{D} = \sum_{\mathcal{U}} \mathrm{ord}_{\mathcal{U}}(f_\alpha).\mathcal{U},$$

where for each $\mathcal{U}$, we choose α so that $O_\alpha \cap \mathcal{U} \neq \emptyset$. Conversely, given a divisor $\mathcal{D} = \sum n_j \mathcal{D}_j$, we can choose a covering $\{O_\alpha\}$ of M such that in each O_α, every hypersurface $\mathcal{D}_j$ has a local function $f_{j\alpha}$. So, we can pose

$$f_\alpha = \prod_j f_{j\alpha}^{n_j} \in H^0\left(\mathcal{U}_\alpha, \mathcal{M}_M^*|_{\mathcal{U}_\alpha}\right), \qquad [4.1]$$

which determines a global section of $\mathcal{M}_M^*/\mathcal{O}_M^*$. Hence, the following isomorphism

$$\mathrm{Div}(M) \simeq H^0\left(M, \mathcal{M}_M^*/\mathcal{O}_M^*\right),$$

and as a result we have the following.

PROPOSITION 4.1.– *Any divisor $\mathcal{D}$ on M is a global section of the quotient sheaf $\mathcal{M}_M^*/\mathcal{O}_M^*$, where $\mathcal{O}_M^*$ is the subsheaf of invertible holomorphic functions. Furthermore, Div(V) and $H^0\left(M, \mathcal{M}_M^*/\mathcal{O}_M^*\right)$ are isomorphic.*

REMARK 4.1.– *Cartier divisors on M are (informally) Weil divisors that can be defined locally by an equation, but some divisors do not have this property. Note that on a non-singular manifold in codimension 1, and therefore in particular if M is smooth, the notions of Weil divisor and Cartier divisor coincide. We do not distinguish these two notions in this work because the complex manifolds or complex tori that will be considered admit a holomorphic embedding into a complex projective space; these are called Abelian varieties, and these are always non-singular.*

4.2. Line bundles

DEFINITION 4.3.– *A (holomorphic) line bundle on a (connected) complex manifold M is equivalent to the data of a complex manifold $\mathcal{L}$ and a holomorphic map (projection) $\pi : \mathcal{L} \longrightarrow M$ such that there exists an open covering $\{O_\alpha\}$ of M and isomorphisms*

$$\varphi_\alpha : \pi^{-1}(O_\alpha) \longrightarrow O_\alpha \times \mathbb{C},$$

called trivializations respecting the vector space structure, i.e. for all alpha and all beta, the composite

$$\varphi_\alpha . \varphi_\beta^{-1}(p, \xi) : (O_\alpha \cap O_\beta) \times \mathbb{C} \longrightarrow (O_\alpha \cap O_\beta) \times \mathbb{C},$$

is given by $(p, \xi) \longmapsto (p, g_{\alpha\beta}(p)\xi)$, where $g_{\alpha\beta}$ are holomorphic functions (called transition functions) on $O_\alpha \cap O_\beta$ that do not vanish.

The transition functions $g_{\alpha\beta}$ satisfy the relations

$$g_{\alpha\beta}.g_{\beta\alpha} = 1 \text{ on } O_\alpha \cap O_\beta, \qquad g_{\alpha\beta}.g_{\beta\gamma}.g_{\gamma.\alpha} = 1 \text{ on } O_\alpha \cap O_\beta \cap O_\gamma,$$

and define a cocycle (of Cech) on M with coefficients in $\mathcal{O}_V^*$.

Given collections $\{g_{\alpha\beta}\}$ and $\{g'_{\alpha\beta}\}$ of transition functions corresponding to two trivializations of the line bundle $\mathcal{L} \longrightarrow M$, we can find a collection of functions $f_\alpha \in H^0(\mathcal{U}_\alpha, \mathcal{O}_M^*|_{\mathcal{U}_\alpha})$ such that we have

$$g'_{\alpha\beta}(z) = \frac{f_\alpha}{f_\beta} g_{\alpha\beta}(z).$$

In other words, the cocycles $\{g_{\alpha\beta}\}$ and $\{g'_{\alpha\beta}\}$ differ by a coboundary, so they are equivalent.

REMARK 4.2.– *A line bundle is often defined as a vector bundle of rank 1. A (holomorphic) vector bundle of rank r on a complex manifold M is a complex manifold E with a map $\pi : E \to M$ such that there exists a covering of M by trivializing open sets O_α and isomorphisms $\varphi_\alpha : \pi^{-1}(O_\alpha) \longrightarrow O_\alpha \times \mathbb{C}^r$, called local trivializations, such that the transition functions*

$$\varphi_\alpha.\varphi_\beta^{-1} : \varphi_\alpha\left(\pi_\alpha^{-1}(O_\alpha \cap O_\beta)\right) \longrightarrow \varphi_\beta\left(\pi_\beta^{-1}(O_\alpha \cap O_\beta)\right),$$

are written $(p, \xi) \longmapsto (p, g_{\alpha\beta}(p)\xi)$, where $g_{\alpha\beta}$ are holomorphic functions with values in $GL(r, \mathbb{C})$. The matrix $(g_{\alpha\beta}(p))$ of order r depending on $p \in M$ is called the transition matrix.

REMARK 4.3.– *In the following, and unless explicitly stated otherwise, when we write line bundles, this means that they are holomorphic bundles. This means that the trivializations $\varphi_\alpha : \pi^{-1}(O_\alpha) \longrightarrow O_\alpha \times \mathbb{C}$, are holomorphic maps, or what amounts to the same thing if and only if the bundle $\mathcal{L}$ can be constructed using holomorphic transition functions $g_{\alpha\beta}$.*

Let $\mathcal{L}$ be a line bundle over the complex manifold M with $\{O_\alpha\}$ being an open covering of M and $g_{\alpha\beta}$ transition functions. The dual $\mathcal{L}^*$ of the line bundle $\mathcal{L}$ is the bundle defined over M trivialized over the same covering $\{O_\alpha\}$ of V and having $\dfrac{1}{g_{\alpha\beta}}$ as transition functions. The tensor product of $\mathcal{L}$ with another line bundle $\mathcal{L}'$ defined on V (with the same cover $\{O_\alpha\}$ of M and $g'_{\alpha\beta}$ of the transition functions) is the bundle $\mathcal{L} \otimes \mathcal{L}'$ defined on M trivialized on the same cover $\{O_\alpha\}$ of M and having $g_{\alpha\beta}g'_{\alpha\beta}$ as transition functions. It should also be noted that if $\mathcal{L}$ is a line bundle and k is an integer, then the bundle $k\mathcal{L}$ defined by the cocycle $g_{\alpha\beta}^k$ is equal to the kth tensor power of $\mathcal{L}$.

The set of line bundles, modulo isomorphism, on M is equipped with a commutative group structure for the tensor product. Recall that this group called the Picard group is denoted by $\mathrm{Pic}(M)$ (see example 3.13). More precisely, these line bundles have a one-to-one correspondence with the elements of the Cech cohomology group $H^1(M, \mathcal{O}_M^*)$, where $\mathcal{O}_M^*$ is the sheaf of holomorphic functions that do not vanish on M. Note that the group operation in $H^1(M, \mathcal{O}_M^*)$ corresponds to the tensor product of line bundles. In summary, we have the following result:

PROPOSITION 4.2.– *The Picard group* $Pic(M)$ *is isomorphic to the cohomology group* $H^1(M, \mathcal{O}_M^*)$.

EXAMPLE 4.1.– *In the special case of complex algebraic curves* $\mathcal{C}$ *(or compact Riemann surfaces which are complex one-dimensional manifolds), the Picard group* $Pic(\mathcal{C})$ *is the group of divisors quotiented by the subgroup of principal divisors.*

4.3. Sections of line bundles

DEFINITION 4.4.– *A (global) section of a line bundle* $\pi : \mathcal{L} \longrightarrow M$ *on a complex manifold* V *is a holomorphic map* $s : V \longrightarrow \mathcal{L}$ *such that:* $\pi o s = id_M$. *The set of global sections is denoted by* $H^0(M, \mathcal{L})$.

Equivalently, we require that there exists an open covering $\{O_\alpha\}$ of M and trivializations such that the section s is defined by a family of functions s_α holomorphic on O_α with $s_\alpha(p) = g_{\alpha\beta}(p)s_\beta(p)$, on $O_\alpha \cap O_\beta$, where $g_{\alpha\beta}$ are the transition functions.

When we speak in the following of a section of a line bundle, it will be (and without specifying it) a holomorphic section.

A section $s : M\backslash\{p\} \longrightarrow \mathcal{L}$ has a pole of order k at p if, in a sufficiently small open trivialization set around p, the representation associated with s is written in the form

$$\sum_{n \geq -k} a_n z^n, \quad 0 < k < \infty, \quad a_{-k} \neq 0.$$

In this case, we will speak of a meromorphic section as well as a divisor of the zeros and poles of s by analogy with what is generally done for meromorphic functions on Riemann surfaces (see, for example, Lesfari (2023)).

EXAMPLE 4.2.– *Consider the Cartesian product* $V \times \mathbb{C}$. *Then, the projection* $M \times \mathbb{C} \longrightarrow M$ *determines a line bundle. The sections of this bundle are the holomorphic functions on* M. *Any line bundle isomorphic to this bundle is said to be trivial. A line bundle is trivial if and only if it has a section that is never zero (see exercise 3.1).*

Let $\mathcal{D}$ be a divisor on the variety M, defined locally by functions $f_\alpha \in \mathcal{M}^*(O_\alpha)$. To this divisor is naturally attached a line bundle denoted by $[\mathcal{D}]$. This is defined by the transition functions

$$g_{\alpha\beta} = \frac{f_\alpha}{f_\beta} \in H^0\left(O_\alpha \cap O_\beta, \mathcal{O}_M^*|_{O_\alpha \cap O_\beta}\right).$$

Obviously, we have $g_{\alpha\beta}g_{\beta\gamma}g_{\gamma\alpha} = 1$, on $O_\alpha \cap O_\beta \cap O_\gamma$. The line bundle $[\mathcal{D}]$ does not depend on the choice of the local equations of the divisor. The map

$$[.] : \text{Div}(V) \longrightarrow \text{Pic}(M), \quad \mathcal{D} \longmapsto [\mathcal{D}],$$

is a group homomorphism with

$$[\mathcal{D} + \mathcal{D}'] = [\mathcal{D}] \otimes [\mathcal{D}'], \quad (\mathcal{D}, \mathcal{D}' \in \text{Div}(M)), \qquad [k\mathcal{D}] = [\mathcal{D}]^k, \quad k \in \mathbb{Z},$$

and for $\pi : \mathcal{L} \longrightarrow \mathcal{D}$, we have $\pi^*([\mathcal{D}]) = [\pi^*(\mathcal{D})]$.

From the exact sequence of sheaves

$$0 \longrightarrow \mathcal{O}_M^* \stackrel{i}{\longrightarrow} \mathcal{M}_M^* \stackrel{j}{\longrightarrow} \mathcal{M}_M^*/\mathcal{O}_M^* \longrightarrow 0,$$

we deduce the exact cohomology sequence

$$0 \to H^0(M, \mathcal{O}_M^*) \stackrel{i^*}{\to} H^0(M, \mathcal{M}_M^*) \stackrel{j^*}{\to} H^0(M, \mathcal{M}_M^*/\mathcal{O}_M^*) \stackrel{\delta}{\to} H^1(M, \mathcal{O}_M^*) \to \cdots$$

In terms of the following identifications:

$$\text{Div}(M) \simeq H^0(M, \mathcal{M}_M^*/\mathcal{O}_M^*), \qquad \text{Pic}(M) \simeq H^1(M, \mathcal{O}_M^*),$$

we have

$$\cdots \longrightarrow \mathcal{M}_M^* \longrightarrow \text{Div}(M) \longrightarrow \text{Pic}(M) \longrightarrow \cdots$$

Therefore, $j^*f = (f)$ for any meromorphic function f. The map δ of the group of divisors to the group of line bundles on M is a homomorphism and $\delta\mathcal{D} = [\mathcal{D}]$ for any divisor $\mathcal{D}$ on M. Hence,

PROPOSITION 4.3.– *$\mathcal{D}$ is a principal divisor if and only if the line bundle $[\mathcal{D}]$ associated with this divisor is trivial.*

For any line bundle $\mathcal{L}$ on M, we consider the set $\mathcal{O}_M(\mathcal{L})(\mathcal{U})$ of sections associated with a line bundle $\mathcal{L}|_O$ where O is an open set of M. We know that for any open covering $\{O_\alpha\}$ of O, a section $s \in \mathcal{O}_M(\mathcal{L})(O)$ is entirely determined by its restrictions $s|_{O_\alpha} \in \mathcal{O}_M(\mathcal{L})(O_\alpha)$. Conversely, if for all α, a section $s_\alpha \in \mathcal{O}_M(\mathcal{L})(O_\alpha)$ is such that: for all α, β, the sections s_α, s_β, have the same restriction to $O_\alpha \cap O_\beta$, then we can find a section $s \in \mathcal{O}_M(\mathcal{L})(O)$ satisfying $s|_{O_\alpha} = s_\alpha$ for all α. The sheaf thus defined is the sheaf associated with $\mathcal{L}$ and we denote it $\mathcal{O}_M(\mathcal{L})$; we sometimes write $H^0(O, \mathcal{O}_M)$ instead of $\mathcal{O}_M(\mathcal{L})$. When $\mathcal{L}$ is trivial, we simply have $\mathcal{O}_M(\mathcal{L}) \simeq \mathcal{O}_M$. For a divisor $\mathcal{D}$ on M, if $\mathcal{L} = [\mathcal{D}]$ is the line bundle associated with this divisor, then we simply write $\mathcal{O}_M(\mathcal{D})$ instead of $\mathcal{O}_M(\mathcal{L})$. The space of sections of the line bundle $[\mathcal{D}]$ is denoted by $H^0(M, \mathcal{O}_M([\mathcal{D}]))$ or simply $H^0(M, [\mathcal{D}])$.

EXAMPLE 4.3.– *An interesting example of a line bundle is the bundle denoted by $\mathcal{O}_{\mathbb{P}^n(\mathbb{C})}(1)$. To define it, we associate with a point $[z]$ of the complex projective space $\mathbb{P}^n(\mathbb{C})$ the line $\mathbb{C}z$ of $\mathbb{C}^{n+1}$. We thus construct a line bundle (tautological bundle) denoted by $\mathcal{O}_{\mathbb{P}^n(\mathbb{C})}(-1)$ which is in fact a line subbundle of the trivial bundle $\mathbb{C}^{n+1} \times \mathbb{P}^n(\mathbb{C})$. By definition, the dual of this line bundle is the bundle $\mathcal{O}_{\mathbb{P}^n(\mathbb{C})}(1)$. The latter admits as global sections the linear forms on $\mathbb{C}^{n+1}$, while the global sections of $\mathcal{O}_{\mathbb{P}^n(\mathbb{C})}(-1)$ are zero. Conversely, any linear form on $\mathbb{C}^{n+1}$ defines a section of the dual bundle $\mathcal{O}_{\mathbb{P}^n(\mathbb{C})}(1)$. The sections of $\mathcal{O}_{\mathbb{P}^n(\mathbb{C})}(1)$ are identified with homogeneous polynomials of degree 1 in $n+1$ variables.*

EXAMPLE 4.4.– *The line bundle $\mathcal{O}_{\mathbb{P}^n(\mathbb{C})}(d)$ is identified with $\mathcal{O}_{\mathbb{P}^n(\mathbb{C})}(1)^{\otimes d}$ for $d \geq 0$ and with $\mathcal{O}_{\mathbb{P}^n(\mathbb{C})}(-1)^{\otimes -d}$ for $d < 0$. This bundle is isomorphic to the line bundle associated with a hypersurface of degree d, i.e. whose sections are defined by homogeneous polynomials of degree d in $n+1$ variables. The transition functions of $\mathcal{O}_{\mathbb{P}^n(\mathbb{C})}(d)$, $d \geq 0$, are*

$$\varphi_{\alpha\beta}(p) = \left(\frac{p_\beta}{p_\alpha}\right)^d,$$

and a section of this bundle is

$$s_\alpha(p) = \frac{Q_d(p)}{p_\alpha^d},$$

where $Q_d(p)$ denotes a polynomial of degree $d \geq 0$ in $n+1$ variables. We have

$$\mathcal{O}_{\mathbb{P}^n(\mathbb{C})}(d_1) \otimes \mathcal{O}_{\mathbb{P}^n(\mathbb{C})}(d_2) \simeq \mathcal{O}_{\mathbb{P}^n(\mathbb{C})}(d_1 + d_2),$$

for all integers d_1 and d_2. The bundle $\mathcal{O}_{\mathbb{P}^n(\mathbb{C})}(d_1)$ has a non-zero global section if and only if $d_1 \geq 0$. So, for $d_1 < 0$, the only section is the zero section. Hence, the bundle $\mathcal{O}_{\mathbb{P}^n(\mathbb{C})}(d_1) \otimes \mathcal{O}_{\mathbb{P}^n(\mathbb{C})}(d_2)$ has a section if and only if $d_1 + d_2 \geq 0$. Therefore, the bundles $\mathcal{O}_{\mathbb{P}^n(\mathbb{C})}(d_1)$ and $\mathcal{O}_{\mathbb{P}^n(\mathbb{C})}(d_2)$ are isomorphic if and only if $d_1 = d_2$. We thus obtain all the line bundles on the space $\mathbb{P}^n(\mathbb{C})$ and its Picard group is isomorphic to $\mathbb{Z}$.

REMARK 4.4.– *We take the global section $\{f_\alpha\}$ of the sheaf $\mathcal{M}_M^*/\mathcal{O}_M^*$ where f_α is given explicitly by formula [4.1]. As we have already explained, we can associate with*

$\{f_\alpha\}$ a divisor $\mathcal{D}$ and there exists a section $s \in H^0(M, \mathcal{O}_M[\mathcal{D}])$ of the line bundle $[\mathcal{D}]$ such that: $s^{-1}(0) = \mathcal{D}$. The sequence

$$0 \longrightarrow \mathcal{O}_M(\mathcal{L} \otimes [-\mathcal{D}]) \longrightarrow \mathcal{O}_M(\mathcal{L}) \xrightarrow{r} \mathcal{O}_{\mathcal{D}}(\mathcal{L}|_{\mathcal{D}}) \longrightarrow 0,$$

(where r denotes the restricted map) is exact.

It should be noted that the space $H^0(M, \mathcal{O}_M^*[\mathcal{D}])$ of sections of the line bundle $[\mathcal{D}]$, is interpreted (see below) as the vector space

$$\mathcal{L}(\mathcal{D}) = \{f \text{ meromorphic on } M : (f) + \mathcal{D} \geq 0\},$$

attached to a divisor $\mathcal{D}$ on the manifold M, it is formed from the meromorphic functions f on M such that: $(f) + \mathcal{D} \geq 0$.

Recall that in the special case of divisors

$$\mathcal{D} = \sum_j n_j p_j, \quad n_j \in \mathbb{Z},$$

on a Riemann surface $\mathcal{C}$, ((p_j) being a locally finite family of points of $\mathcal{C}$), saying that $f \in \mathcal{L}(\mathcal{D}) = \{f \text{ meromorphic on } \mathcal{C} : (f) + \mathcal{D} \geq 0\}$, means that if for p_j in the divisor $\mathcal{D}$, the coefficient n_j is less than 0, then f must have a zero of multiplicity at least $|n_j|$ in p_j; otherwise, f can have a pole of multiplicity at most n_j at p_j. In particular, if $\mathcal{D}$ is positive, then the space $\mathcal{L}(\mathcal{D})$ consists of meromorphic functions that can only have poles at points p_j with multiplicity not exceeding n_j. It is clear that $\mathcal{L}(\mathcal{D})$ forms a vector space over $\mathbb{C}$.

In the case of a divisor $\mathcal{D}$ on a complex manifold M, of the form

$$\mathcal{D} = \sum_j n_j \mathcal{D}_j, \quad n_j \in \mathbb{Z},$$

where the $\mathcal{D}_j$ are irreducible analytic subvarieties of codimension 1, the space $\mathcal{L}(\mathcal{D})$ represents the set of holomorphic functions on $M \backslash \mathcal{D}$, having at most one pole of order n_j on $\mathcal{D}_j$ when $n_j > 0$ and a zero of order at least equal to $-n_j$ when $n_j \leq 0$. Two divisors $\mathcal{D}$ and $\mathcal{D}'$ are linearly equivalent (we denote by $\mathcal{D} \sim \mathcal{D}'$) if $\mathcal{D} - \mathcal{D}'$ is principal, i.e. $\mathcal{D} - \mathcal{D}' = (f)$, where f is a meromorphic function.

We denote by $|\mathcal{D}|$ the complete linear system of the divisor $\mathcal{D}$; it is the set of positive divisors linearly equivalent to $\mathcal{D}$, i.e.

$$|\mathcal{D}| = \{\mathcal{E} \in \text{Div}(M) : \mathcal{E} \geq 0, \mathcal{E} \sim \mathcal{D}\}.$$

For any line bundle $\mathcal{L}$ on a compact complex manifold M, the space $H^0(V, \mathcal{O}_M(\mathcal{L}))$ of sections of $\mathcal{L}$ is finite-dimensional (in general, if the manifold M is not compact, then this space is infinite-dimensional). We denote the projective space associated with $H^0(M, \mathcal{O}_M(\mathcal{L}))$, by $\mathbb{P}\left(H^0(M, \mathcal{O}_M(\mathcal{L}))\right)$. Note that the map

$$\mathbb{P}\left(H^0(M, \mathcal{O}_M(\mathcal{L}))\right) \simeq \mathbb{P}(\mathcal{L}(\mathcal{D})) \longrightarrow |\mathcal{D}|, \quad \dot{f} \longmapsto \mathcal{D} + (f),$$

ensures that the system $|\mathcal{D}|$ is parameterized by the space $\mathbb{P}(\mathcal{L}(\mathcal{D})) \cong \mathbb{P}^N$ where $N = \dim \mathcal{L}(\mathcal{D} - 1$ (the notation $\dot{f}$ denotes here the class of f).

Any subspace of $\mathbb{P}(H^0(M, \mathcal{O}_M(\mathcal{L})))$ is called a linear system of divisors. This is complete if it is of the form $|\mathcal{D}|$ for at least one divisor $\mathcal{D}$. We associate with the line bundle $\mathcal{L}$ a holomorphic map from an open set O of M to the dual projective space $\mathbb{P}(H^0(M, \mathcal{O}_M(\mathcal{L}))^*)$. A point $p \in M$ is called a base point of $\mathcal{L}$ if $s(p) = 0$ for any section $s \in H^0(M, \mathcal{O}_M(\mathcal{L}))$ and the set of base points is called a base locus. The latter is an analytic submanifold. The base locus of a linear system of the form $\{\mathcal{D}_\alpha\}$ is the common intersection $\bigcap_\alpha \mathcal{D}_\alpha$ and a fixed component of the bundle $\{\mathcal{D}_\alpha\}$ is a divisor $\mathcal{S}$ such that: $\mathcal{D}_\alpha - \mathcal{S} \geq 0$ for all α. Another way to describe the basis points of $|\mathcal{D}|$ is to consider a basis $(f_1, ..., f_N)$ of the space $\mathcal{L}(\mathcal{D})$ (we show in the following proposition that there is an isomorphic vector spaces between $\mathcal{L}(\mathcal{D})$ and $H^0(M, \mathcal{O}_M(\mathcal{D}))$ and we will see in Chapter 8 how to explicitly determine such a basis when dealing with concrete problems), the morphism

$$M \longrightarrow \mathbb{P}^N, \quad z \longmapsto [f_1(z) : ... : f_N(z)],$$

where $N = \dim \mathcal{L}(\mathcal{D}) - 1$ (which is indeed defined modulo the automorphisms of the projective space $\mathbb{P}^N$ and outside the poles of the functions f_j and their common zeros), and the intersection of the supports of the divisors in $|\mathcal{D}|$. Then, the base points of $|\mathcal{D}|$ are the points in this intersection. Moreover, such an intersection may be empty; in this case, we will say that $|\mathcal{D}|$ has no base point.

PROPOSITION 4.4.– *We have the following isomorphism:*

$$|\mathcal{D}| \simeq \mathbb{P}(\mathcal{L}(\mathcal{D})) \simeq \mathbb{P}\left(H^0(M, \mathcal{O}_M(\mathcal{D}))\right).$$

PROOF.– Let us first show that there is an isomorphism vector space between $\mathcal{L}(\mathcal{D})$ and $H^0(M, \mathcal{O}_M(\mathcal{D}))$. Let s_0 be a global meromorphic section of $[\mathcal{D}]$ with $\mathcal{D} = (s_0)$ the divisor of s_0. This can be defined by taking local trivializations. Therefore, for any global section $s \in H^0(M, \mathcal{O}_M(\mathcal{D}))$, we can consider the quotient $f_s = \dfrac{s}{s_0}$ as a rational function, so as a meromorphic function on M with

$$(f_s) + \mathcal{D} = (s) - (s_0) + \mathcal{D} = (s) \geq 0,$$

i.e. $f_s \in \mathcal{L}(\mathcal{D})$. Moreover, for any function $f \in \mathcal{L}(\mathcal{D})$, the section $s = f.s_0$ of $[\mathcal{D}]$ is holomorphic. Therefore, multiplication by s_0 determines a (non-canonical) isomorphism,

$$\mathcal{L}(\mathcal{D}) \xrightarrow{\otimes s_0} H^0(M, \mathcal{O}_M(\mathcal{D})).$$

For every $\mathcal{D}' \in |\mathcal{D}|$, there exists a function $f \in \mathcal{L}(\mathcal{D})$ such that: $\mathcal{D}' - \mathcal{D} = (f)$. We have $\deg \mathcal{D}' = \deg \mathcal{D}$ and $\dim \mathcal{L}(\mathcal{D}') = \dim \mathcal{L}(\mathcal{D})$, for two equivalent divisors. Two functions $f, g \in \mathcal{L}(\mathcal{D})$ can only satisfy the relation,

$$(f) + \mathcal{D} = (g) + \mathcal{D},$$

if $\frac{f}{g}$ is a non-zero constant of $\mathbb{C}$. This proves that the set $|\mathcal{D}|$ is identified with the projective space $\mathbb{P}(\mathcal{L}(\mathcal{D}))$ of dimension $\dim \mathcal{L}(\mathcal{D}) - 1$. This completes the proof of the proposition. ■

EXAMPLE 4.5.– *Consider the case where M is the complex projective space $\mathbb{P}^n(\mathbb{C})$ of dimension n and let $\mathcal{D} = \mathcal{H}$ be an arbitrary hyperplane of $\mathbb{P}^n(\mathbb{C})$. The transition functions for the line bundle $[\mathcal{H}]$ are determined by*

$$(p, \xi) \longmapsto \left(p, \frac{p_k}{p_j}.\xi\right),$$

in $O_j \cap O_k$ where O_j is an open covering of the projective space $\mathbb{P}^n(\mathbb{C})$. Here, $H^0\left(\mathbb{P}^n(\mathbb{C}), \mathcal{O}_{\mathbb{P}^n(\mathbb{C})}[\mathcal{H}]\right)$ is the vector space of all linear forms

$$(p_0, p_1, ..., p_n) \longmapsto \lambda_0 p_0 + \cdots + \lambda_n p_n,$$

on $\mathbb{C}^{n+1}$.

EXAMPLE 4.6.– *Let ω_1 and ω_2 be two holomorphic n-differential forms on M where n is the dimension of the manifold M. We can choose a meromorphic function f on M such that: $f = \frac{\omega_1}{\omega_2}$. Therefore,*

$$(\omega_1) - (\omega_2) = (f),$$

and so $(\omega_1) \sim (\omega_2)$, i.e. the zero loci of ω_1 and ω_2 are linearly equivalent. Conversely, given a holomorphic n-differential form on M, any positive divisor linearly equivalent to (ω) is the locus of zeros of a holomorphic n-differential form on M. So, all holomorphic n-differential forms on M define the only and the same line bundle on M. This is called the canonical fiber of M and is denoted by K_M, while the divisor (ω) where ω is a holomorphic n-differential form on M is called a canonical divisor. The sections of K_M are the holomorphic n-differential forms on M and hence $\mathcal{O}_M(K_M) = \Omega^n_M$. Therefore, for any line bundle over M, we have an isomorphism sheaf,

$$\mathcal{O}_M(\mathcal{L}) \simeq \Omega^n_M(\mathcal{L} \otimes K_M^{-1}).$$

The Euler–Poincaré characteristic of a line bundle $\mathcal{L}$ on a compact complex manifold M is defined by

$$\chi(\mathcal{L}) = \sum_{k=0}^{n}(-1)^k \dim H^k(M, \mathcal{O}_V(\mathcal{L})).$$

This is a topological invariant. In the case of a divisor $\mathcal{D}$, we use the same notation.

4.4. Tori and Riemann form

Let E be a real vector space of finite dimension n. A lattice Λ in E is by definition a discrete subgroup of rank n. Recall that Λ is discrete if and only if there exist r-vectors $e_1, ..., e_r$ linearly independent of $\mathbb{R}$ where $r \leq n$ is an integer (called the rank of Λ) and such that

$$\Lambda = \mathbb{Z}e_1 \oplus \cdots \oplus \mathbb{Z}e_r = \left\{ \sum_{j=1}^{r} k_j e_j : k_j \in \mathbb{Z} \right\}.$$

Thus, Λ is a subgroup generated by a basis of the space E.

A real torus is a topological (Lie) group diffeomorphic to the quotient E/Λ; it is equipped with the quotient topology. We show that E/Λ is diffeomorphic to a product of circles,

$$E/\Lambda \simeq \mathbb{R}^n/\mathbb{Z}^n = (\mathbb{R}/\mathbb{Z})^n = \left(S^1\right)^n,$$

where S^1 is the unit circle. Note that in particular, Λ is a lattice if and only if E/Λ is compact.

Let us now assume that E is a complex vector space of complex dimension n (so, of real dimension $2n$) and that Λ is a lattice of E (considered as a real vector space, therefore of rank $2n$).

A complex torus E/Λ of dimension n is the quotient of a complex vector space E of dimension n by a lattice Λ, i.e. by a discrete and free $\mathbb{Z}$-module, of rank $2n$. We show that E/Λ is a complex manifold of complex dimension n, and it is equipped with an Abelian Lie group structure.

Let E/Λ be a complex torus, N be a non-zero integer, and

$$\psi : E/\Lambda \longrightarrow \mathbb{P}^N(\mathbb{C}).$$

Then, we say that ψ is an embedding if ψ is injective and if the differential $d\psi_z$ is injective for all $z \in E/\Lambda$.

Let E be a complex vector space. A function $\varphi : E \times E \longrightarrow \mathbb{C}$, is a sesquilinear form if it is linear in its first argument and semilinear in its second argument, which means that

$$\varphi(x_1 + x_2, y) = \varphi(x_1, y) + \varphi(x_2, y), \quad \varphi(x, y_1 + y_2) = \varphi(x, y_1) + \varphi(x, y_2),$$

$$\varphi(\alpha x, y) = \alpha\varphi(x, y), \qquad \varphi(x, \beta y) = \overline{\beta}\varphi(x, y),$$

for all $x, y, x_1, x_2, y_1, y_2 \in E$ and all $\alpha, \beta \in \mathbb{C}$. A function $\varphi : E \times E \longrightarrow \mathbb{C}$, is a hermitian form if it is sesquilinear and if

$$\varphi(y, x) = \overline{\varphi(x, y)},$$

for all $x, y \in E$.

Let $\mathbb{T} = E/\Lambda$ be a complex torus where Λ is a lattice of E. We will denote h_1 the real part of h and h_2 its imaginary part; we write

$$h(x, y) = h_1(x, y) + ih_2(x, y).$$

Since

$$h(y, x) = \overline{h(x, y)} = h_1(x, y) - ih_2(x, y),$$

then

$$h_1(y, x) = h_1(x, y), \qquad h_2(y, x) = -h_2(x, y),$$

i.e. the form h_1 is $\mathbb{R}$-bilinear symmetric while h_2 is an alternating (or antisymmetric) form on E. Note that $h(ix, y) = -ih(x, y)$, which shows that $h_1(ix, y) = h_2(x, y)$ and $h_2(ix, y) = -h_1(x, y)$. Similarly, we have $h(x, iy) = ih(x, y)$, hence $h_1(x, iy) = -h_2(x, y)$ and $h_2(x, iy) = h_1(x, y)$. We also have $h(ix, iy) = h(x, y)$, which implies that $h_1(ix, iy) = h_1(x, y)$ and $h_2(ix, iy) = ih_2(x, y)$. We have $h_2(x, ix) > 0$ if $x \neq 0$. Taking these expressions into account, we can therefore write

$$h(x, y) = h_1(x, y) + ih_2(x, y) = h_1(x, y) - ih_2(x, iy) = h_2(x, iy) + ih_2(x, y).$$

The form h_2 is called a Riemann form. If h_2 verifies $h_2(\Lambda, \Lambda) \subset \mathbb{Z}$ (i.e. the value $h_2(x, y)$ is an integer for all $x, y \in \Lambda$), then we say that h is a polarization for the torus $\mathbb{T}$ (and that $\mathbb{T}$ is polarized or polarizable). Note that

$$\det h_2 = (-1)^n \delta^2,$$

because rg $\Lambda = 2n$, h_2 is alternating and $h_2(\Lambda, \Lambda) \subset \mathbb{Z}$. We say that δ is the degree of polarization h. A polarization of degree 1 is called a principal polarization.

DEFINITION 4.5.– *A Riemann form on a lattice Λ of a complex vector space E is a non-degenerate positive definite hermitian form $h(x, y)$ on E such that its imaginary part $h_2 =$ Im h takes integer values on $\Lambda \times \Lambda$, (i.e. $h_2(x, y) \in \mathbb{Z}$ for all $x, y \in \Lambda$).*

4.5. Line bundles on complex tori

Let $\mathbb{T}^n = \mathbb{C}^n/\Lambda$ be a complex torus where Λ is a lattice of $\mathbb{C}^n$ and $\tau : \mathcal{L} \longrightarrow \mathbb{T}^n$, a line bundle (also denoted by $\mathcal{L}$). Consider the pullback

$$\pi^*\mathcal{L} = \{(u, v) \in \mathbb{C}^n \times \mathcal{L} : \pi(u) = \tau(v)\},$$

with the regular map $\pi : \mathbb{C}^n \longrightarrow \mathbb{T}^n$. Note that the space $\mathbb{C}^n$ is contractible (recall that a topological space E is contractible if it is homotopically equivalent to a point. In other words, if Id_E is homotopic to a constant map: $E \longrightarrow E$. Here, $E = \mathbb{C}^n$ and the map $f : \mathbb{C}^n \times [0, 1] \longrightarrow \mathbb{C}^n$, $(z, t) \longmapsto tz$, is a continuous deformation of $Id_{\mathbb{C}^n}$ to the zero constant function, then $\mathbb{C}^n$ is contractible). So, since $\mathbb{C}^n$ is contractible and moreover any bundle over a contractible space is trivial, we deduce that the bundle $\pi^*\mathcal{L}$ is trivial and we have a global trivialization:

$$\begin{aligned}
\varphi &: \pi^*\mathcal{L} \longrightarrow \mathbb{C}^n \times \mathbb{C},\\
\varphi_z &: (\pi^*\mathcal{L})_z \longrightarrow \mathbb{C}, \quad z \in \mathbb{C}^n\\
\varphi_{z+\lambda} &: (\pi^*\mathcal{L})_{z+\lambda} \longrightarrow \mathbb{C}, \quad z \in \mathbb{C}^n, \quad \lambda \in \Lambda
\end{aligned}$$

We have $(\varphi_z) = (\varphi_{z+\lambda})$ and

$$\mathbb{C} \overset{\varphi_z}{\longleftarrow} (\varphi_z) = (\varphi_{z+\lambda}) \overset{\varphi_{z+\lambda}}{\longrightarrow} \mathbb{C},$$

provides a linear automorphism. Therefore, the multiples for $\mathcal{L}$ are given by

$$e_\lambda(z) = \varphi_{z+\lambda}^{-1} . \varphi_z \in \mathbb{C},$$

and satisfy for all $\lambda, \mu \in \Lambda$ the relations

$$e_\mu(z + \lambda)e_\lambda(z) = e_\lambda(z + \mu)e_\mu(z) = e_{\lambda+\mu}(z).$$

Conversely, given a set of multiples $e_\lambda(z)$ satisfying these relations, we can define the line bundle $\mathcal{L} \longrightarrow \mathbb{T}^n$ by taking

$$\mathcal{L} \simeq \mathbb{C}^n \times \mathbb{C}/\{(z, \xi) \sim (z + \lambda, e_\lambda(z).\xi)\}.$$

PROPOSITION 4.5.– *Let $\mathcal{L}$ be a line bundle over a torus $\mathbb{T}^n = \mathbb{C}^n/\Lambda$. Then, there exists a basis $(\lambda_1, ..., \lambda_{2n})$ of Λ and a trivialization of $\pi^*\mathcal{L}$ such that: $e_{\lambda_\alpha}(z) \equiv 1$, $\alpha = 1, ..., n$.*

PROOF.– We assume that $\lambda_1, ..., \lambda_n$ are linearly independent on $\mathbb{C}$, hence

$$\mathbb{C}^n/\mathbb{Z}\{\lambda_1, ..., \lambda_n\} \simeq (\mathbb{C}^*)^n.$$

Let $\pi_1^* \mathcal{L}$ be the pullback of the bundle $\mathcal{L}$ with $\pi_1 : (\mathbb{C}^*)^n \longrightarrow \mathbb{T}^n$. On $\mathbb{T}^n$, we have an exact sequence of sheaves of groups:

$$0 \longrightarrow \mathbb{Z} \longrightarrow \mathcal{O}_{\mathbb{T}^n} \overset{\exp}{\longrightarrow} \mathcal{O}^*_{\mathbb{T}^n} \longrightarrow 0,$$

where $\mathcal{O}_{\mathbb{T}^n}$ denotes the sheaf of holomorphic functions and $\mathcal{O}^*_{\mathbb{T}^n}$ denotes the sheaf of invertible holomorphic functions on $\mathbb{T}^n$ and where $exp(z) = e^{2\pi i z}$. Since

$$H^1((\mathbb{C}^*)^n, \mathcal{O}_{\mathbb{T}^n}) = H^2((\mathbb{C}^*)^n, \mathcal{O}_{\mathbb{T}^n}) = 0,$$

we immediately deduce that:

$$H^1((\mathbb{C}^*)^n, \mathcal{O}^*_{\mathbb{T}^n}) \overset{c_1}{\longrightarrow} H^2((\mathbb{C}^*)^n, \mathbb{Z}),$$

the class $c_1(\mathcal{L})$ is the (first) Chern class of the line bundle $\mathcal{L}$. For the latter, we can choose a basis $(\lambda_1, ..., \lambda_{2n})$ of Λ such that in the corresponding dual basis $(x_1, ..., x_{2n})$ we have,

$$c_1(\mathcal{L}) = \sum_{\alpha=1}^{n} \delta_\alpha dx_\alpha \wedge dx_{n+\alpha}.$$

Hence, $c_1(\pi_1^* \mathcal{L}) = \pi_1^*(c_1 \mathcal{L}) = 0$ (because the coordinates $x_{n+\alpha}$ are well-defined on $(\mathbb{C}^*)^n$), and consequently $\pi_1^* \mathcal{L}$ is trivial. Taking a trivialization $\pi_1^* \mathcal{L} \longrightarrow (\mathbb{C}^*)^n \times \mathbb{C}$, and extending it to $\pi^* \mathcal{L} \longrightarrow \mathbb{C}^n$, we obtain $e_{\lambda_\alpha}(z) \equiv 1$, $\alpha = 1, ..., n$. The proof ends. ■

PROPOSITION 4.6.– *Any line bundle $\mathcal{L}$ on a complex torus $\mathbb{T}^n = \mathbb{C}^n / \Lambda$ with the multiples $e_\lambda(z)$ verifying*

$$e_{\lambda_\alpha}(z) \equiv 1, \qquad e_{\lambda_{n+\alpha}}(z) = e^{-2\pi i z_\alpha}, \quad \alpha = 1, ..., n \tag{4.2}$$

admits the Chern class $c_1(\mathcal{L}) = [\omega]$.

PROOF.– We can choose a basis $(\lambda_1, ..., \lambda_{2n})$ of Λ such that in the dual basis $(x_1, ..., x_{2n})$, the period matrix is written in the form:

$$\Omega = (\Delta_\delta, Z), \qquad \Delta_\delta = \text{diag}\,(\delta_1, ..., \delta_n), \quad \delta_\alpha \in \mathbb{N}^*,$$

with $\delta_\alpha | \delta_{\alpha+1}$, $\alpha = 1, ..., n$, $Z^\top = Z$, $\text{Im}\, Z > 0$. The Hodge form is written as,

$$\omega = \sum_{\alpha=1}^{n} \delta_\alpha dx_\alpha \wedge dx_{n+\alpha}.$$

Let $e_\lambda(z)$ be the multiples of the line bundle $\mathcal{L}$ on $\mathbb{T}^n$, satisfying [4.2]. Since $Z_{\alpha\beta} = Z_{\beta\alpha}$, we deduce that these multiples satisfy the relations:

$$e_{\lambda_\alpha}(z + \lambda_\beta) e_{\lambda_\beta}(z) = e_{\lambda_\beta}(z + \lambda_\alpha) e_{\lambda_\alpha}(z).$$

Let us write Z in the form $Z = X + iY$. Since $Y = \text{Im } Z > 0$, then we can set $W = (w_{\alpha\beta}) = Y^{-1}$. Therefore, the function defined by

$$h(z) = e^{\frac{\pi}{2}\sum W_{\alpha\beta}(z_\alpha - \overline{z}_\alpha)(z_\beta - \overline{z}_\beta - 2iY_{\beta\beta})},$$

satisfies the relations:

$$h(z + \lambda_\alpha) = h(z), \qquad h(z + \lambda_{n+\alpha}) = \left|e^{2\pi i z_\alpha}\right|^2 h(z). \tag{4.3}$$

Such a function determines a metric on the line bundle $\mathcal{L}$. A direct calculation shows that the curvature form $\Theta_{\mathcal{L}}$ associated with this metric is

$$\Theta_{\mathcal{L}} = \partial\overline{\partial}\log\frac{1}{h} = \pi\sum_{\alpha,\beta} W_{\alpha\alpha} dz_\alpha \wedge d\overline{z}_\beta.$$

Since

$$dz_\alpha = \delta_\alpha dx_\alpha + \sum_\beta z_{\alpha\beta} dx_{n+\beta}, \qquad d\overline{z}_\beta = \delta_\alpha dx_\alpha + \sum_\beta \overline{z}_{\alpha\beta} dx_{n+\beta},$$

so,

$$\Theta_{\mathcal{L}} = -2\pi i \sum_\alpha \delta_\alpha dx_\alpha \wedge dx_{n+\alpha} = -2\pi i\omega,$$

which shows that

$$c_1(\mathcal{L}) = \left[\frac{i}{2\pi}\Theta\right] = [\omega],$$

and the proof ends. ■

PROPOSITION 4.7.– *Let $\mathcal{L}$ be a line bundle over a torus $\mathbb{T}^n = \mathbb{C}^n/\Lambda$. Then, any line bundle with the same Chern class as $\mathcal{L}$ is a translate of $\mathcal{L}$. Furthermore, $\mathcal{L}$ is fixed by $\prod_{\alpha=1}^n \delta_\alpha^2$ translations where $\delta_1, ..., \delta_n$ are the elementary divisors associated with the polarization $c_1(\mathcal{L})$.*

PROOF.– We first consider the case of a line bundle $\widetilde{\mathcal{L}} \longrightarrow \mathbb{T}^n$ whose Chern class is zero. From the following exact sequences of sheaves:

$$\begin{array}{ccccccccc}
0 & \longrightarrow & \mathbb{Z} & \longrightarrow & \mathcal{O} & \xrightarrow{\exp} & \mathcal{O}^* & \longrightarrow & 0 \\
 & & \| & & \uparrow & & \uparrow & & \\
0 & \longrightarrow & \mathbb{Z} & \longrightarrow & \mathbb{C} & \longrightarrow & \mathbb{C}^* & \longrightarrow & 0
\end{array}$$

on $\mathbb{T}^n$ (or on any compact Kähler manifold, definition 5.6), we deduce the following commutative diagram:

$$\begin{array}{ccccc}
H^1(\mathbb{T}^n, \mathcal{O}) & \longrightarrow & H^1(\mathbb{T}^n, \mathcal{O}^*) & \xrightarrow{c_1} & H^2(\mathbb{T}^n, \mathbb{Z}) \\
f \uparrow & & g \uparrow & & \Vert \\
H^1(\mathbb{T}^n, \mathbb{C}) & \longrightarrow & H^1(\mathbb{T}^n, \mathbb{C}^*) & \longrightarrow & H^2(\mathbb{T}^n, \mathbb{Z})
\end{array}$$

where

$$H^1(\mathbb{T}^n, \mathbb{C}) = H^{1,0}(\mathbb{T}^n) \oplus H^{0,1}(\mathbb{T}^n).$$

Since f is surjective, then the lines are exact, the kernel of c_1 lies in the image of g and consequently the bundle $\widetilde{\mathcal{L}}$ is expressed using transition or passage (constant) functions. The trivialization of $\pi^*\widetilde{\mathcal{L}}$ has constant multiples. From proposition 4.6, we know that the multiples $e_\lambda(z)$ of $\mathcal{L}$ satisfy relations [4.2], i.e. $e_{\lambda_\alpha}(z) \equiv 1$, $e_{\lambda_{n+\alpha}}(z) = e^{-2\pi i z_\alpha}$, $\alpha = 1, ..., n$, so for all $\nu \in \mathbb{T}^n$, under a translation by ν,

$$\tau_\nu : \mathbb{T}^n \longrightarrow \mathbb{T}^n, \quad z \longmapsto z + \nu,$$

the multiples of $\tau_\nu^*\mathcal{L}$ are

$$e'_{\lambda_\alpha}(z) = e_{\lambda_\alpha}(z+\nu) \equiv 1, \qquad e'_{\lambda_{n+\alpha}}(z) = e_{\lambda_{n+\alpha}}(z+\nu) = e^{-2\pi i(z_\alpha+\nu_\alpha)}.$$

Let $\mathcal{L}' = \tau_\nu^*\mathcal{L}$. The constant multiples of $\widetilde{\mathcal{L}} = \mathcal{L}^{-1} \otimes \mathcal{L}'$ are given by $e_{\lambda_\alpha} \equiv 1$, $e_{\lambda_{n+\alpha}} = e^{-2\pi i \mu_\alpha}$, $\alpha = 1, ..., n$ with $c_1(\mathcal{L}') = c_1(\mathcal{L})$, which completes the proof of the proposition. ■

PROPOSITION 4.8.– *Let $\mathcal{L}$ be a positive line bundle on $\mathbb{T}^n$. Then, $\mathcal{L}$ is fixed by $\prod_{\alpha=1}^n \delta_\alpha^2$, translations (of the form $\{\tau_s : s \in \mathbb{Z}\{\delta_\alpha^{-1}\lambda_\alpha, \delta_\alpha^{-1}\lambda_{\alpha+n}\}\}$) where $\delta_1, ..., \delta_\alpha$ are the elementary divisors associated with the polarization $c_1(\mathcal{L})$.*

PROOF.– As in the previous proposition, we consider $\tau_\nu^*\mathcal{L}$, $\nu \in \mathbb{T}^n$ and

$$\tau_\nu : \mathbb{T}^n \longrightarrow \mathbb{T}^n, \quad \tau_\nu \longmapsto z + \nu,$$

as a translation, and the problem consists of showing that

$$\varphi_{\mathcal{L}} : \mathbb{T}^n \longrightarrow \text{Pic}^\circ(\mathbb{T}^n) = \frac{H^1(\mathbb{T}^n, \mathcal{O}_{\mathbb{T}^n})}{H^1(\mathbb{T}^n, \mathbb{Z})}, \quad \nu \longmapsto \mathcal{L}^{-1} \otimes \tau_\nu^*\mathcal{L},$$

is an isogeny of degree $\prod_{\alpha=1}^{n} \delta_\alpha^2$. We will explicitly describe $\varphi_{\mathcal{L}}$ in terms of the bases (λ_α) of Λ and (e_α) of $\mathbb{C}^n$ of the dual basis (y_α^*) of Λ^*. Similar reasoning to the previous proposition shows that multiples of

$$\varphi_{\mathcal{L}}\left(\sum_{\alpha=1}^{n} c_\alpha \lambda_\alpha\right) = \varphi_{\mathcal{L}}\left(\sum_{\alpha=1}^{n} c_\alpha \delta_\alpha e_\alpha\right),$$

are $e_{\lambda_\alpha} \equiv 1$, $e_{\lambda_{n+\alpha}} = e^{-2\pi i \delta_\alpha c_\alpha}$. In other words,

$$\varphi_{\mathcal{L}}\left(\sum_{\alpha=1}^{n} c_\alpha \lambda_\alpha\right) = -\sum_{\alpha=1}^{n} c_\alpha \delta_\alpha x_{n+\alpha}^* = \sum_{\alpha=1}^{n} c_\alpha \delta_\alpha y_\alpha^*,$$

where $x_{n+\alpha}^*$ are coordinates on $\mathbb{C}^n$. Similarly, multiples of

$$\varphi_{\mathcal{L}}\left(\sum_{\alpha=1}^{n} c_\alpha \lambda_{n+\alpha}\right) = \varphi_{\mathcal{L}}\left(\sum_{\alpha=1}^{n} c_\alpha Z_{\alpha\beta} e_\beta\right),$$

are $e_{\lambda_\alpha} \equiv 1$, $e_{\lambda_{n+\beta}} = e^{-2\pi i Z_{\alpha\beta}}$, or what amounts to the same thing

$$\varphi_{\mathcal{L}}\left(\sum_{\alpha=1}^{n} c_\alpha \lambda_{n+\alpha}\right) = \sum_{\alpha=1}^{n} c_\alpha \delta_\alpha x_\alpha^* = \sum_{\alpha=1}^{n} c_\alpha \delta_\alpha y_{n+\alpha}^*.$$

The map $\varphi_{\mathcal{L}}$ applies the subgroup of the complex torus $\mathbb{T}^n$ generated by $\{\delta_\alpha^{-1}\lambda_\alpha, \delta_\alpha^{-1}\lambda_{n+\alpha}\}$ to the trivial line bundle. The kernel of $\varphi_{\mathcal{L}}$ coincides with this subgroup, and the result follows. ■

Let $\mathbb{T} = E/\Lambda$ be a complex torus where Λ is a lattice of a complex vector space E. Let $h : E \times E \longrightarrow \mathbb{C}$ be a Riemann form (see definition 4.5) on Λ.

DEFINITION 4.6.– *A semicharacter α for the Riemann form h is a map defined by*

$$\alpha : \Lambda \longrightarrow \{z \in \mathbb{C} : |z| = 1\} \equiv U,$$

such that:

$$\alpha(\lambda_1 + \lambda_2) = \alpha(\lambda_1)\alpha(\lambda_2)e^{\pi i h(\lambda_1,\lambda_2)}, \quad (\lambda_1, \lambda_2 \in \Lambda).$$

The manipulation of line bundles on a complex torus can be done more simply thanks to the Appell–Humbert theorem, for which we will, before stating it, describe some essential steps which intervene during its proof. We first establish a canonical isomorphism

$$\text{Pic}(\mathbb{T}) \simeq H^1(\mathbb{T}, \mathcal{O}_{\mathbb{T}}^*) \simeq H^1(\Lambda, \mathcal{O}^*(E)).$$

Therefore, with the above notations, for any (h, α), the element (called the autonomy factor)

$$l_\lambda(z) = \alpha(\lambda) e^{\pi i h(z,\lambda) + \frac{\pi}{2} h(\lambda,\lambda)},$$

determines a line bundle that we denote by $\mathcal{L}(h, \alpha)$. From the exponential sequence

$$0 \longrightarrow \mathbb{Z} \longrightarrow \mathcal{O}_{\mathbb{T}} \longrightarrow \mathcal{O}^*_{\mathbb{T}} \longrightarrow 0,$$

we obtain the following diagram:

$$\begin{array}{ccc}
H^1\left(\Lambda, \mathcal{O}^*(E)\right) & \longrightarrow & H^2(\Lambda, \mathbb{Z}) \\
\downarrow \simeq & & \downarrow \simeq \\
H^1\left(\mathbb{T}, \mathcal{O}^*_{\mathbb{T}}\right) & \longrightarrow & H^2(\mathbb{T}, \mathbb{Z})
\end{array}$$

and we show that it is commutative. The image of any line bundle in the group of alternating forms $H^2(\Lambda, \mathbb{Z}) \simeq \bigwedge^2 \mathrm{Hom}(\Lambda, \mathbb{Z})$ provides an alternating form h. We show that the image of any line bundle in $H^2(\Lambda, \mathbb{Z})$ is a Riemann form. Finally, taking into account the following commutative diagram:

$$\begin{array}{ccccccccc}
0 & \longrightarrow & \mathrm{Hom}(\Lambda, U) & \longrightarrow & \{(h, \alpha)\} & \longrightarrow & \{\text{Riemann forms}\} & \longrightarrow & 0 \\
 & & \downarrow \simeq & & \downarrow \simeq & & \downarrow \simeq & & \\
0 & \longrightarrow & \mathrm{Pic}^0(\mathbb{T}) & \xrightarrow{c_1} & \mathrm{Pic}(\mathbb{T}) & \longrightarrow & NS(\mathbb{T}) & \longrightarrow & 0
\end{array}$$

where $NS(\mathbb{T})$ is the Néron–Severi group, i.e. the kernel of

$$H^2(\mathbb{T}, \mathbb{Z}) \longrightarrow H^2(\mathbb{T}, \mathcal{O}_{\mathbb{T}});$$

this is the set of Chern classes c_1 of the elements of the Picard group $\mathrm{Pic}(\mathbb{T})$. The two horizontal sequences are exact, and it must be shown that the left and right vertical arrows are isomorphisms. For surjectivity, we have the following diagram:

$$\begin{array}{ccccc}
H^1(\Lambda, \mathbb{C}) & \longrightarrow & H^1(\Lambda, \mathcal{O}_E) & \longrightarrow & \ker\left[H^1(\Lambda, \mathcal{O}^*_E) \longrightarrow H^2(\Lambda, \mathbb{Z})\right] \\
\downarrow \simeq & & \downarrow \simeq & & \downarrow \simeq \\
H^1(\mathbb{T}, \mathbb{C}) & \longrightarrow & H^1(\Lambda, \mathcal{O}_{\mathbb{T}}) & \longrightarrow & \ker\left[H^1(\Lambda, \mathcal{O}^*_{\mathbb{T}}) \xrightarrow{c_1} H^2(\Lambda, \mathbb{Z})\right]
\end{array}$$

where

$$\ker\left[H^1(\Lambda, \mathcal{O}^*_{\mathbb{T}}) \xrightarrow{c_1} H^2(\Lambda, \mathbb{Z})\right] = \mathrm{Pic}^0(\mathbb{T}).$$

For further details, see Mumford (1975b, p. 19) or Birkenhake and Lange (2004, p. 32).

The Appel–Humbert theorem is stated as follows.

THEOREM 4.1 (APPEL–HUMBERT).– *Any line bundle over a complex torus is isomorphic to a bundle $\mathcal{L}(h, \alpha)$ where h is a Riemann form and α is a semi-character for h. The pair (h, α) is uniquely determined.*

4.6. Exercises

EXERCISE 4.1.– *Determine the canonical bundle $K_{\mathbb{P}^n(\mathbb{C})}$ of the complex projective space $\mathbb{P}^n(\mathbb{C})$. What can we conclude from this?*

Solution: Let $Z_0, ..., Z_n$ be homogeneous coordinates in $\mathbb{P}^n(\mathbb{C})$ and

$$x_1 = \frac{Z_1}{Z_0}, \quad x_2 = \frac{Z_2}{Z_0}, \quad \dots \quad , x_n = \frac{Z_n}{Z_0},$$

be affine coordinates in $\mathbb{P}^n(\mathbb{C})\backslash\{Z_0 = 0\}$. Let us determine the divisor of the n-meromorphic form

$$\omega = \frac{dx_1}{x_1} \wedge ... \wedge \frac{dx_n}{x_n}.$$

To do this, let us consider on $\mathbb{P}^n(\mathbb{C})\backslash\{Z_n = 0\}$ the following affine coordinates:

$$u_0 = \frac{Z_0}{Z_n}, \quad u_1 = \frac{Z_1}{Z_n}, \quad \dots \quad , u_{n-1} = \frac{Z_{n-1}}{Z_n}.$$

We have

$$u_0 = \frac{1}{x_n}, \quad u_1 = \frac{x_1}{x_n}, \quad \dots \quad , u_{n-1} = \frac{x_{n-1}}{x_n},$$

and so,

$$\omega = \frac{1}{u_0^{n+1}} du_0 \wedge du_1 \wedge ... \wedge du_{n-1}.$$

Hence, the divisor of ω is

$$(\omega) = -(n+1)H_\infty,$$

where H_∞ is a hyperplane in $\mathbb{P}^n(\mathbb{C})$ and

$$K_{\mathbb{P}^n(\mathbb{C})} = [(\omega)] = [-(n+1)H_\infty].$$

We conclude that $K_{\mathbb{P}^n(\mathbb{C})}$ is isomorphic to $\mathcal{O}(-n-1)$.

EXERCISE 4.2.– *Let Λ be a lattice of $\mathbb{C}^n$, h be an $\mathbb{R}$-bilinear form of $\mathbb{C}^n$ with values in $\mathbb{R}$ and Ω be a matrix whose columns are $u_1, ..., u_{2n}$ where $(u_1, ..., u_{2n})$ is a basis of Λ with respect to $\mathbb{Z}$. We assume that:*

(1) $h(\Lambda, \Lambda) \subset \mathbb{Z}$, i.e. the value $h(x, y)$ is an integer for all $x, y \in \Lambda$.

(2) $h(y, x) = -h(x, y)$, i.e. h is an alternate form.

(3) $h(x, iy)$ is a positive symmetric and non-degenerate form.

Show that these conditions are equivalent to the following:

(a) The coefficients of the matrix $\mathcal{M}$ are integers, where $\mathcal{M}$ is a matrix of order $2n$ such that

$$h(\Omega A, \Omega B) = A^\top \mathcal{M} B, \quad (A, B \in \mathbb{R}^{2n}).$$

($A^\top$ denotes the transpose of the vector A).

(b) $\mathcal{M}^\top = -\mathcal{M}$.

(c) $\Omega \left(\mathcal{M}^{-1}\right)^\top \Omega^\top = 0$.

(d) $i\Omega \left(\mathcal{M}^{-1}\right)^\top \overline{\Omega}^\top$ is a non-degenerate positive hermitian matrix.

(The form $h(x, y)$ is called the Riemann form on the torus $\mathbb{C}^n/\Lambda$; see also definition 4.5.)

EXERCISE 4.3.– *Let Λ be a free $\mathbb{Z}$-module of rank $2n$, $n \in \mathbb{N}^*$ and h be a non-degenerate alternating bilinear form on Λ and with integer values. Show that there exists a basis of Λ in which the matrix associated with h is written in the form,*

$$\begin{pmatrix} 0 & \Delta_\delta \\ -\Delta_\delta & 0 \end{pmatrix}, \qquad \Delta_\delta = diag\,(\delta_1, ..., \delta_n),$$

where Δ_δ is the diagonal matrix with coefficients $\delta_1, \ldots, \delta_n \in \mathbb{N}^*$ (called elementary divisors) and such that $\delta_j \mid \delta_{j+1}$ for all $j = 1, ..., n-1$. (The determinant of this matrix is the square of the integer $\delta_1 \cdots \delta_n$ called its Pfaffian.) We say that the n-tuple $(\delta_1, \ldots, \delta_n)$ is the type of the form h.

Hint: The solution consists of reasoning by recurrence on the rank of Λ; see, for example, Debarre (1999).

EXERCISE 4.4.– *Show that the Picard group of the projective space $\mathbb{P}^n$ is isomorphic to the group of relative integers $\mathbb{Z}$.*

Solution: Any line bundle over $\mathbb{P}^n$ is of the form $\mathcal{O}(d)$ for $d \in \mathbb{Z}$. The line bundles on $\mathbb{P}^n$ are precisely the tensor powers $\mathcal{O}(d)$ for $d \in \mathbb{Z}$. Furthermore, we know that: $\mathcal{O}(d_1) \otimes \mathcal{O}(d_2) \simeq \mathcal{O}(d_1 + d_2)$. Hence, the Picard group of $\mathbb{P}^n$ is isomorphic to $\mathbb{Z}$, generated by the class of the hyperplane bundle $\mathcal{O}(1)$. We have $\text{Pic}(\mathbb{P}^n(\mathbb{C})) \simeq H^2(\mathbb{P}^n(\mathbb{C}), \mathbb{Z}) \simeq \mathbb{Z}$.

EXERCISE 4.5.– *Let* $\mathbb{T}_1 = E_1/\Lambda_1$ *and* $\mathbb{T}_2 = E_2/\Lambda_2$ *be two complex tori and* $f : \mathbb{T}_1 \longrightarrow \mathbb{T}_2$ *be a holomorphic map. Show that there exists a (unique)* $\mathbb{C}$*-linear map* $g : E_1 \longrightarrow E_2$ *such that* $g(\Lambda_1) \subset \Lambda_2$ *and inducing a homomorphism* h *with* $f(z) = h(z)f(0)$.

Solution: Without restricting generality, we can assume that $f(0)$ is zero in $\mathbb{T}_2$; it suffices to replace f by $h = f - f(0)$ (in other words, it suffices to take a translation of f by $f(0)$), and thus bring f back to a map h sending 0 to 0. Therefore, for all $z \in \mathbb{T}_1$, $f(z) = h(z) + f(0)$. Then, we lift the composition $E_1 \longrightarrow \mathbb{T}_1 \stackrel{f}{\longrightarrow} \mathbb{T}_2$, to the universal covering E_2 of $\mathbb{T}_2$. We denote this lifting g and choose it such that: $g(0) = 0$.

$$\begin{array}{ccc} E_1 & \stackrel{g}{\longrightarrow} & E_2 \\ \downarrow & \searrow & \downarrow \\ \mathbb{T}_1 & \stackrel{f}{\longrightarrow} & \mathbb{T}_2 \end{array}$$

Let $z \in E_1$ and $\mu \in \Lambda_1$ be given. The points z and $z + \mu$ have the same image in $\mathbb{T}_1$, so $g(z+\mu) - g(z) \in \Lambda_2$. The mapping $z \longmapsto g(z+\mu) - g(z) \in \Lambda_2$ maps the connected space E_1 into the discrete space Λ_2. It is constant, and since $g(0) = 0$, then $g(z+\mu) - g(z) + g(\mu)$, $(g(\mu) \equiv c_\mu)$. The partial derivatives of g are holomorphic and $2m$-periodic (where $m \equiv \dim \mathbb{T}_1$), therefore bounded. Therefore, these partial derivatives are constant by the Liouville theorem and hence g is $\mathbb{C}$-linear.

EXERCISE 4.6.– *Recall that a line bundle* $\mathcal{L}$ *on a complex manifold* M *is trivial if it is isomorphic to* $\mathbb{C} \times M$ *(see example 4.2). Show that the bundle* $\mathcal{L}$ *is trivial if and only if it has a section that vanishes nowhere.*

Solution: Let $\varphi : \mathcal{L} \longrightarrow M \times \mathbb{C}$ be a trivialization, it is obvious that $\varphi^{-1}(p, 1)$ is a section that does not vanish anywhere. Conversely, let $s : M \longrightarrow \mathcal{L}$ be such that $\pi o s = id_M$ where $\pi : \mathcal{L} \longrightarrow M$, i.e. a section $s \in H^0(M, \mathcal{L})$ and we assume that this section does not vanish anywhere. Hence, the morphism (a trivialization) defined by

$$M \times \mathbb{C} \longrightarrow \mathcal{L}, \quad (p, \tau) \longmapsto \tau.s(p),$$

is an isomorphism and its inverse is provided by

$$u \longmapsto \left(\pi(u), \frac{u}{s\pi(u)}\right).$$

EXERCISE 4.7.– *Show that any line bundle on $\mathbb{C}^n$ is trivial.*

Solution: Just use the exact exponential sequence:

$$0 \longrightarrow \mathbb{Z} \xrightarrow{i} \mathcal{O}_{\mathbb{C}^n} \xrightarrow{\exp} \mathcal{O}^*_{\mathbb{C}^n} \longrightarrow 0,$$

which provides the exact cohomology sequence

$$\ldots \longrightarrow H^1\left(\mathbb{C}^n, \mathcal{O}_{\mathbb{C}^n}\right) \xrightarrow{\exp^*} H^1\left(\mathbb{C}^n, \mathcal{O}^*_{\mathbb{C}^n}\right) \xrightarrow{c_1} H^2(\mathbb{C}^n, \mathbb{Z}) \longrightarrow \cdots$$

and show that:

$$H^k\left(\mathbb{C}^n, \mathbb{Z}\right) = H^k\left(\mathbb{C}^n, \mathcal{O}_{\mathbb{C}^n}\right) = H^k\left(\mathbb{C}^n, \mathcal{O}^*_{\mathbb{C}^n}\right) = 0,$$

for all $k > 0$, while taking into account that there exists a group isomorphism: $Pic(\mathbb{C}^n) \cong H^1\left(\mathbb{C}^n, \mathcal{O}^*_{\mathbb{C}^n}\right)$.

EXERCISE 4.8.– *Show that any line bundle on a compact Riemann surface or complex algebraic curve C has a non-zero meromorphic section.*

Hint: See, for example, Dolbeault (1990, p. 174).

EXERCISE 4.9.– *Let $\mathbb{T} = E/\Lambda$ be a complex torus where Λ is a lattice of a complex vector space E. Show that the image of any line bundle in $H^2(\Lambda, \mathbb{Z})$ is a Riemann form.*

Hint: See Scheidemann (2023).

EXERCISE 4.10.– *Let E be a complex vector space of dimension k. Consider two line bundles $\mathcal{L}(h_1, \alpha_1)$ and $\mathcal{L}(h_2, \alpha_2)$ where α_1, α_2 are semicharacters for Riemann forms h_1, h_2 on E, respectively. Determine $\mathcal{L}(h_1, \alpha_1) \otimes \mathcal{L}(h_2, \alpha_2)$ as well as $\mathcal{L}(h_1, \alpha_1)^k$.*

Solution: We easily obtain $\mathcal{L}(h_1, \alpha_1) \otimes \mathcal{L}(h_2, \alpha_2) = \mathcal{L}(h_1 + h_2, \alpha_1\alpha_2)$. Similarly, we have $\mathcal{L}(h_1, \alpha_1)^k = \mathcal{L}(kh_1, \alpha_1^k)$.

5

Some Fundamental Theorems

5.1. Preliminaries and various notions

5.1.1. *Projective variety*

Let $\mathcal{L}$ be a line bundle over a compact complex manifold M and $s_0, ..., s_N$ be sections of $\mathcal{L}$ that are assumed not to be all zero. Recall that since M is compact, the dimension of the $\mathbb{C}$-vector space $H^0(M, \mathcal{O}_M(\mathcal{L}))$ of sections of $\mathcal{L}$ over M is finite. Let $\mathcal{B}$ denote the base locus

$$\mathcal{B} \equiv \{s \in H^0(M, \mathcal{O}_M(\mathcal{L})) : s(p) = 0, p \in M\}.$$

The latter is an analytic subvariety; it can be put in the form

$$\mathcal{B} = D(s_0) \cap ... \cap D(s_N),$$

where $D(s_j)$ is an element of $\mathrm{Div}(M)$.

PROPOSITION 5.1.– *Let* $(s_0, ..., s_N)$ *be a basis of the space* $H^0(M, \mathcal{O}_M(\mathcal{L}))$. *The following map,*

$$\psi_{\mathcal{L}} : M \backslash \mathcal{B} \longrightarrow \mathbb{P}^N(\mathbb{C}), \quad p \longmapsto [s_0(p) : ... : s_N(p)],$$

is holomorphic and has values in the complex projective space $\mathbb{P}^N(\mathbb{C})$.

PROOF.– Since $\mathcal{B} \subset M$ is a closed subset, $V \backslash \mathcal{B}$ is an open submanifold of M, and therefore this notion of holomorphy does make sense. To show that the map $\psi_{\mathcal{L}}$ is holomorphic and has values in $\mathbb{P}^N(\mathbb{C})$, it suffices to locally choose a holomorphic trivialization of the line bundle $\mathcal{L}$. Such a choice allows us to locally view the sections $s_0, ..., s_N$ as holomorphic functions. Then, a change of trivialization multiplies these functions by the same invertible function, which gives rise to the same map with values in $\mathbb{P}^N(\mathbb{C})$. Explicitly, let $p \in M\backslash\mathcal{B}$ and denote by φ a trivialization of $\mathcal{L}$ on an open neighborhood of $p \in M$. Therefore,

$$\psi_{\mathcal{L}}(p) = [s_0(p) : ... : s_N(p)],$$

denotes the point $[\varphi(s_0(p)) : ... : \varphi(s_N(p))] \in \mathbb{P}^N(\mathbb{C})$. The definition of the map $\psi_{\mathcal{L}}$ is independent of the trivialization φ because for any other trivialization, it is of the form $\lambda\varphi$, $\lambda \in \mathbb{C}^*$ and

$$[\lambda(p)\varphi(s_0(p)) : ... : \lambda(p)\varphi(s_N(p))] = [\varphi(s_0(p)) : ... : \varphi(s_N(p))].$$

The local description of the map $\psi_{\mathcal{L}}$ shows that it is well defined and holomorphic on $V\backslash\mathcal{B}$, which completes the proof. ■

REMARK 5.1.– *a) Note that instead of choosing $(s_0, ..., s_N)$ as a basis, we can consider any other collection of sections $s_0, ..., s_N$. We define a holomorphic map,*

$$\psi_{\mathcal{L}} : M\backslash\mathcal{B} \longrightarrow \mathbb{P}^N(\mathbb{C}) \simeq \mathbb{P}\left(H^0(M, \mathcal{O}_M(\mathcal{L}))^*\right), p \longmapsto [s_0(p) : ... : s_N(p)],$$

by associating with a point $p \in M\backslash\mathcal{B}$ the linear map

$$H^0(M, \mathcal{O}_M(\mathcal{L}) \longrightarrow \mathbb{C}, \quad s \longmapsto s(p),$$

or what amounts to the same thing, by sending a point p of an open O_α of $V\backslash\mathcal{B}$ onto the point of homogeneous coordinates $[s_0(p) : ... : s_N(p)]$ of the projective space $\mathbb{P}(\mathbb{C}) \simeq \mathbb{P}\left(H^0(V, \mathcal{O}_M(\mathcal{L}))^*\right)$, defined by the hyperplane consisting of sections vanishing at p. Note that the point p does not depend on the open set O_α because in another open set, all these coordinates are multiplied by the same non-zero scalar $g_{\alpha\beta}(p)$ and it is well defined since we have excluded the zero sections.

b) When studying Kodaira embedding theorem, we will work with a meromorphic map of the form

$$\psi_{\mathcal{L}} : M \longrightarrow \mathbb{P}^N(\mathbb{C}) \simeq \mathbb{P}\left(H^0(M, \mathcal{O}_M(\mathcal{L}))^*\right),$$

more precisely, $\psi_{\mathcal{L}^k}$ with $\mathcal{L}^k$ a tensor power, $k \in \mathbb{N}^*$.

Let $\mathcal{D} = \sum_j n_j \mathcal{D}_j$, $n_j \in \mathbb{Z}$, be a divisor on the complex manifold M, where the $\mathcal{D}_j$ are irreducible analytic submanifolds of codimension 1. Consider again the set $\mathcal{L}(\mathcal{D})$ of holomorphic functions on $V\backslash\mathcal{D}$, having at most one pole of order n_j on $\mathcal{D}_j$ when $n_j > 0$ and a zero of order at least equal to $-n_j$ when $n_j \leq 0$ and let $(1, f_1, ..., f_N)$ be a basis of this space. If $\mathcal{L}$ is the line bundle associated with the divisor $\mathcal{D}$ (which we denote by $\mathcal{L} \equiv [\mathcal{D}]$), then the restriction of $\psi_{\mathcal{D}} \equiv \psi_{\mathcal{L}}$ to $M\backslash\mathcal{D}$ can be constructed as above by writing

$$\psi_{\mathcal{L}} : M\backslash\mathcal{D} \longrightarrow \mathbb{P}^N(\mathbb{C}), \quad p \longmapsto [1, f_1(p) : ... : f_N(p)]. \qquad [5.1]$$

DEFINITION 5.1.– *We say that the complex manifold M is projective or equivalently M admits a holomorphic embedding in a projective space if it admits a line bundle without a base point which determines a holomorphic embedding of M in $\mathbb{P}^N(\mathbb{C})$.*

A projective variety is a closed analytic subvariety of a complex projective space. Furthermore, it is compact, and every regular point has a neighborhood that can be parameterized using analytic functions.

5.1.2. *Tangent and cotangent bundles*

Let M be a complex manifold and $z_1, ..., z_n$, $n = \dim_{\mathbb{C}} V$, be holomorphic local coordinates on M. In fact, at each point of M, we define local coordinates z_j and $\overline{z}_j$ as well as the tangent bundle $\mathbb{C}\left[\frac{\partial}{\partial z_j}, \frac{\partial}{\partial \overline{z}_j}\right]$ which we denote by TM. This one admits the decomposition

$$TM = T'M \oplus T''M,$$

where $T'M$ (respectively $T''M$) is the holomorphic (respectively antiholomorphic) part of TM generated by $\frac{\partial}{\partial z_j}$ (respectively $\frac{\partial}{\partial \overline{z}_j}$). Similarly, the cotangent bundle T^*M (dual of TM) admits the decomposition

$$T^*M = T'^*M \oplus T''^*M,$$

into holomorphic and antiholomorphic parts. The points of this bundle are the linear forms on the fibers of $TM \longrightarrow M$. A p-differential form is a section of the bundle $\Lambda^p T^*M$ (the exterior powers of T^*M). The points of the latter are the alternating p-multilinear forms on T^*M.

5.1.3. *Dolbeault cohomology group*

Let M be a complex n-manifold. Note that the $dz_1, ..., dz_n, d\overline{z}_1, ..., d\overline{z}_n$ form a local basis of the space of differential forms. A differential form on M of type $(p, q))$ is given locally by,

$$\omega = \sum_{\substack{1 \leq j_1 < ... < j_p \leq n \\ 1 \leq k_1 < ... < k_q \leq n}} f_{j_1...j_p k_1...k_q} dz_{j_1} \wedge ... \wedge dz_{j_p} \wedge d\overline{z}_{k_1} \wedge ... \wedge d\overline{z}_{k_q},$$

where $f_{j_1...j_p k_1...k_q}$ are complex-valued functions of class C^∞. These are the forms generated locally by the $dz_I \wedge d\overline{z}_J$, with I and J being subsets of $\{1, ..., n\}$ such that $\text{card}(I) = p$ and $\text{card}(J) = q$. This definition does not depend on the choice of holomorphic local coordinates. Obviously, a form of type $(0, 0)$ is a function.

The set of differential forms of type (p, q) on M is a complex vector bundle denoted by $A^{p,q}(M)$. It should be noted that only $A^{p,0}(M)$ is holomorphic. The differential of ω is

$$\begin{aligned} d\omega &= \sum_{\substack{1\le j_1<\ldots<j_p\le n\\ 1\le k_1<\ldots<k_q\le n}} df_{j_1\ldots j_p k_1\ldots k_q} \wedge dz_{j_1} \wedge \ldots \wedge dz_{j_p} \wedge d\overline{z}_{k_1} \wedge \ldots \wedge d\overline{z}_{k_q}, \\ &= \sum_{\substack{1\le j_1<\ldots<j_p\le n\\ 1\le k_1<\ldots<k_q\le n}} \partial f_{j_1\ldots j_p k_1\ldots k_q} \wedge dz_{j_1} \wedge \ldots \wedge dz_{j_p} \wedge d\overline{z}_{k_1} \wedge \ldots \wedge d\overline{z}_{k_q} \\ &\quad + \sum_{\substack{1\le j_1<\ldots<j_p\le n\\ 1\le k_1<\ldots<k_q\le n}} \overline{\partial} f_{j_1\ldots j_p k_1\ldots k_q} \wedge dz_{j_1} \wedge \ldots \wedge dz_{j_p} \wedge d\overline{z}_{k_1} \wedge \ldots \wedge d\overline{z}_{k_q}, \\ &= \partial\omega + \overline{\partial}\omega, \end{aligned}$$

where

$$\partial : A^{p,q}(M) \longrightarrow A^{p+1,q}(M), \quad \partial \equiv \frac{\partial}{\partial z} dz,$$

and

$$\overline{\partial} : A^{p,q}(M) \longrightarrow A^{p,q+1}(M), \quad \overline{\partial} \equiv \frac{\partial}{\partial \overline{z}} d\overline{z}.$$

Or in condensed form, we have

$$\omega = \sum_{I,J} f_{IJ} dz_I \wedge d\overline{z}_J, \qquad d\omega = \sum_{I,J} df_{IJ} \wedge dz_I \wedge d\overline{z}_J,$$

where

$$df_{IJ} = \sum_k \frac{f_{IJ}}{\partial z_k} dz_k + \sum_k \frac{f_{IJ}}{\partial \overline{z}_k} d\overline{z}_k = \partial f_{IJ} + \overline{\partial} f_{IJ},$$

and thus

$$\partial\omega = \sum_{I,J} \partial f_{IJ} \wedge dz_I \wedge d\overline{z}_J, \quad \overline{\partial}\omega = \sum_{I,J} \overline{\partial} f_{IJ} \wedge dz_I \wedge d\overline{z}_J.$$

The Dolbeault operators $\partial, \overline{\partial}$ are defined in terms of

$$d : A^{p,q}(M) \longrightarrow A^{p+q+1}(M),$$

($A^r(M)$ being the complex valued r-forms, it is a sheaf) and the projection maps as

$$\partial = pr_{p+1,q} o d, \qquad \overline{\partial} = pr_{p,q+1} o d,$$

it is clear from the above that: $d = \partial + \overline{\partial}$.

Since $\partial f_{j_1 \ldots j_p k_1 \ldots k_q}$ is of type $(1,0)$ and $\overline{\partial} f_{j_1 \ldots j_p k_1 \ldots k_q}$ is of type $(0,1)$, then the component $\partial\omega$ of $d\omega$ is of type $(p+1,q)$ and $\overline{\partial}\omega$ is of type $(p,q+1)$. We therefore have the following decomposition: $d = \partial + \overline{\partial}$ and the operators $\partial, \overline{\partial}$ are independent of the choice of holomorphic local coordinates. The relation $d^2 = 0$ means here that

$$\partial^2 = \overline{\partial}^2 = \partial\overline{\partial} + \overline{\partial}\partial = 0.$$

Indeed, from the relations $d^2 = 0$ and $d = \partial + \overline{\partial}$, we derive

$$0 = d^2 = \partial^2 + \overline{\partial}^2 + \partial\overline{\partial} + \overline{\partial}\partial.$$

Note that if ω is a form of type (p,q), then $\partial^2\omega$ is of type $(p+2,q)$, $\overline{\partial}^2\omega$ is of type $(p,q+2)$ and $(\partial\overline{\partial} + \overline{\partial}\partial)$ is of type $(p+1,q+1)$. The result follows from the fact that the sum of these three forms is zero. The operator $\overline{\partial}$ satisfies Leibniz's rule,

$$\overline{\partial}(\omega \wedge \lambda) = \overline{\partial}\omega \wedge \lambda + (-1)^{deg\omega} \omega \wedge \overline{\partial}\lambda.$$

Noting that $\partial\omega = \overline{\overline{\partial}\overline{\omega}}$, we also obtain

$$\partial(\omega \wedge \lambda) = \partial\omega \wedge \lambda + (-1)^{deg\omega} \omega \wedge \partial\lambda.$$

Moreover, the differential form ω is said to be $\overline{\partial}$-closed if $\overline{\partial}\omega = 0$ and $\overline{\partial}$-exact if there exists a form λ such that: $\omega = \overline{\partial}\lambda$. Any $\overline{\partial}$-exact form is $\overline{\partial}$-closed. The analogue of the Poincaré lemma for the operator $\overline{\partial}$ can be stated as follows: if ω is a form of type (p,q) of class C^1, $q > 0$ with $\overline{\partial}\omega = 0$, then we can find locally on M a λ form of class C^1 of type $(p, q-1)$ such that: $\omega = \overline{\partial}\lambda$. Recall that the Poincaré lemma implies the nullity of cohomology groups in the case of open star sets. We have just seen that: $\overline{\partial}^2 = 0$, so we have for each p a cochain complex (Dolbeault complex):

$$0 \longrightarrow A^{p,0}(M) \xrightarrow{\overline{\partial}} A^{p,1}(M) \xrightarrow{\overline{\partial}} \cdots \xrightarrow{\overline{\partial}} A^{p,n}(M) \longrightarrow 0.$$

Hence, we have the following definition:

DEFINITION 5.2.– *We define the Dolbeault cohomology groups $H^{p,q}(M)$ of a n-complex manifold M as the cohomology groups of the above complex, to be the following complex vector spaces:*

$$H^{p,q}(M) = Z^{p,q}(M)/\overline{\partial}A^{p,q-1}(M) = \frac{\ker\{\overline{\partial} : A^{p,q}(M) \longrightarrow A^{p,q+1}(M)\}}{Im\{\overline{\partial} : A^{p,q-1}(M) \longrightarrow A^{p,q}(M)\}}.$$

($Z^{p,q}(M)$ is the kernel of $\overline{\partial}$ on $A^{p,q}$, i.e. the sheaf of $\overline{\partial}$-closed forms of type (p,q). Here, p,q are integers between 0 and n and we adopt the convention $A^{p,-1}(M)=0$).

In other words,

$$H^{p,q}(M)=\frac{\{\text{forms of type (p,q) } \overline{\partial}\text{-closed on } M\}}{\{\text{forms of type (p,q) } \overline{\partial}\text{-exact on } M\}}.$$

i.e. $H^{p,q}(M)$ is the set of classes representable by a closed form on M of type (p,q). Moreover, we have a natural morphism

$$\bigoplus_{p+q=k} H^{p,q}(M)\longrightarrow H^k(M,\mathbb{C}).$$

In general, this morphism is neither injective nor surjective.

5.1.4. *First Chern class*

Let M be a complex manifold and $\mathcal{L}$ be a line bundle over M. Consider the exact exponential sequence

$$0\longrightarrow\mathbb{Z}\xrightarrow{i}\mathcal{O}_M\xrightarrow{\exp}\mathcal{O}_M^*\longrightarrow 0,$$

where $\mathcal{O}_M$ is the sheaf of holomorphic functions on M, $\mathcal{O}_M^*$ is the sheaf of non-vanishing holomorphic functions on M, i is the trivial inclusion and exp is the exponential map $\exp f = e^{2\pi\sqrt{-1}f}$. This sequence means that an invertible holomorphic function is locally the exponential of a holomorphic function defined up to a constant multiple of i. From this sequence, we deduce a connection morphism (first Chern class), $c_1: H^1(M,\mathcal{O}_M^*)\longrightarrow H^2(M,\mathbb{Z})$ which associates with any line bundle a cohomology class in $H^2(M,\mathbb{Z})$. More precisely, we deduce the long exact cohomology sequence

$$0\longrightarrow H^0(M,\mathbb{Z})\longrightarrow H^0(M,\mathcal{O}_M)\longrightarrow H^0(M,\mathcal{O}_M^*)\longrightarrow H^1(M,\mathbb{Z})\longrightarrow$$

$$H^1(M,\mathcal{O}_M)\xrightarrow{\exp^*}H^1(M,\mathcal{O}_M^*)\xrightarrow{c_1}H^2(M,\mathbb{Z})\xrightarrow{i_*}H^2(M,\mathcal{O}_M)\longrightarrow\cdots$$

We know from proposition 4.2 that the Picard group $Pic(M)$ is isomorphic to the cohomology group $H^1(M,\mathcal{O}_M^*)$.

DEFINITION 5.3.– *The image of the class of $\mathcal{L}$ in $Pic(M)$ by the morphism*

$$c_1: Pic(M)\simeq H^1(M,\mathcal{O}_M^*)\longrightarrow H^2(M,\mathbb{Z}),$$

is called the first Chern class of $\mathcal{L}$. We denote it by $c_1(\mathcal{L})$ and we also denote by $Pic^0(M)$ the kernel of c_1.

Therefore, to every divisor on M corresponds a positive line bundle $\mathcal{L} \in H^1(M, \mathcal{O}_M^*)$ (this means that its Chern class $c_1(\mathcal{L}) \in H^2(M, \mathbb{R})$ can be represented by a closed $(1,1)$-positive real form). Furthermore, for a complex Riemann surface or algebraic curve $\mathcal{C}$, we have $H^2(\mathcal{C}, \mathbb{Z}) \simeq \mathbb{Z}$ and for the bundle $\mathcal{L} = \mathcal{O}_{\mathcal{C}}(\mathcal{D})$, the first Chern class $c_1(\mathcal{L})$ is identified with the degree of the divisor $\mathcal{D}$.

5.1.5. *Hodge forms, hermitian metrics, harmonic space, Hodge star operator and Hodge theorem*

Let M be a compact complex manifold of dimension n (with the hermitian structure h). A differential form ψ satisfying the Laplace equation: $\Delta\psi = 0$ is called a harmonic form. For any $0 \leq p, q \leq n$, define the linear (over $\mathbb{R}$) isomorphism (Hodge star operator) $*$ of a (p,q)-form to be a $(n-p, n-q)$-form: $* : A^{p,q}(M) \longrightarrow A^{n-p,n-q}(M)$ by requiring for any $\psi, \eta \in A^{p,q}(M)$,

$$\psi \wedge *\eta = h(\psi, \eta)\frac{\omega^n}{n!}.$$

In order to write down $*$ locally and explicitly, let $I = (i_1, ..., i_p)$ in $\{1, ..., n\}$ and $J = (j_1, ..., j_q)$ in $\{1, ..., n\}$, and write $\psi = \sum_{I,J} \psi_{IJ}\varphi_I \wedge \overline{\varphi}_J$, $|I| = p$, $|J| = q$. Then, $*$ is given by

$$*\psi = 2^{p+q-n} \sum_{I,J} \varepsilon_{IJ}\psi_{\overline{I}\overline{J}}\varphi_{I^0} \wedge \overline{\varphi}_{J^0}$$

where $I^0 = \{1, ..., n\}\backslash I$ and ε_{IJ} denote the sign of the permutation:

$$(1, ..., n, 1', ..., n') \longrightarrow (i_1, ..., i_p, j_1, ..., j_q, i_1^0, ..., i_{n-p}^0, j_1^0, ..., j_{n-q}^0).$$

It is straight forward to check that $**\eta = (-1)^{p+q}\eta$ for any (p,q)-form η and we have $*\left(\frac{\omega^k}{k!}\right) = \frac{\omega^{n-k}}{(n-k)!}$. Consider the hermitian inner product (the L^2 norm) on each $A^{p,q}(M)$ defined by

$$(\psi, \eta) = \int_M \psi \wedge *\eta$$

where $\psi, \eta \in A^{p,q}(M)$. For any (p,q)-forms ψ and η, we have $(*\psi, *\eta) = (\psi, \eta)$. In terms of star, the operator:

$$\overline{\partial}^* = (-1)^{p+q} *^{-1} \overline{\partial}* = - * \overline{\partial}* : A^{p,q}(M) \longrightarrow A^{p,q-1}(M),$$

is adjoint to $\overline{\partial}$. Indeed, for any $\psi \in A^{p,q-1}(M)$ and any $\eta \in A^{p,q}(M)$, we have

$$\begin{aligned}\overline{\partial}\psi \wedge *\eta - \psi \wedge *\overline{\partial}^*\eta &= \overline{\partial}\psi \wedge *\eta + (-1)^{p+q-1}\psi \wedge \overline{\partial} * \eta \\ &= \overline{\partial}(\psi \wedge *\eta) = d(\psi \wedge *\eta)\end{aligned}$$

($\overline{\partial} = d$ on forms of type $(n, n-1)$). Hence,

$$(\overline{\partial}\psi, \eta) = \int_M \overline{\partial}\psi \wedge *\eta = \int_M \psi \wedge *\overline{\partial}^*\eta + d(\psi \wedge *\eta) = \int_M \psi \wedge *\overline{\partial}^*\eta = (\psi, \overline{\partial}^*\eta)$$

since by Stokes's theorem $\int_M d(\psi \wedge *\eta) = 0$.

The $\overline{\partial}$-laplacian $\Delta_{\overline{\partial}} : A^{p,q}(M) \longrightarrow A^{p,q}(M)$ is then defined by

$$\Delta_{\overline{\partial}} = (\overline{\partial} + \overline{\partial}^*) = \overline{\partial}\overline{\partial}^* + \overline{\partial}^*\overline{\partial}.$$

Note that $\overline{\partial}^*$ is the adjoint of $\overline{\partial}$ and $\Delta_{\overline{\partial}}$ self-adjoint under the L^2-norm. It has a non-negative eigenvalue:

$$(\Delta_{\overline{\partial}}\psi, \psi) = (\overline{\partial}\psi, \overline{\partial}\psi) + (\overline{\partial}^*\psi, \overline{\partial}^*\psi) \geq 0,$$

for every $\psi \in A^{p,q}(M)$, and is an elliptic differential operator. A (p,q)-form ψ is $\overline{\partial}$-harmonic if $\Delta_{\overline{\partial}}\psi = 0$, or equivalently if both ψ and $*\psi$ are $\overline{\partial}$-closed (i.e. $\overline{\partial}\psi = \overline{\partial}^*\psi = 0$). The space of harmonic forms of type (p,q), $0 \leq p, q \leq n$ is denoted by $\mathcal{H}^{p,q}(M)$ and called the harmonic space. Hodge's theorem (Griffiths and Harris 1978) is stated as follows.

THEOREM 5.1 (HODGE).– *On a compact complex manifold M of dimension n (with the hermitian structure h), the space $\mathcal{H}^{p,q}(M)$ is finite dimensional for any $0 \leq p, q \leq n$, so the orthogonal projection $\mathcal{H} : A^{p,q}(M) \longrightarrow \mathcal{H}^{p,q}(M)$ is well defined. In addition, there exists a unique continuous linear operator (Green operator) $G : A^{p,q}(M) \longrightarrow A^{p,q}(M)$ satisfying*

$$G(\mathcal{H}^{p,q}(M)) = 0, \qquad I = \mathcal{H} + \Delta_{\overline{\partial}}G,$$

where I is the identity map of $A^{p,q}(M)$, and $\mathcal{H}$ is the orthogonal projection from $A^{p,q}(M)$ onto $\mathcal{H}^{p,q}(M)$. Furthermore, $[G, \overline{\partial}] = [G, \overline{\partial}^*] = 0$, i.e. G commutes with $\overline{\partial}$ and $\overline{\partial}^*$.

Consequently, there exist orthogonal decompositions:

$$A^{p,q}(M) = \mathcal{H}^{p,q}(M) \oplus \overline{\partial}A^{p,q-1}(M) \oplus \overline{\partial}^* A^{p,q+1}(M),$$

$$\ker \overline{\partial}^{p,q} = \mathcal{H}^{p,q}(M) \oplus \overline{\partial}A^{p,q-1}(M),$$

since given a (p,q)-form ψ, we have (the so-called decomposition on forms):

$$\psi = (\mathcal{H} + \Delta_{\overline{\partial}G})\psi = \mathcal{H}(\psi) + \overline{\partial}(\overline{\partial}^* G\psi) + \overline{\partial}^*(\overline{\partial}G\psi).$$

If ψ is closed, this means that ψ can be uniquely expressed as the sum of a harmonic form and a $\overline{\partial}$-exact form. Thus, there is a 1–1 correspondence between harmonic space and the Dolbeault cohomology group. Also, we have an isomorphism:

$$H^{p,q}_{\overline{\partial}}(M) \simeq \mathcal{H}^{p,q}(M),$$

which shows the finite dimensionality of $H^{p,q}_{\overline{\partial}}(M)$ for compact M (the compactness assumption in the above theorem is crucial).

DEFINITION 5.4.– *We call a Hodge form a 2-form ω of type $(1,1)$, positive and closed such that its cohomology class is entire, i.e. $[\omega] \in H^2(M, \mathbb{Z})$.*

DEFINITION 5.5.– *A hermitian metric h on a complex manifold M is a positive definite hermitian form of class $\mathcal{C}^\infty$ on the tangent bundle TM. A hermitian manifold (M, h) is a complex manifold M together with a hermitian metric h on M.*

In local complex coordinates $z_1, ..., z_n$, this form is written as

$$h(z) = \sum_{j,l=1}^{n} h_{jl}(z) dz_j \otimes d\overline{z}_l,$$

where (h_{jl}) denotes a positive hermitian metric with coefficients $\mathcal{C}^\infty$ and $n = \dim M$. On a hermitian manifold (M, h), we have a Riemannian metric $g = \text{Re}\, h$ and to the form h, we associate a form ω of type $(1,1)$ by taking the imaginary part of h; more precisely,

$$\omega = -\text{Im}\, h = \frac{i}{2} \sum_{j,l=1}^{n} h_{jl} dz_j \wedge d\overline{z}_l, \quad i = \sqrt{-1}.$$

5.1.6. *Kähler metric, Kähler form, Kähler manifold*

DEFINITION 5.6.– *A Kähler manifold is a complex manifold equipped with a hermitian metric whose imaginary part, which is a 2-form ω of type $(1,1)$ relative to the complex structure, is closed. Such a metric is called a Kähler metric, and the ω form is called a Kähler form.*

Let M be a compact Kähler manifold of dimension n, equipped with its Kähler metric ω. Note that: $\omega^n = \omega \wedge ... \wedge \omega$ (n-factors) is a hermitian volume element (noted $Vol._\omega(M)$) of M. Let $z_n = x_n + iy_n$. A direct calculation gives

$$\begin{aligned}\frac{\omega^n}{n!} = \frac{1}{n!}\omega \wedge ... \wedge \omega &= \alpha i^n n! \det(h_{jl}) dz_1 \wedge d\overline{z}_1 \wedge ... \wedge dz_n \wedge d\overline{z}_n, \\ &= \alpha n! \det(h_{jl}) dx_1 \wedge dy_1 \wedge ... \wedge dx_n \wedge dy_n,\end{aligned}$$

where α is a positive constant. Hence, the volume form $dv = \frac{\omega^n}{n!}$ is positive and coincides with the hermitian volume element of M. If M is compact, then

$$\int_M \omega^n = n!\mathrm{Vol}._\omega(M) > 0.$$

So, if the cohomology class $[\omega] \in H^2(M, \mathbb{R})$, then $[\omega]^n \neq 0$ and furthermore,

$$H^{2k}(M, \mathbb{R}) \neq 0, \quad 0 \leq k \leq n.$$

EXAMPLE 5.1.– *Every Riemann surface is Kähler because every* 2*-form is closed. Every hermitian metric on* M*, where* $\dim M = 1$*, is Kähler.*

EXAMPLE 5.2.– $\mathbb{C}^n$ *is Kähler with the standard metric* $\sum_{j=1}^{n} dz_j \otimes d\overline{z}_j$.

EXAMPLE 5.3.– *Any (analytic) submanifold* N *of a Kähler manifold* M *is Kähler for the induced metric (the hermitian metric* h *on* M *induces a hermitian metric on* N *whose associated form is the pull-back of the associated form of* h*). In particular, any projective variety is Kähler.*

EXAMPLE 5.4.– *Let* $\mathbb{T}^n = \mathbb{C}^n/\Lambda$ *be a complex torus of dimension* n*, where* $\Lambda \simeq H_1(\mathbb{T}^n, \mathbb{Z})$*, is a lattice isomorphic to* $\mathbb{Z}^{2n}$*. The torus* $\mathbb{T}^n$ *is a Kähler manifold.*

EXAMPLE 5.5.– *Let* M *be a projective complex variety. Then,* M *is a Kähler manifold and all its complex de Rham cohomology groups decompose as follows:*

$$H^k(M, \mathbb{C}) = \bigoplus_{p+q=k} H^{p,q}(M),$$

for $0 \leq k \leq n$ *with* $\overline{H^{p,q}(M)} = H^{q,p}(M)$.

EXAMPLE 5.6.– *The complex projective space* $\mathbb{P}^n(\mathbb{C})$ *is a Kähler manifold. This important example consists of using the Fubini–Study metric (this metric will be specified in the solution to exercise 5.6 concerning this problem).*

An interesting result concerning Kähler varieties was obtained by Moishezon. First, recall that according to a Siegel theorem, any complex manifold of dimension n admits at most n algebraically independent meromorphic functions (Zheng 2000). So-called Moishezon varieties (i.e. compact complex analytic varieties that become projective after a finite number of blow-ups of smooth centers) possess the maximum number of algebraically independent meromorphic functions. More precisely, we have the following fundamental result obtained by Moishezon (1967):

THEOREM 5.2 (MOISHEZON THEOREM).– *A compact complex analytic variety of dimension* n *is Moishezon if and only if it admits* n *algebraically independent meromorphic functions.*

Equivalently, the Moishezon theorem says that a compact Kähler variety of dimension n is projective if and only if it has n algebraically independent meromorphic functions. One consequence of the Moishezon theorem is the Chow–Kodaira theorem, which states that a smooth compact complex surface is Moishezon if and only if it is projective. In fact, we can only blow up points in two dimensions. But in general, if M is a compact manifold and if $\widetilde{M}$ denotes the manifold M blowing up at point p, then M is projective if and only if $\widetilde{M}$ is. Even in dimension three, we can find non-projective Moishezon varieties, as shown in the following example constructed by Hironaka: it is a variety containing a compact curve C homologous to 0. The presence of this curve shows that M is not Kähler and therefore not projective. It is enough to reason by the absurd by assuming that M admits a Kähler metric ω. The latter is written in local coordinates $(z_1, ..., zn)$ of M in the form:

$$h = \sum_{j,l} h_{jl} dz_j \otimes d\overline{z}_l,$$

where (h_{jl}) is a positive definite hermitian matrix at each point and we associate it with the real closed differential form $\omega = -\text{Im}\, h$, of type $(1,1)$. We have shown previously that: $\int_C \omega^n > 0$ where the form $\omega^n = \omega \wedge ... \wedge \omega$ (n-factors) is a hermitian volume element of M. Now, C is homologous to 0 and ω is closed, so $\int_C \omega^n = 0$ which is absurd.

5.2. The Kodaira–Nakano vanishing theorem

Let M be a compact complex manifold of dimension n. A line bundle $\mathcal{L}$ is said to be positive over M if its Chern class $c_1(\mathcal{L}) \in H^2(M, \mathbb{R})$ can be represented by a closed positive $(1,1)$-real form. The Kodaira–Nakano vanishing theorem is an assertion that certain cohomology groups of positive line bundles vanish and it is stated as follows: the higher cohomology groups $H^q(M, \Omega^p(\mathcal{L})) = 0$ for all $q > 0$ provided that $\mathcal{L} \longrightarrow M$ is positive and $p + q > n$. This is a generalization of the original statement of Kodaira which said that $H^q(M, \omega_M \otimes \mathcal{L}) = 0$ under the same assumption. To prove the Kodaira–Nakano vanishing theorem, we use isomorphism between sheaf cohomology and the space of harmonic forms.

THEOREM 5.3 (KODAIRA–NAKANO VANISHING THEOREM).– *If $\mathcal{L}$ is a positive line bundle over M, then*

$$H^q(M, \Omega^p(\mathcal{L})) = 0, \textit{ for } p + q > n.$$

Dualizing this statement, if $\mathcal{L}$ is negative, then

$$H^q(M, \Omega^p(\mathcal{L})) = 0, \textit{ for } p + q < 0.$$

PROOF.– Since the bundle $\mathcal{L}$ is positive, we can construct a metric on $\mathcal{L}$ whose associated curvature form $\frac{i}{\sqrt{2\pi}}\Theta$ corresponds to a Kähler metric. Furthermore, we know that $H^q(M, \Omega^p(\mathcal{L})) \simeq \mathcal{H}^{p,q}(\mathcal{L})$, so it suffices to look at harmonic forms. Let $\eta \in \mathcal{H}^{p,q}(\mathcal{L})$, so $\overline{\partial}\eta = 0$ and let $D = D' + \overline{\partial}$ be the connection metric on $\mathcal{L}$. Hence,

$$\Theta\eta = D^2\eta = (\overline{\partial}D' + D'\partial)\eta = \overline{\partial}D'\eta.$$

Let

$$L : \mathcal{A}_M^{p,q}(\mathcal{L}) \longrightarrow \mathcal{A}_M^{p+1,q+1}(\mathcal{L}),$$

be the operator defined by

$$L(\eta \otimes s) = \omega \wedge \otimes s,$$

where ω is the form associated with the Kähler metric. Let $\Lambda = L^*$ be the adjoint operator of L. We easily show that:

$$[\Lambda, \overline{\partial}] = \frac{i}{2}D'^*, \qquad [\Lambda, L]\eta = (n - p - q)\eta,$$

for all $\eta \in H^0(M, \mathcal{A}_M^{p,q}(\mathcal{L}))$. In addition, we have

$$2i\langle\Lambda\Theta\eta, \eta\rangle = 2i\langle\Lambda\overline{\partial}D'\eta, \eta\rangle = 2i\langle(\overline{\partial}\Lambda - \frac{i}{2}D'^*\eta)D'\eta, \eta\rangle = \langle D'^*D'\eta, \eta\rangle,$$

hence,

$$2i\langle\Lambda\Theta\eta, \eta\rangle = \langle D'\eta, D'\eta\rangle \geq 0,$$

because $\langle\overline{\partial}\Lambda D'\eta, \eta\rangle = \langle\Lambda D'\eta, \overline{\partial}^*\eta\rangle = 0$. Similarly, we have

$$-2i\langle\Theta\Lambda\eta, \eta\rangle = -2i\langle D'\overline{\partial}D'\eta, \eta\rangle = -2i\langle D'(\overline{\partial}\Lambda + \frac{i}{2}D'^*\eta)\eta, \eta\rangle = \langle D'D'^*\eta, \eta\rangle,$$

hence,

$$-2i\langle\Theta\Lambda\eta, \eta\rangle\langle D'^*\eta, D'^*\eta\rangle \geq 0.$$

From the two inequalities above, we derive $2i\langle[\Lambda, \Theta]\eta, \eta\rangle \geq 0$. We recall that the external product

$$\wedge : \mathcal{A}_M^{p,q}(\mathcal{L}^*) \otimes \mathcal{A}_M^{p',q'}(\mathcal{L}^*) \longrightarrow \mathcal{A}_M^{p+p',q+q'}(\mathcal{L}^*),$$

can be defined by setting

$$(\eta \otimes s) \wedge (\eta' \otimes s') = \langle s, s'\rangle\eta \wedge \eta'.$$

Since $\Theta = -2\pi i L$, then

$$2i\langle[\Lambda, \Theta]\eta, \eta\rangle = 4\pi\langle[\Lambda, L]\eta, \eta\rangle = 4\pi(n - p - q)\langle\eta, \eta\rangle \geq 0.$$

Therefore, for $p+q>n$, we have $\langle\eta,\eta\rangle=\|\eta\|^2=0$, which implies that $\eta=0$. In other words, for $p+q>n$ we have $\mathcal{H}^{p,q}(\mathcal{L})=0$. Similarly, we show using Kodaira–Serre duality that we have $H^q(M,\Omega^p(\mathcal{L}))=0$, for $p+q<0$. This completes the proof. ■

5.3. The Lefschetz theorem on hyperplane sections

THEOREM 5.4 (LEFSCHETZ THEOREM ON HYPERPLANE SECTIONS).– *Let M be an n-dimensional compact complex manifold and $V\subset M$ be a smooth hypersurface with $\mathcal{L}=[V]$ positive. Then, the map*

$$H^q(M,\mathbb{Q})\longrightarrow H^q(V,\mathbb{Q}),$$

induced by the inclusion $j:V\hookrightarrow M$ is an isomorphism for $q\leq n-2$ and is injective when $q=n-1$.

PROOF.– It suffices to prove the theorem for cohomology with complex coefficients. According to the Hodge decomposition, the cohomologies $H^k(M,\mathbb{C})$ as well as the corresponding cohomologies $H^k(V,\mathbb{C})$ have the following decomposition:

$$H^k(M,\mathbb{C})=\bigoplus_{p+q=k}H^{p,q}(M),$$

where $H^{p,q}(M)\simeq H^q(M,\Omega^p_M)$, according to the Dolbeault theorem. So, it suffices to show that the map $H^q(M,\Omega^p_M)\xrightarrow{j^*}H^q(V,\Omega^p_V)$ is an isomorphism for $p+q\leq n-2$ and injective for $p+q=n-1$. By hypothesis, $\mathcal{L}=[V]$ is positive, i.e. $[-V]$ is negative and by the Kodaira vanishing theorem, we have

$$H^q(M,\Omega^p_M([-V]))=0,\quad p+q<n,$$
$$H^q(M,\Omega^p_V([-V]))=0,\quad p+q<n-1.$$

To make the connection with our case, we first decompose the following restriction $\Omega^p_M\longrightarrow\Omega^p_M$ as a composition:

$$\Omega^p_M\xrightarrow{k}\Omega^p_M|_V\xrightarrow{j}\Omega^p_V.$$

Note that the kernel of k is the sheaf of holomorphic p-forms vanishing on V. Hence, we have an exact sequence of sheaves on M,

$$0\longrightarrow\Omega^p_M([-V])\longrightarrow\Omega^p_M\xrightarrow{k}\Omega^p_M|_V\longrightarrow 0.\qquad [5.2]$$

The map j is part of an exact sequence of sheaves. Indeed, for any point $x\in V$, we have

$$0\longrightarrow N^*_{V,x}\longrightarrow T'^*_x(M)\longrightarrow T'^*_x(V)\longrightarrow 0,$$

where $N^*_{V,x}$ is the conormal bundle (i.e. the dual bundle of the normal bundle; it is a subbundle of the cotangent bundle). Since $N^*_{V,x} \hookrightarrow T^*_x(M)$ has dimensions equal to 1, we have

$$0 \longrightarrow N^*_{V,x} \otimes \Lambda^{p-1} T'^*_x(V) \longrightarrow T\Lambda^p T'^*_x(M) \longrightarrow \Lambda^p T'^*_x(V) \longrightarrow 0,$$

and therefore provides an exact sequence of sheaves

$$0 \longrightarrow \Omega^{p-1}_V(N^*_V) \longrightarrow \Omega^p_M|_V \xrightarrow{j} \Omega^p_V \longrightarrow 0.$$

According to the addition formula, we know that: $N^*_V = [-V]|_V$, so

$$0 \longrightarrow \Omega^{p-1}_V([-V]) \longrightarrow \Omega^p_M|_V \xrightarrow{j} \Omega^p_V \longrightarrow 0. \qquad [5.3]$$

By hypothesis, $[-V]$ is negative on M and the same applies to $[-V]|_V$. Therefore, according to the Kodaira vanishing theorem and taking into account the fact that $H^q(M, \Omega^p_M|_V) \simeq H^q(V, \Omega^p_V|_V)$, the cohomology sequences associated with the sequences [5.2] and [5.3] provide

$$\begin{array}{c} H^q(M, \Omega^p_M([-V])) \longrightarrow H^q(M, \Omega^p_M) \longrightarrow H^q(M, \Omega^p_M|_V) \simeq H^q(V, \Omega^p_M|_V) \\ \downarrow \\ H^{q+1}(M, \Omega^p_M([-V])) \end{array}$$

and

$$H^q(V, \Omega^{p-1}_V([-V])) \longrightarrow H^q(V, \Omega^p_M|_V) \longrightarrow H^q(V, \Omega^p_V) \longrightarrow H^{q+1}(V, \Omega^{p-1}_V([-V])).$$

Therefore, when $p + q \leq n - 2$, the map

$$H^q(M, \Omega^p_M) \longrightarrow H^q(M, \Omega^p_M|_V) \simeq H^q(V, \Omega^p_M|_V) \longrightarrow H^q(V, \Omega^p_V),$$

is an isomorphism and it is injective when $p + q = n - 1$. The proof ends. ■

The name hyperplane sections comes from the fact that if M is a subvariety of $\mathbb{P}^N$, $V = M \cap H$ for a hyperplane H verifies the hypothesis of the theorem. In fact, the theorem holds for $\mathbb{Z}$-coefficient cohomology groups, and even for homotopy groups. The proof for more general cases can be sketched out using the so-called Morse functions (see, for example, Griffiths and Harris (1978, p. 158)).

5.4. The Lefschetz theorem on $(1, 1)$-classes

Before studying the theorem in question, we first prove two results. We start by sketching out a different kind of vanishing theorem.

THEOREM 5.5.– *Let M be a compact complex manifold, $\mathcal{L} \longrightarrow M$ be a positive line bundle, and $E \longrightarrow M$ be a holomorphic vector bundle. Then, for sufficiently large μ and $q > 0$, we have*

$$H^q(M, \mathcal{O}_M(\mathcal{L}^\mu \otimes E)) = 0.$$

PROOF.– Using the Kodaira–Serre duality and Dolbeault theorem, we reduce the problem to

$$\mathcal{H}^{0,p}(\mathcal{L}^{-\mu} \otimes E^* \otimes K_M) = 0,$$

for sufficiently large μ and $p < n = \dim M$. By hypothesis E is any holomorphic vector bundle, we can therefore replace $E^* \otimes K_M$ with E. Let us choose a metric on the bundle $\mathcal{L}$ equipped with a connection such that the form $\omega = \frac{i}{2\pi}\Theta$ is positive where Θ is the curvature associated with this metric. Hence, the curvature form of $\mathcal{L}^{-\mu} \otimes E$ is

$$\Theta_{\mathcal{L}^{-\mu}\otimes E} = 2\pi i \mu\omega \otimes 1_E + \Theta_E.$$

Let $\eta \in \mathcal{H}^{0,p}(\mathcal{L}^{-\mu} \otimes E)$ be a harmonic form. We denote by Θ the curvature form $\Theta_{\mathcal{L}^{-\mu}\otimes E}$ and use reasoning similar to that used in the proof of the Kodaira vanishing theorem. We obtain $2i\langle \Lambda\Theta\eta, \eta\rangle \geq 0, 2i\langle \Theta\Lambda\eta, \eta\rangle \leq 0$, from which $2i\langle [\Lambda, \Theta]\eta, \eta\rangle \geq 0$. Or

$$\Theta = \Theta_{\mathcal{L}^{-\mu}\otimes E} = 2\pi i \mu\omega + \Theta_E,$$

hence

$$2i\langle [\Lambda, \Theta]\eta, \eta\rangle = 2i\langle [\Lambda, \Theta_E]\eta, \eta\rangle - 4\pi\mu(n-p)\|\eta\|^2,$$

and

$$4\pi\mu(n-p)\|\eta\|^2 \leq 2i\langle [\Lambda, \Theta_E]\eta, \eta\rangle \leq C\|\eta\|^2,$$

where C is a constant. Therefore, for $p < n$, we can choose sufficiently large μ (e.g. $\mu > \frac{C}{2\pi}$) so that $4\pi\mu(n-p) > C$, which implies that $\eta = 0$ and the theorem is proven. ■

REMARK 5.2.– *Note that in the case where the bundle E is of rank 1, i.e. a line bundle, the previous theorem follows immediately from the Kodaira vanishing theorem. Indeed, it suffices to choose sufficiently large μ so that the bundle $\mathcal{L}^\mu \otimes E \otimes K_M^*$ is positive. Since*

$$c_1(\mathcal{L}^\mu \otimes E) = \mu c_1(\mathcal{L}) + c_1(E),$$

then for sufficiently large μ and $q > 0$,

$$H^q(M, \mathcal{O}(\mathcal{L}^\mu \otimes E)) = H^q(M, \Omega^n(\mathcal{L}^\mu \otimes E \otimes K_M^*)) = 0.$$

We now use the vanishing theorem above to prove the following important proposition.

PROPOSITION 5.2.– *Any line bundle $\mathcal{L}$ over a submanifold $M \subset \mathbb{P}^N(\mathbb{C})$ is written in the form $\mathcal{L} = [\mathcal{D}]$ where $\mathcal{D}$ is a divisor.*

PROOF.– We will show that any line bundle $\mathcal{L}$ over M has a non-trivial meromorphic global section. Let H be the restriction to M of a hyperplane bundle over $\mathbb{P}^N(\mathbb{C})$. Let us show that for sufficiently large μ, $\mathcal{L} + \mu H$ has a non-trivial holomorphic global section. We proceed by descending induction on $n = \dim M$. We assume that this is true for any submanifold $M \subset \mathbb{P}^N(\mathbb{C})$ of dimensions strictly less than n, i.e. that $H^0(M, \mathcal{O}_M(\mathcal{L}+\mu H)) \neq 0$, for sufficiently large μ. According to the Bertini theorem (see exercise 5.5), M is smooth via the choice of a hyperplane $\mathbb{P}^{N-1}(\mathbb{C})$ (i.e. $M = \mathbb{P}^{N-1}(\mathbb{C}) \cap M$). Hence, $H = [M]$ and we can consider the exact sequence of sheaves

$$0 \longrightarrow \mathcal{O}_M(\mathcal{L} + (\mu - 1)H) \overset{\otimes s}{\longrightarrow} \mathcal{O}_M(\mathcal{L} + \mu H) \longrightarrow \mathcal{O}_M(\mathcal{L} + \mu H) \longrightarrow 0,$$

where s is the section of H which vanishes on M and the second map is the restriction. We deduce from the sequence above

$$H^0(M, \mathcal{O}_M(\mathcal{L}+\mu H)) \longrightarrow H^0(M, \mathcal{O}_V(\mathcal{L}+\mu H)) \longrightarrow H^0(M, \mathcal{O}_M(\mathcal{L}+(\mu-1)H)).$$

For sufficiently large μ, we have by hypothesis $H^0(V, \mathcal{O}_M(\mathcal{L} + \mu H)) \neq 0$ and by the previous theorem, we have $H^0(M, \mathcal{O}_M(\mathcal{L} + (\mu - 1)H)) = 0$. Therefore, $H^0(M, \mathcal{O}_M(\mathcal{L} + \mu H)) \neq 0$ and the proposition is proven. ■

We now prove the Lefschetz theorem for classes of type (1,1).

THEOREM 5.6 (LEFSCHETZ THEOREM FOR CLASSES OF TYPE (1,1)).– *For a submanifold $M \subset \mathbb{P}^N(\mathbb{C})$, any homology class γ in $H^{1,1}(M) \cap H^2(M, \mathbb{Z})$ is analytic. In other words, γ is the Poincaré dual $\eta_{\mathcal{D}}$ (see section 3.7) of the fundamental class of a divisor $\mathcal{D}$ over M.*

PROOF.– Consider the following diagram (see Chapter 3 for notations):

$$\begin{array}{ccccc}
H^1(M, \mathcal{O}^*_M) & \overset{c_1}{\longrightarrow} & H^2(M, \mathbb{Z}) & \overset{i_*}{\longrightarrow} & H^2(M, \mathcal{O}_M) \simeq H^{0,2}(M) \\
 & & \downarrow & & \\
H^{1,1}(M) \cap H^2(M, \mathbb{Z}) & \subset & H^2(M, \mathbb{C}) & & \downarrow \\
 & & \downarrow & & \\
 & & H^1(\mathcal{Z}^1_d) & & H^1(\mathcal{Z}^{0,1}_{\bar\partial}) \\
 & & \downarrow & & \downarrow \\
\dfrac{H^0(\mathcal{Z}^2_d)}{dH^0(\mathcal{Z}^1_d)} & = & H^2_{DR}(M, \mathbb{C}) & \overset{\pi^{(0,2)}}{\longrightarrow} & H^{0,2}_{\bar\partial}(M) = \dfrac{H^0(\mathcal{Z}^{0,2}_{\bar\partial})}{dH^0(\mathcal{Z}^{0,1}_{\bar\partial})}
\end{array}$$

We can see that this is indeed a commutative diagram:

$$\begin{array}{ccc} H^2(M,\mathbb{Z}) & \xrightarrow{i_*} & H^2(M,\mathcal{O}_M) \\ \downarrow & & \downarrow \\ H^2(M,\mathbb{C}) & \longrightarrow & H^{0,2}(M) \end{array}$$

The cohomology sequence from above follows from the exact sequence:

$$0 \longrightarrow \mathbb{Z} \longrightarrow \mathcal{O}_M \xrightarrow{exp} \mathcal{O}_M^* \longrightarrow 0.$$

The vertical arrows on the left express the de Rham isomorphism and those on the right the Dolbeault isomorphism. Let $z = (z_{\alpha\beta\gamma}) \in \mathcal{Z}^2(M,\mathbb{Z}) \subset \mathcal{Z}^2(M,\mathbb{C})$, a Cech cochain, and determine the image of z under the de Rham isomorphism. Locally, we can write $z_{\alpha\beta\gamma}$ in the form: $z_{\alpha\beta\gamma} = f_{\alpha\beta} + f_{\beta\gamma} + f_{\gamma\alpha}$ on $\mathcal{U}_\alpha \cap \mathcal{U}_\beta \cap \mathcal{U}_\gamma$. Since $z_{\alpha\beta\gamma}$ is constant, then $df_{\alpha\beta} + df_{\beta\gamma} + df_{\gamma\alpha} = 0$, so $(df_{\alpha\beta})$ is a closed cochain over $\mathcal{Z}^1(M, \mathcal{Z}_d^1)$ which represents the image of z in $H^1(\mathcal{Z}_d^1)$. And we can find a 1-form ω_α such that: $df_{\alpha\beta} = \omega_\beta - \omega_\alpha$ in $\mathcal{U}_\alpha \cap \mathcal{U}_\beta$ and $(d\omega_\alpha)$ represents the image of z in $H^2_{DR}(M,\mathbb{C})$. The image of z under the Dolbeault isomorphism is represented by $(\overline{\partial} f_{\alpha\beta}) \in \mathcal{Z}^1(\mathcal{Z}^{0,1}_{\overline{\partial}})$. Locally, we have

$$\overline{\partial} f_{\alpha\beta} = \omega_\beta^{0,1} - \omega_\alpha^{0,1},$$

where $\omega_\alpha^{0,1} = \pi^{(0,1)}\omega_\alpha$ and $\overline{\partial}\omega_\alpha^{0,1} = (d\omega_\alpha)^{0,2}$ represents the image of z in $H^{0,2}_{\overline{\partial}}(M)$. Therefore, taking into account the projection $\pi^{(0,2)}$, the de Rham isomorphism is the same as the Dolbeault isomorphism. Therefore, we have $i_*(\gamma) = 0$ and we can find a line bundle $\mathcal{L} \in H^1(M,\mathcal{O}_M^*)$ with $\gamma = c_1(\mathcal{L})$ being the Chern class. For a divisor $\mathcal{D} = \sum n_j V_j$ on M with fundamental class $\sum n_j(V_j)$, we have, according to proposition 5.2, $\mathcal{L} = [\mathcal{D}]$ and

$$\gamma = c_1([\mathcal{D}]) = \sum n_j \eta_{V_j} = \eta_{\mathcal{D}},$$

the Poincaré dual, which completes the proof. ■

Note that the fundamental class of any subvariety of dimension $(n-p)$ of a manifold of dimension n is dual to a (p,p)-form, because only the integral of an $(n-p, n-p)$-form over this subvariety can be non-zero. We deduce that the converse of theorem 5.6 is verified in a fairly trivial way

5.5. The Kodaira embedding theorem

The archetypal embedding theorem in a complex projective space is the Kodaira embedding theorem for compact complex manifolds. The purpose of this section is to provide a proof of this theorem We will also examine the different embeddings of the manifold into a complex projective space $\mathbb{P}^N(\mathbb{C})$. The following Kodaira embedding theorem characterizes compact complex manifolds admitting a holomorphic embedding into a projective space $\mathbb{P}^N(\mathbb{C})$.

THEOREM 5.7 (KODAIRA EMBEDDING THEOREM).– *Let M be a compact complex manifold and let $\mathcal{L} \longrightarrow M$ be a positive line bundle over M. Then, there exists k_0 such that for $k \geq k_0$, the map*

$$\psi_{\mathcal{L}^k} : M \longrightarrow \mathbb{P}^N(\mathbb{C}),$$

is well defined and is an embedding of M into the projective space $\mathbb{P}^N(\mathbb{C})$. In other words, for a divisor $\mathcal{D}$ if $\mathcal{L} \equiv [\mathcal{D}]$ is positive, then for $k \in \mathbb{N}$, the map [5.1] defined by the functions of the space $\mathcal{L}(k\mathcal{D})$ embeds M into $\mathbb{P}^N(\mathbb{C})$ for large enough N (=$\dim \mathcal{L}(k\mathcal{D} - 1)$).

REMARK 5.3.– *The proof of this theorem proceeds in several steps, some of which are analytic and others cohomological techniques, but before that it is convenient to introduce some preliminaries. To show that $\psi_{\mathcal{L}}$ is an embedding, we will first express this in terms of cohomology and thus see the possibility of using the Kodaira vanishing theorem. For the map $\psi_{\mathcal{L}}$ to be well defined on M (a morphism), it is necessary and sufficient that for every $p \in M$, there exists at least one section $s \in H^0(M, \mathcal{O}_M(\mathcal{L}))$ that is never zero; or what amounts to the same thing, it is necessary and sufficient that the base locus $\mathcal{B}$ is non-empty. The evaluation of the sections of $\mathcal{L}$ at the point $p \in M$, $p \neq 0$, determines a linear map*

$$H^0(M, \mathcal{O}_V(\mathcal{L})) \xrightarrow{r_p} \mathcal{L}_p.$$

When p is a base point of $\mathcal{L}$, then all sections of $\mathcal{L}$ vanish at p and therefore the map r_p is identically zero. By fixing an isomorphism of $\mathcal{L}_p$ with $\mathbb{C}$, the map r_p is interpreted as a linear form and depends on the choice of the isomorphism $\mathcal{L}_p \simeq \mathbb{C}$ only by a multiplicative factor. At any point p of M that is not a base point of $\mathcal{L}$, we obtain an element of the projective space $\mathbb{P}(H^0(M, \mathcal{O}_V(\mathcal{L}))^*)$. For line bundles that admit at least one non-zero section, this process determines a meromorphic map

$$\psi_{\mathcal{L}} : M \longrightarrow \mathbb{P}(H^0(M, \mathcal{O}_M(\mathcal{L}))^*),$$

whose indeterminacy points are contained in the base locus $\mathcal{B}$. Therefore, equivalently, to show that $\psi_{\mathcal{L}}$ is an embedding, it is necessary and sufficient that the following restricted map

$$\mathbb{C}^{N+1} = H^0(M, \mathcal{O}_M(\mathcal{L})) \xrightarrow{r_p} \mathcal{L}_p = \mathbb{C},$$

be surjective. If this is the case, then the map $\psi_{\mathcal{L}}$ is an embedding if the following two conditions are satisfied:

– The first condition*: the map $\psi_{\mathcal{L}}$ is injective. In other words, $\psi_{\mathcal{L}}$ separates the points, i.e. for $p, q \in M$ with $p \neq q$, we can find a section $s \in H^0(M, \mathcal{O}_V(\mathcal{L}))$ satisfying $s(p) = 0$ and $s(q) \neq 0$. Equivalently, $\psi_{\mathcal{L}}$ is injective if and only if for $p, q \in M$ with $p \neq q$, the restricted map*

$$\mathbb{C}^{N+1} = H^0(M, \mathcal{O}_M(\mathcal{L})) \xrightarrow{r_{p,q}} \mathcal{L}_p \oplus \mathcal{L}_q = \mathbb{C}^2, \qquad [5.4]$$

is surjective (or again if and only if $\ker r_{p,q} = \ker r_p \cap \ker r_q$ *is of codimension two). In this case, the complete linear system* $|\mathcal{L}|$ *determines an injective morphism and it is base point-free.*

– The second condition*: the differential*

$$D\psi_{\mathcal{L}}(p) : T_pM \longrightarrow T_{\psi_{\mathcal{L}}(p)}\mathbb{P}^N(\mathbb{C}), \quad p \in \mathcal{U}_\alpha,$$

is injective if and only if $s_0(p) \neq 0$ *and*

$$D\psi_{\mathcal{L}}(p) = \{ds_1(p), ..., ds_(p)\} : T_pM \longrightarrow \mathbb{C}^N, \quad p \in \mathcal{U}_\alpha,$$

is an embedding. We can assume that the image of $\psi_{\mathcal{L}}(p)$ *is* $\psi_{\mathcal{L}}(p) = (1, 0, ..., 0)$*; it suffices to make a change of coordinates* s_j *by adding* λs_0, $\lambda \in \mathbb{C}^*$ *to* $s_j(p)$*. So, the second condition is equivalent to the fact that:* $\bigcap_{j\geq 1} ds_j(p) \neq 0$ *or if you like* $\psi_{\mathcal{L}}$ *admits a differential that is never zero. Equivalently (it suffices to use a linear transformation) for any* $v* \in T_p^*(M)$*, we can find a section* $s \in H^0(M, \mathcal{O}_V(\mathcal{L}))$ *satisfying* $s_\alpha(p) = 0$, $ds_\alpha(p) = v^*$*, where* $s_\alpha = \varphi_\alpha^* s$ *and* φ_α *denote a trivialization of* $\mathcal{L}$ *around* p*. We can also formulate this second condition in another way, independent of the choice of the neighborhood* $\mathcal{U}_p$ *of* p*. We denote by* $\mathcal{J}_p$ *the sheaf of holomorphic functions vanishing at* p *and by* $\mathcal{J}_p(\mathcal{L})$ *the sheaf of sections of* $\mathcal{L}$ *vanishing at* p*. Let* s *be a section of* $\mathcal{J}_p(\mathcal{L})$ *defined near* p *and* φ_α, φ_β *be trivializations of* $\mathcal{L}$ *in the neighborhoods* $\mathcal{U}_\alpha$ *and* $\mathcal{U}_\beta$ *of* p*. By writing* $s_\alpha = \varphi_\alpha^* s$, $s_\beta = \varphi_\beta^* s$, $s_\alpha = g_{\alpha\beta} s_\beta$*, where* $g_{\alpha\beta}$ *is the transition function of* $\mathcal{L}$*, the differentials* ds_α *and* ds_β *corresponding to the change from* $\mathcal{U}_\alpha$ *to* $\mathcal{U}_\beta$ *are related by the relation:* $d(s_\alpha) = d(s_\beta)g_{\alpha\beta}$ *at* p*. Hence, we have defined a map* $d_p : \mathcal{J}_p(\mathcal{L}) \longrightarrow T_p'^* \otimes \mathcal{L}_p$*, which is well defined and therefore the second condition above can be formulated by saying that for all* p $\in M$*, the map*

$$H^0(M, \mathcal{J}_p(\mathcal{L})) \xrightarrow{d_{p,q}} T_p'^* \otimes \mathcal{L}_p, \qquad [5.5]$$

is surjective. This is a limiting case of [5.4] when $q \longrightarrow p$*.*

Proof of theorem 5.7: We now propose to demonstrate the Kodaira embedding theorem for which these remarks are preliminary.

1) We have two exact sequences of sheaves

$$0 \longrightarrow \mathcal{J}_{p,q}(\mathcal{L}) \longrightarrow \mathcal{O}_V(\mathcal{L}) \xrightarrow{r_{p,q}} \mathcal{L}_p \oplus \mathcal{L}_q \longrightarrow 0,$$

and

$$0 \longrightarrow \mathcal{J}_p^2(\mathcal{L}) \longrightarrow \mathcal{J}_p(\mathcal{L}) \xrightarrow{d_p} T_p'^* \otimes \mathcal{L}_p \longrightarrow 0.$$

So, to show that maps [5.4] and [5.5] are surjective, it suffices to prove that

$$H^1(M, \mathcal{J}_{p,q}(\mathcal{L})) = 0, \qquad [5.6]$$

and

$$H^1(M, \mathcal{J}_p^2(\mathcal{L})) = 0. \qquad [5.7]$$

Note that in the theorem, having $\mathcal{L}^k$ ($k \in \mathbb{N}^*$) instead of $\mathcal{L}$ does not pose a problem because this construction can be repeated by replacing $\mathcal{L}$ by its tensorial power $\mathcal{L}^k$ and we use the fact that

$$H^1(M, \mathcal{O}_M(\mathcal{L}^k)) = 0, \quad k \geq k_1.$$

We mentioned above that we will see the possibility of using vanishing theorems. Now, our problem here (except in the case where $\dim M = 1$) is due to the fact that we are dealing with the sheaves $\mathcal{J}_{p,q}$, $\mathcal{J}_p^2$ and these are not line bundles; they are in fact coherent sheaves. We can certainly use a divisor and proceed by induction, but the problem is that we do not know if M contains such divisors. To get around these difficulties, we will use the blow-up method.

2) Let $\widetilde{M_p}$ be the blow-up of M at point p and $\pi : \widetilde{M_p} \longrightarrow M$, a holomorphic projection. The preimage $\mathcal{P} \equiv \pi^{-1}(p)$ (which we also denote by $\mathcal{P}_p$ if there is a risk of confusion) is a divisor in M, biholomorphic to $\mathbb{P}^{n-1}(\mathbb{C})$. The map

$$\pi : \widetilde{M_p} \backslash \mathcal{P} \longrightarrow M \backslash \{p\},$$

is a diffeomorphism. We construct the variety $\widetilde{M_p}$ in the following way: let $z_1, ..., z_n \in \Sigma_p$ be holomorphic coordinates in p where $n = \dim M$ and Σ_p be an open neighborhood of p. Let

$$\widetilde{\Sigma}_p = \left\{(z, l) \in \Sigma_p \times \mathbb{P}^{n-1}(\mathbb{C}) : z_i l_j - z_j l_i = 0, \forall i, j\right\},$$

be the blow-up of Σ_p at p with $z = (z_1, ..., z_n) \in \Sigma_p$ and $l = [l_1 : ... : l_n]$ be homogeneous coordinates in $\mathbb{P}^{n-1}(\mathbb{C})$. Considering $l \in \mathbb{P}^{n-1}(\mathbb{C})$ as a straight line in $\mathbb{C}^n$, the condition $z_i l_j = z_j l_i$ means that this straight line contains the point z. It is easy to check that the manifold $\widetilde{\Sigma}_p$ is smooth and that it projects holomorphically onto Σ_p by the projection to the first factor; i.e.

$$\pi : \widetilde{\Sigma}_p \subset \Sigma_p \times \mathbb{P}^{n-1}(\mathbb{C}) \longrightarrow \Sigma_p, \quad (z, l) \longmapsto z.$$

Obviously, π is an isomorphism if $z \neq 0$ and $\mathcal{P} = \pi^{-1}(0) = \mathbb{P}^{n-1}(\mathbb{C})$ is a smooth complex hypersurface (in fact, $\mathcal{P} = \mathbb{P}(T_p M)$). We call $\mathcal{P}$ the exceptional divisor, whose ideal sheaf $\mathcal{J}_{\mathcal{P}}$ is an invertible sheaf denoted by $\mathcal{O}(-\mathcal{P})$. Finally, the blow-up $\widetilde{\mathcal{P}}_p$ of M at p is obtained by gluing $\widetilde{\Sigma}_p$ and $\mathcal{P} \backslash \{p\}$ following the open set

$$\widetilde{\Sigma}_p \backslash \mathcal{P} \simeq \Sigma_p \backslash \{p\}.$$

3) Let $\widetilde{M}$ be the blow-up of the manifold M (of dimension n) at the points p, q and $\pi : \widetilde{M} \longrightarrow M$ a holomorphic projection. The preimages $\mathcal{P}_p = \pi^{-1}(p)$ and $\mathcal{P}_q = \pi^{-1}(q)$ are exceptional divisors. Let

$$\mathcal{P} = \mathcal{P}_p + \mathcal{P}_q, \qquad \widetilde{\mathcal{L}} = \pi^* \mathcal{L}.$$

For $n = 1$, we take $\widetilde{M} = M$, $\pi = Id$; in fact, in this case, we do not need to practice the blow-up and the result follows directly from the Kodaira vanishing theorem. We can therefore assume $n \geq 2$. The pull-back map of holomorphic forms of maximum degree

$$H^0(M, \mathcal{O}_M(\mathcal{L}^k)) \xrightarrow{\pi^*} H^0(\widetilde{M}, \mathcal{O}_{\widetilde{M}}(\widetilde{\mathcal{L}}^k)),$$

according to Hartog theorem (for any global section $\widetilde{s}$ of $\widetilde{\mathcal{L}}^k$, the section s of $\mathcal{L}^k$ on $M\backslash\{p, q\}$ extends to a global section $s \in H^0(M, \mathcal{O}_M(\mathcal{L})))$. By definition, $\widetilde{\mathcal{L}}^k$ being trivial along $\mathcal{P}_p$ and $\mathcal{P}_q$, we deduce that we have

$$H^0(\mathcal{P}, \mathcal{O}_{\mathcal{P}}(\widetilde{\mathcal{L}}^k)) = \mathcal{L}_p^k \oplus \mathcal{L}_q^k.$$

These actions give the following commutative diagram:

$$\begin{array}{ccc} H^0(M, \mathcal{O}_M(\mathcal{L}^k)) & \xrightarrow{\quad r_{p,q} \quad} & \mathcal{L}_p^k \oplus \mathcal{L}_q^k \\ {\scriptstyle \pi^*}\big\downarrow & & \big\| \\ H^0(\widetilde{M}, \mathcal{O}_{\widetilde{M}}(\widetilde{\mathcal{L}}^k)) & \xrightarrow{\quad r_{\mathcal{P}} \quad} & H^0(\mathcal{P}, \mathcal{O}_{\mathcal{P}}(\widetilde{\mathcal{L}}^k)) \end{array}$$

where $r_{\mathcal{P}}$ denotes the restriction to $\mathcal{P}$. To prove [5.4] is the same as showing the surjectivity of $r_{\mathcal{P}}$. Considering the exact sequence on the blow-up $\widetilde{M}$,

$$0 \longrightarrow \mathcal{O}_{\widetilde{M}}(\mathcal{L}^k \otimes [-\mathcal{P}]) \longrightarrow \mathcal{O}_{\widetilde{M}}(\widetilde{\mathcal{L}}^k) \xrightarrow{r_{\mathcal{P}}} \mathcal{O}_{\mathcal{P}}(\widetilde{\mathcal{L}}^k_{|\mathcal{P}}) \longrightarrow 0,$$

and the associated long exact sequence, we can see that it is sufficient to show that for sufficiently large k, we have

$$H^1(\widetilde{M}, \mathcal{O}_{\widetilde{M}}(\widetilde{\mathcal{L}}^k \otimes [-\mathcal{P}])) = 0.$$

4) We choose k_1 such that: $\mathcal{L}^{k_1} \otimes K_M^*$ is positive on M. Let us show that we can choose k_2 such that: $\widetilde{\mathcal{L}}^k \otimes [-\mathcal{P}]^n$ is positive for $k \geq k_2$ (the choice of $k_1 + k_2$ will come later, towards the end of the proof of the theorem). Let $\mathcal{U}$ be a neighborhood of $p \in M$ and let $z : \mathcal{U} \longrightarrow \Sigma_p$ be a local coordinate around p. Let $\Theta_{[\mathcal{P}]}$ be the curvature of the line bundle $[\mathcal{P}]$ and let

$$\Omega_{[\mathcal{P}]} = \frac{i}{2\pi} \Theta_{[\mathcal{P}]}.$$

We can find a metric h on $\widetilde{M}$ such that

$$\Omega_{[-\mathcal{P}]} = -\Omega_{[\mathcal{P}]},$$

is zero on $\widetilde{M}\backslash\widetilde{\mathcal{U}}_{2\delta'}$, greater than or equal to zero on $\widetilde{\mathcal{U}}_\delta$ and strictly greater than zero on $T_p\mathcal{P} \subset T'_p\widetilde{M}$ for all $p \in \mathcal{P}$. Here, $\mathcal{U}_\delta$ denotes the ball $\|z\| < \delta$ with $\delta \in \mathbb{R}^*_+$. To determine such a metric, we proceed as follows: let h_1 be the metric on $[\mathcal{P}]_{\widetilde{\mathcal{U}}}$ given by $|(l_1, ..., l_n)|^2 = \|l\|^2$. Let $s \in H^(\widetilde{M}, \mathcal{O}_{\widetilde{M}}(\mathcal{P}))$ be a global section of $[\mathcal{P}]$ on $\widetilde{M}$ with $(s) = \mathcal{P}$. Let h_2 on $[\mathcal{P}]_{\widetilde{M}\backslash\mathcal{P}}$ be defined by $|s(z)| = 1$. For $\delta > 0$, we consider the ball $\|z\| < \delta$ with center 0 and radius $\delta \in \mathbb{R}^*_+$. We set $\widetilde{\mathcal{U}}_\delta = \pi^{-1}(\mathcal{U}_\delta)$. Let $\{\rho_1, \rho_2\}$ be a partition of unity for $\{\widetilde{\mathcal{U}}_{2\delta}, \widetilde{M}\backslash\widetilde{\mathcal{U}}_\delta\}$ and let $h = \rho_1 h_1 + \rho_2 h_2$.

– On $\widetilde{M}\backslash\widetilde{\mathcal{U}}_{2\delta}$, we have $\rho_2 = 1$, hence $\|s\|^2 = 1$, and therefore

$$\Omega_{[\mathcal{P}]} = -\frac{i}{2\pi}\log\frac{1}{\|s\|^2} = 0.$$

– Likewise, on $\widetilde{M}\backslash\mathcal{P} \simeq \mathcal{U}_\delta\backslash\{p\}$, we write $s(z, l) = z$ and therefore

$$\Omega_{[\mathcal{P}]} = -\frac{i}{2\pi}\log\frac{1}{\| z \|^2} = \frac{i}{2\pi}\log \| z \|^2 .$$

In other words,

$$-\Omega_{[\varepsilon]} = \pi'^*\omega,$$

where ω is the $(1, 1)$-form associated with the Fubini–Study metric (for more information on this metric, see exercise 5.7) on the space $\mathbb{P}^{n-1}(\mathbb{C})$, with

$$\pi' : \widetilde{\mathcal{U}}_\delta \longrightarrow \mathbb{P}^{n-1}(\mathbb{C}), \quad (z, l) \longmapsto l.$$

So, we have $-\Omega_{[\mathcal{P}]} \geq 0$ on $\widetilde{\mathcal{U}}_\delta\backslash\mathcal{P}$. By continuity, $-\Omega_{[\mathcal{P}]} = \pi'^*\omega$ on $\widetilde{\mathcal{U}}_\delta$ and consequently $\Omega_{[\mathcal{P}]}|_\mathcal{P} = \omega > 0$. Now, we will see that there exists k_2 such that: $\widetilde{\mathcal{L}}^k \otimes [-\mathcal{P}]^n$ is positive for $k \geq k_2$. Indeed, we have $\Omega_{\widetilde{\mathcal{L}}} = \pi^*\Omega_{[\mathcal{L}]}$ and from the above $\Omega_{\widetilde{\mathcal{L}}}$ is positive except along $\mathcal{P}$. Furthermore, we have $\langle\Omega_{\widetilde{\mathcal{L}}}; v, \overline{v}\rangle \geq 0$ if and only if $v \in T_p(\widetilde{M})$, $p \in \mathcal{P}$, $\langle\Omega_{\widetilde{\mathcal{L}}}; v, \overline{v}\rangle = 0$, if and only if $\pi_*(v) = 0$. This means that the form $\Omega_{\widetilde{\mathcal{L}}}$ is everywhere greater than or equal to zero and it is strictly greater than zero on $\widetilde{M}\backslash\mathcal{P}$ and on $T'_p\widetilde{M}/T'_p\mathcal{P}$, $p \in \mathcal{P}$. The form

$$\Omega_{\widetilde{\mathcal{L}}^k\otimes[-\mathcal{P}]^n} = \Omega_{\widetilde{\mathcal{L}}^k} + \Omega_{[-\mathcal{P}]^n} = k\Omega_{\widetilde{\mathcal{L}}} + n\Omega_{[-\mathcal{P}]},$$

is positive everywhere on $\widetilde{U}_\delta$ and $\widetilde{M}\backslash\widetilde{U}_{2\delta}$. Taking into account compactness, then if $v \notin T_p\mathcal{P}$ then $\langle\Omega_{\widetilde{\mathcal{L}}}; v, \overline{v}\rangle$ is bounded. Therefore, $\widetilde{\mathcal{L}}^k \otimes [-\mathcal{P}]^n$ is positive for k large enough or what amounts to the same thing there exists k_2 such that: $\widetilde{\mathcal{L}}^k \otimes [-\mathcal{P}]^n$ is positive for $k \geq k_2$.

5) Let us show that the canonical bundle $K_{\widetilde{M}}$ of $\widetilde{M}$ is given by the formula

$$\widetilde{M} = \widetilde{K_M} \otimes [\mathcal{P}]^{n-1}, \quad \widetilde{K_M} = \pi^* K_M.$$

Indeed, in this proof, we will use the equations $z_i l_j - z_j l_i = 0$, which are involved in the definition of the blow-up (see *2)*). Since this is a local calculation, then we can, without restricting the generality, take $M = \mathbb{C}^n$. Let us choose an open covering

$$\widetilde{M} = \bigcup \mathcal{U}_\alpha, \quad \mathcal{U}_\alpha \equiv \{(z, l) : z_\alpha \neq 0\}, \quad \alpha = 1, ..., n,$$

and coordinates

$$\varphi_\alpha : \mathcal{U}_\alpha \simeq \mathbb{C}^n \subset \mathbb{C}^{n+1}, \quad (z, l) \longmapsto \left(\frac{z_1}{z_\alpha}, ..., 1, ..., \frac{z_n}{z_\alpha}, z_\alpha\right).$$

Therefore, $\mathbb{C}^n \subset \mathbb{C}^{n+1}$ is the affine subspace defined by $u_\alpha = 1$. The transition functions are given by

$$\varphi_{\alpha\beta}(u_1, ..., u_{n+1}) = \left(\frac{u_1}{u_\alpha}, ..., \frac{u_n}{u_\alpha}, ..., \frac{u_{n+1}}{u_\alpha}\right).$$

Consequently, the determinant of the change matrix of charts is

$$\det J(\varphi_{\alpha\beta}) = \gamma_{\alpha\beta}.u_\alpha,$$

where $\{\gamma_{\alpha\beta}\}$ is the corresponding cocycle on $\mathbb{P}^{n-1}(\mathbb{C})$, i.e. $\gamma_{\alpha\beta} = \frac{1}{u_\alpha^n}$. So, $K_{\widetilde{M}}$ is given by the cocycle

$$\frac{1}{u_\alpha^{n-1}} = \left(\frac{z_\beta}{z_\alpha}\right)^{n-1}.$$

Furthermore, the divisor $\mathcal{P} \subset \widetilde{M}$ is defined as the locus of the zeros of the functions $l_1, ..., l_n$. Using $z_i l_j - z_j l_i = 0$, this can be described on the open subset $\mathcal{U}_\alpha$ by the single equation $l_\alpha = 0$. So, the cocycle associated with $\mathcal{P}$ and which describes the line bundle $\mathcal{O}(\mathcal{P})$ is $\{\mathcal{U}_{\alpha\beta}, \frac{1}{u_\alpha}\}$. We therefore deduce from this calculation that for a canonical form defined in the neighborhood of $p \subset M$ and non-zero at p, its pull-back by π is a canonical form on the blow-up $\widetilde{M}$ and which vanishes along the exceptional divisor $\mathcal{P}$ to order $n-1$. We therefore have a line bundle isomorphism

$$\pi^* K_M \longrightarrow K_{\widetilde{M}} \otimes [\mathcal{P}]^{1-n}, \quad \lambda \longmapsto s^{1-n}\pi^*(\lambda),$$

where $s \in H^0(\widetilde{M}, \mathcal{O}_M(\mathcal{P}))$ is a section.

6) Let $k_0 \equiv k_1 + k_2$, and note that the choice of k_0 is independent of that of p because the manifold M is compact. For $k \geq k_0$, we have by virtue of the previous step,

$$\begin{aligned} \mathcal{O}_{\widetilde{M}}\left(\widetilde{\mathcal{L}}^k \otimes [-\mathcal{P}]\right) &= \Omega^n_{\widetilde{M}}\left(\widetilde{\mathcal{L}}^k \otimes [-\mathcal{P}] \otimes K^*_{\widetilde{M}}\right), \\ &= \Omega^n_{\widetilde{M}}\left((\widetilde{\mathcal{L}}^{k_1} \otimes \widetilde{K}^*_M) \otimes (\widetilde{\mathcal{L}}^{k'} \otimes [-\mathcal{P}]^n)\right), \quad k' \geq k_2 \end{aligned}$$

We chose above k_1 so that $\mathcal{L}^{k_1} \otimes K^*_M$ is positive on M, and $\widetilde{\mathcal{L}}^{k_1} \otimes \widetilde{K}^*_M$ is only semi-positive on M but this is corrected by the fact that by our choice $\widetilde{\mathcal{L}}^{k_1} \otimes [-\mathcal{P}]^n$ is positive on $\widetilde{M}$. Therefore, $(\widetilde{\mathcal{L}}^k_1 \otimes \widetilde{K}^*_M) \otimes (\widetilde{\mathcal{L}}^{k'} \otimes [-\mathcal{P}]^n)$ is positive on $\widetilde{M}$ and by the Kodaira–Nakano vanishing theorem, we have for $k \geq k_0$,

$$H^1\left(\widetilde{M}, \mathcal{O}_{\widetilde{M}}(\widetilde{\mathcal{L}}^k \otimes [-\mathcal{P}])\right) = H^1\left(\widetilde{M}, \Omega^n_{\widetilde{M}}((\widetilde{\mathcal{L}}^{k_1} \otimes \widetilde{K}^*_M) \otimes (\widetilde{\mathcal{L}}^{k'} \otimes [-\mathcal{P}]^n)\right) = 0.$$

Therefore, [5.6] is satisfied or what amounts to the same thing the map [5.4] is surjective.

7) Let us now prove [5.7], i.e. that the map [5.5] is surjective. Let $\widetilde{M}$ be the blow-up of M at point p and $\pi : \widetilde{M} \longrightarrow M$ be a holomorphic projection. The preimage $\mathcal{P} = \pi^{-1}(p)$ is an exceptional divisor. The pull-back map

$$H^0\left(M, \mathcal{O}_M(\mathcal{L}^k)\right) \xrightarrow{\pi^*} H^0\left(\widetilde{M}, \mathcal{O}_{\widetilde{M}}(\widetilde{\mathcal{L}}^k)\right),$$

is an isomorphism according to the Hartog theorem. If $s \in H^0\left(M, \mathcal{O}_M(\mathcal{L}^k)\right)$, then $s(p) = 0$ if and only if $\pi^* s$ vanishes on $\mathcal{P}$. Therefore, π^* is limited to providing an isomorphism

$$H^0\left(M, \mathcal{J}_p(\mathcal{L}^k)\right) \xrightarrow{\pi^*} H^0\left(\widetilde{M}, \mathcal{O}_{\widetilde{M}}(\widetilde{\mathcal{L}}^k \otimes [-\mathcal{P}])\right).$$

As before, we have

$$H^0\left(\mathcal{P}, \mathcal{O}_{\mathcal{P}}(\widetilde{\mathcal{L}}^k \otimes [-\mathcal{P}])\right) = \mathcal{L}^k_p \otimes H^0\left(\mathcal{P}, \mathcal{O}_{\mathcal{P}}([-\mathcal{P}])\right) \simeq \mathcal{L}^k_p \otimes T'^*_p.$$

The following diagram:

$$\begin{array}{ccc} H^0\left(M, \mathcal{J}_p(\mathcal{L}^k)\right) & \xrightarrow{\quad d_p \quad} & T'^*_p \oplus \mathcal{L}^k_p \\ {\scriptstyle \pi^*}\downarrow \simeq & & \Big\| \\ H^0\left(\widetilde{M}, \mathcal{O}_{\widetilde{M}}(\widetilde{\mathcal{L}}^k \otimes [-\mathcal{P}])\right) & \xrightarrow{\quad r_{\mathcal{P}} \quad} & H^0\left(\mathcal{P}, \mathcal{O}_{\mathcal{P}}(\widetilde{\mathcal{L}}^k \otimes [-\mathcal{P}])\right) \end{array}$$

is commutative and it is therefore sufficient to show that the map $r_{\mathcal{P}}$ is surjective for large enough k. By analogy with what was done in the previous step, we use the exact sequence on $\widetilde{\mathcal{P}}$,

$$0 \longrightarrow \mathcal{O}_{\widetilde{\mathcal{P}}}\left(\widetilde{\mathcal{L}}^k \otimes [-\mathcal{P}]^2\right) \longrightarrow \mathcal{O}_{\widetilde{\mathcal{P}}}\left(\widetilde{\mathcal{L}}^k \otimes [-\mathcal{P}]\right) \xrightarrow{r_{\mathcal{P}}} \mathcal{O}_{\mathcal{P}}\left(\widetilde{\mathcal{L}}^k \otimes [-\mathcal{P}]\right) \longrightarrow 0.$$

Then, we choose k_1 such that: $\mathcal{L}^{k_1} \otimes K_M^*$ is positive on M and according to [5.4] and [5.5], there exists k_2 such that: $\widetilde{\mathcal{L}}^{k'} \otimes [-\mathcal{P}]^{n+1}$ is positive on $\widetilde{M}$ for $k' \geq k_2$. For $k \geq k_0 \equiv k_1 + k_2$, we have

$$\mathcal{O}_{\widetilde{M}}\left(\widetilde{\mathcal{L}}^k \otimes [-\mathcal{P}]^2\right) = \Omega^n_{\widetilde{M}}\left((\mathcal{L}^{k_1} \otimes \widetilde{K}^*_M) \oplus (\widetilde{\mathcal{L}}^{k'} \otimes [-\mathcal{P}]^{n+1})\right),$$

with $k' \geq k_2$. Note that here too (since M is compact), the choice of k_0 does not depend on p. According to the Kodaira vanishing theorem, we have

$$H^1\left(\widetilde{M}, \mathcal{O}_{\widetilde{M}}(\widetilde{\mathcal{L}}^k \otimes [-\mathcal{P}]^2)\right) = 0,$$

for $k \geq k_0$. Consequently, [5.7] is satisfied, so the map [5.5] is surjective and the proof is complete. ■

DEFINITION 5.7.– *A line bundle $\mathcal{L}$ is said to be very ample if its global sections provide an embedding of the manifold M into a projective space or in other words if the map $\psi_{\mathcal{L}}$ embeds the manifold M into a projective space. A line bundle $\mathcal{L}$ is said to be ample if a power $\mathcal{L}^k = \mathcal{L}^{\otimes k}$, $k \in \mathbb{N}^*$, is very ample. A divisor $\mathcal{D}$ is said to be ample (respectively very ample) if the line bundle $[\mathcal{D}]$ associated with this divisor is ample (respectively very ample).*

With this terminology, we can translate the Kodaira embedding theorem as follows:

THEOREM 5.8.– *Let $\mathcal{L} \to M$ be a line bundle over a complex manifold M. Then, $\mathcal{L}$ is ample if, and only if, its first Chern class $c_1(\mathcal{L})$ is positive.*

Regarding definition 5.7, another way to describe ample and very ample divisors is as follows: consider a basis $(f_1, ..., f_N)$ of the space $\mathcal{L}(\mathcal{D})$ and the morphism

$$\psi : M \longrightarrow \mathbb{P}^N(\mathbb{C}), \quad z \longmapsto [f_1(z) : ...f_N(z)],$$

where $N = \dim \mathcal{L}(\mathcal{D}) - 1$ (which is well defined, as we have already pointed out, modulo the automorphisms of the space $\mathbb{P}^N(\mathbb{C})$ and outside the poles of the functions f_j and their common zeros). In these terms, the divisor $\mathcal{D}$ is said to be very ample if the above morphism is an embedding and it is said to be ample if one of its (positive) multiples is very ample.

We will discuss the Kodaira embedding theorem, its relationship to Hodge theory and Kähler manifolds and other equivalent versions of the embedding theorem in complex projective space.

According to Kodaira embedding theorem, there exists a positive divisor if and only if M has a Hodge form ω. For integer k, $[k\omega] \in H^2(M, \mathbb{Z})$. Using the Lefschetz theorem for classes of type $(1, 1)$ which ensures that for a submanifold $M \subset \mathbb{P}^N(\mathbb{C})$, any cohomology class in $H^{1,1}(M) \cap H^2(M, \mathbb{Z})$ is analytic (this theorem classifies the cohomology classes of a projective complex variety which are the Chern classes of line bundles), this assures us of the existence of a line bundle $\mathcal{L}$ such that its Chern class is $c_1(\mathcal{L}) = k\omega$. So, $\mathcal{L}$ is positive and we can therefore reformulate the Kodaira embedding theorem as follows:

THEOREM 5.9.– *A compact complex manifold is projective if and only if it admits a Hodge form.*

Compact Kähler manifolds form a remarkable class of complex analytic manifolds. Generally, we are interested in the class of Kähler varieties by previleging projective varieties. One reason is that the latter contain many complex subvarieties, whereas Kähler manifolds generally do not. It is known that non-Kähler compact complex manifolds can be found, but it is very difficult to construct or decide whether a complex manifold is Kähler or not. Analytic complex projective manifolds are special examples of compact Kähler manifolds. A Kähler form with an entire cohomology class is simply a Hodge form. Another equivalent way of formulating the Kodaira embedding theorem is therefore as follows:

THEOREM 5.10.– *Let M be a compact complex manifold. Then, M is projective if and only if it admits a Kähler form whose cohomology class is entire.*

5.6. Exercises

EXERCISE 5.1.– *Explain why on $\mathbb{C}^n$ there are no interesting compact submanifolds.*

Solution: This is because the only complex and compact submanifold of $\mathbb{C}^n$ is a point since if $\psi : M \longrightarrow \mathbb{C}^n$ is a holomorphic embedding where M is a compact complex manifold, then ψ is constant by virtue of Liouville theorem.

EXERCISE 5.2.– *We have seen previously that the autonomy factor determines a line bundle denoted by $\mathcal{L}(h, \alpha)$. Show that the latter is ample if and only if the form h is positive definite.*

Hint: See Mumford (1975b, p. 30) or Birkenhake and Lange (2004, p. 85).

EXERCISE 5.3.– *Show that a complex p-form $\omega \in A^{p,0}(M)$ is holomorphic if and only if $\overline{\partial}\omega = 0$.*

Solution: In local coordinates for $\sum_{I,J} f_{I,J} dz_I \wedge d\overline{z}_J$, we have:

$$0 = \overline{\partial}\omega = \sum_{I,J} \overline{\partial} f_{I,J} \wedge dz_I \wedge d\overline{z}_J,$$

if and only if $\overline{\partial} f_{I,J} = 0$ for all I, J, i.e. all $f_{I,J}$ are holomorphic.

EXERCISE 5.4.– *Using the notations of theorem 5.1 and its consequences, show that the linear isomorphism $H^{p,q}(M) \simeq \mathcal{H}^{p,q}(M)$, $[\psi] \longmapsto \mathcal{H}\psi$, is not a ring homomorphism.*

Hint: The wedge product of two harmonic forms is not harmonic in general.

EXERCISE 5.5 (BERTINI THEOREM).– *Let M be a projective manifold (i.e. a closed analytic submanifold of a smooth complex projective space $\mathbb{P}^n(\mathbb{C})$). Show that every generic hyperplane section of M is smooth.*

Solution: Recall that a hyperplane section of $M \subset \mathbb{P}^n(\mathbb{C})$ is the intersection of M with a hyperplane of $\mathbb{P}^n(\mathbb{C})$. The idea of the solution is to consider the dual projective space $(\mathbb{P}^n(\mathbb{C}))^*$ consisting of the hyperplanes of $\mathbb{P}^n(\mathbb{C})$; this space parameterizes the hyperplane sections of M. The projection of the set $I = \{(p, H) \in M \times (\mathbb{P}^n(\mathbb{C}))^* : T_pM \subset H\}$, onto M realizes I as a projective bundle onto M and therefore the fibers of this projection are projective spaces of dimension equal to $n - 1$ and hence I is irreducible. Consequently, I is a manifold of dimension $n - 1$ and the projection of I onto $(\mathbb{P}^n(\mathbb{C}))^*$ is not surjective. The "good" hyperplanes are parameterized by the complement of this manifold.

EXERCISE 5.6.– *Determine a Kähler metric on the complex torus.*

Solution: Let Λ be a lattice of $\mathbb{C}^n$ and $\mathbb{T}^n = \mathbb{C}^n/\Lambda$ be a complex torus. This is a compact complex manifold. Any hermitian form

$$\omega = i \sum f_{jk} dz_j \wedge d\overline{z}_k, \quad i = \sqrt{-1},$$

with constant coefficients on $\mathbb{C}^n$ and positive definite determines a Kähler metric on $\mathbb{T}^n$.

EXERCISE 5.7.– *a) Show that the complex projective space $\mathbb{P}^n(\mathbb{C})$ is a Kähler variety for the Fubini–Study metric (this metric will be defined and specified during the solution).*

b) Show that any complex manifold admitting a holomorphic embedding in the space $\mathbb{P}^n(\mathbb{C})$ is a Kähler variety.

Solution: *a)* Indeed, on the projective space $\mathbb{P}^n(\mathbb{C})$ of dimension n, we have a canonical hermitian metric, the Fubini–Study metric (see below) which makes $\mathbb{P}^n(\mathbb{C})$

a Kähler variety. Let $[Z_0 : ... : Z_n] \in \mathbb{P}^n(\mathbb{C})$ be homogeneous coordinates and let $\mathcal{U}_j = \{[Z_0 : ... : Z_n] \in \mathbb{P}^n(\mathbb{C}) : Z_j \neq 0\}$ be the set of lines for which $Z_j \neq 0$. We also consider the map,

$$\varphi_j : \mathcal{U}_j \longrightarrow \mathbb{C}^n, [Z_0 : ... : Z_n] \longmapsto \left(\frac{Z_0}{Z_j}, ..., \frac{Z_{j-1}}{Z_j}, \frac{Z_{j+1}}{Z_j}, ..., \frac{Z_n}{Z_j}\right) \equiv (z_1, ..., z_n),$$

with $z_k = \frac{Z_{k-1}}{Z_j}$ if $k \leq j$ and $z_k = \frac{Z_k}{Z_j}$ if $k > j$, where $1 \leq k \leq n$. We consider in $\mathcal{U}_j$ the $(1,1)$-differential form,

$$\omega_j = \frac{i}{2\pi}\partial\overline{\partial}\log\sum_{k=0}^{n}\left|\frac{Z_k}{Z_j}\right|^2 = \frac{i}{2\pi}\partial\overline{\partial}\log\left(1+\sum_{k=1}^{n}|z_k|^2\right), \quad i=\sqrt{-1}.$$

Note that $\omega_j|_{\mathcal{U}_j\cap\mathcal{U}_k} = \omega_k|_{\mathcal{U}_j\cap\mathcal{U}_k}$ (the 2-forms ω_j must stick together); in other words, these forms provide a global form ω over all $\mathbb{P}^n(\mathbb{C})$ (called the Fubini–Study form). Indeed, we have

$$\log\sum_{k=0}^{n}\left|\frac{Z_k}{Z_j}\right|^2 = \log\left|\frac{Z_l}{Z_j}\right|^2 + \log\sum_{k=0}^{n}\left|\frac{Z_k}{Z_l}\right|^2,$$

and it suffices to show that $\partial\overline{\partial}\log\left|\frac{Z_l}{Z_j}\right|^2 = 0$, on $\mathcal{U}_j\cap\mathcal{U}_l$. Now, $\frac{Z_l}{Z_j} = z_l$, is the l-th coordinate on $\mathcal{U}_j$, so

$$\partial\overline{\partial}\log|z_l|^2 = \partial\left(\frac{1}{z_l\overline{z}_l}\overline{\partial}(z_l\overline{z}_l)\right) = \partial\left(\frac{z_l d\overline{z}_l}{z_l\overline{z}_l}\right) = \partial\left(\frac{d\overline{z}_l}{\overline{z}_l}\right) = 0.$$

Let us show that ω is a closed $(1,1)$-real form. Indeed, this form is real because from the relation: $\overline{\partial\overline{\partial}} = \overline{\partial}\partial = -\partial\overline{\partial}$, we deduce that $\omega_j = \overline{\omega}_j$, and since $d(\partial\overline{\partial}\lambda) = \partial^2\overline{\partial}\lambda - \overline{\partial}^2\partial\lambda = 0$, then ω is closed. It remains to prove that this metric ω is positive definite. Indeed, we verify this on each $\mathcal{U}_j$. We have

$$\begin{aligned}\omega_j &= \frac{i}{2\pi}\partial\overline{\partial}\log\left(1+\sum_{k=1}^{n}|z_k|^2\right) = \frac{i}{2\pi}\partial\left(\frac{\overline{\partial}\left(1+\sum_{k=1}^{n}|z_k|^2\right)}{1+\sum|z_k|^2}\right),\\ &= \frac{i}{2\pi}\partial\left(\frac{\sum z_k d\overline{z}_k}{1+\sum|z_k|^2}\right),\\ &= \frac{i}{2\pi}\partial\left(\frac{(1+\sum|z_k|^2)\sum dz_k\wedge d\overline{z}_k - (\sum\overline{z}_k dz_k)\wedge(\sum z_k d\overline{z}_k)}{(1+\sum|z_k|^2)^2}\right).\end{aligned}$$

Since ω is invariant under the action of the transitive group $U(n+1)$ on the space $\mathbb{P}^n(\mathbb{C})$, it suffices to see that it is positive at a point, for example, the origin $[1 : 0 : ... : 0] \in \mathbb{P}^n(\mathbb{C})$. This point belongs to the affine chart $\mathcal{U}_0$ and obviously this can be replaced by any $\mathcal{U}_j$, $1 \leq j \leq n$, in which we have

$$\omega_j = \frac{i}{2\pi} \sum dz_k \wedge d\overline{z}_k,$$

which is positive, so ω is positive everywhere. This can also be shown directly using the Cauchy–Schwarz inequality: $|\langle \zeta, \eta \rangle| \leq \|\zeta\|.\|\eta\|$, where $\langle , \rangle$ is the canonical hermitian product on $\mathbb{C}^n$ and $\|\zeta\| = \sqrt{\langle \zeta, \zeta \rangle}$. Indeed, let us first note that the above relation can be written in the form:

$$\omega_j = \frac{i}{2\pi} \frac{1}{(1 + \sum |z_k|^2)^2} \cdot \sum f_{kl} dz_k \wedge dz_l, \qquad f_{kl} = \left(1 + \sum |z_k|^2\right) \delta_{kl} - \overline{z}_k z_l.$$

Let us show that the matrix (f_{kl}) is positive definite. For $\zeta \neq 0$, we have

$$\zeta^{\mathsf{T}} (f_{kl}) \overline{\zeta} = \|\zeta\|^2 + \|z\|^2 \|\zeta\|^2 - \zeta^{\mathsf{T}} \overline{z} z^{\mathsf{T}} \overline{\zeta}.$$

Now $\zeta^{\mathsf{T}} \overline{z} z^{\mathsf{T}} \overline{\zeta} = \langle \zeta, z \rangle \langle z, \zeta \rangle = \overline{\langle z, \zeta \rangle} \langle z, \zeta \rangle = \|\langle z, \zeta \rangle\|^2$, so according to the Cauchy–Schwarz inequality, we have

$$\zeta^{\mathsf{T}} (f_{kl}) \overline{\zeta} = \|\zeta\|^2 + \|z\|^2 \|\zeta\|^2 - \zeta^{\mathsf{T}} (f_{kl}) \overline{\zeta} = \|\zeta\|^2 + \|z\|^2 \|\zeta\|^2 - \zeta^{\mathsf{T}} \overline{z} z^{\mathsf{T}} \overline{\zeta} \geq \|\zeta\|^2.$$

Since $\zeta \neq 0$, then $\zeta^{\mathsf{T}} (f_{kl}) \overline{\zeta} > 0$. Let $\sigma : \mathbb{C}^{n+1} \backslash \{0\} \longrightarrow \mathbb{P}^n(\mathbb{C})$, the standard projection. On $\sigma^{-1}(\mathcal{U}_j) = \{(Z_0 : ... : Z_n) : Z_j \neq 0\}$, we have

$$\sigma^* \omega = \frac{i}{2\pi} \partial \overline{\partial} \log \left(\sum_{k=0}^{n} \left| \frac{Z_k}{Z_j} \right|^2 \right) = \frac{i}{2\pi} \partial \overline{\partial} \left(\log(\|Z\|^2) - \log(|Z_j|^2) \right).$$

We saw above that $\partial \overline{\partial} \log(|Z_j|^2) = 0$, so $\sigma^* \omega = \frac{i}{2\pi} \partial \overline{\partial} \log(\|Z\|^2)$. The Kähler metric thus defined on $\mathbb{P}^n(\mathbb{C})$ is called the Fubini–Study metric. The only non-zero entire cohomology groups of $\mathbb{P}^n(\mathbb{C})$ being

$$H^{2k}(\mathbb{P}^1(\mathbb{C}), \mathbb{Z}) = H^{2k}(S^2, \mathbb{Z}) \simeq \mathbb{Z}, \quad 0 \leq k \leq n,$$

then $[\omega] \in H^2(\mathbb{P}^1(\mathbb{C}), \mathbb{Z})$ is a generator of the cohomology ring $H^\bullet(\mathbb{P}^n(\mathbb{C}), \mathbb{Z})$.

b) Just use the previous question.

EXERCISE 5.8.– *Let M be a compact Kähler manifold and ω be an exact (p, q)-form on M. Show that for some $(p-1, q-1)$-form λ, we have $\omega = \partial \overline{\partial} \lambda$.*

Solution: Let $G : A^{p,q}(M) \longrightarrow A^{p,q}(M)$ be the Green operator (defined in theorem 5.1). It is not difficult to show (see, for example, Griffiths and Harris 1978) that the laplacian Δ_∂, the $\overline{\partial}$-laplacian $\Delta_{\overline{\partial}}$ and $\frac{1}{2}\Delta_d$ coincide on the manifold M and their associated Green operators satisfy $G_\partial = G_{\overline{\partial}} = 2G_d$. So, $\partial, \overline{\partial}, d$ and their adjoints commute with $G_{\overline{\partial}}$. Since the ω-form is exact, then its harmonic projection is zero and by Hodge decomposition, we have $\omega = \overline{\partial}\,\overline{\partial}^* G_{\overline{\partial}}\omega$, where $\overline{\partial}^*$ is the adjoint operator $\overline{\partial}^* : A^{p,q}(M) \longrightarrow A^{p,q-1}(M)$. The form ω being of type (p,q), then $\partial(\overline{\partial}^* G_{\overline{\partial}}\omega) = \pm\overline{\partial}^* G_{\overline{\partial}}(\partial\omega] = 0$. The harmonic projection of $\partial(\overline{\partial}^* G_{\overline{\partial}}\omega)$ is zero and according to the Hodge decomposition, we have $\partial(\overline{\partial}^* G_{\overline{\partial}}\omega) = \partial\partial^* G_\partial(\partial(\overline{\partial}^* G_{\overline{\partial}}\omega)) = \partial(\partial^*\overline{\partial}^* G_{\overline{\partial}}\omega) = \partial\lambda$.

EXERCISE 5.9.– *Let $\mathcal{L}_1$ and $\mathcal{L}_2$ be line bundles. Show that:*

a) If $\mathcal{L}_1$ and $\mathcal{L}_2$ are ample, then $\mathcal{L}_1 \otimes \mathcal{L}_2$ is also ample.

b) If $\mathcal{L}_1$ is ample and $\mathcal{L}_2$ is generated by global sections, then $\mathcal{L}_1 \otimes \mathcal{L}_2$ is ample.

c) If $\mathcal{L}_1$ is very ample and $\mathcal{L}_2$ is generated by global sections, then $\mathcal{L}_1 \otimes \mathcal{L}_2$ is very ample.

d) If $\mathcal{L}_1$ is ample, then there is an $n_0 > 0$ such that $\mathcal{L}_2 \otimes \mathcal{L}_1^{\otimes n}$ is very ample for $n > n_0$.

Hint: See Hartshorne (1977, Chapter II, section 7, exercise 7.5).

EXERCISE 5.10.– *Let $M = \mathbb{C}^2\backslash\{0\}/\mathbb{Z}$ be the Hopf surface defined by the quotient of $\mathbb{C}^2\backslash\{0\}$ by $\mathbb{Z}$. Show that such a surface is not Kähler.*

Solution: Note that M is the space of orbits in $\mathbb{C}^2\backslash\{0\}$ of the group $\mathbb{Z}$ operating by $(k,z) \longmapsto \lambda^k z$, $\lambda \neq 0$. Since $\mathbb{C}^2\backslash\{0\}$ is homeomorphic to $S^3 \times \mathbb{R}_+^*$, then $M \simeq S^3 \times S^1$. Hence, $H^2(M,\mathbb{R}) = 0$, and consequently the Hopf surface is non-Kähler.

EXERCISE 5.11.– *a) A $K3$ surface is a compact, simply connected (smooth complex) surface whose canonical bundle is trivial. In other words, a $K3$ surface is a compact, simply connected (smooth complex) surface that has a holomorphic 2-form that is everywhere non-zero. Give some examples of $K3$ surfaces.*

b) Show that every $K3$ surface is Kähler.

c) Deduce that a compact complex surface is Kähler if and only if its first Betti number[1] is even.

1 Recall that, in general, the dimension of the cohomology group $H^k(M,.)$ (which is a vector space) is called the k-th Betti number of the manifold M.

Hint: *a)* Simple examples of such surfaces include smooth quartic hypersurfaces of $\mathbb{P}^3(\mathbb{C})$, double coverings of the projective plane $\mathbb{P}^2(\mathbb{C})$ ramified along a smooth sextic curve and Kummer surfaces (recall that a Kummer surface is a surface obtained by quotienting a torus $\mathbb{C}^2/\Lambda$ by the involution $(x, y) \longmapsto (-x, -y)$).

b) See, for example, Ueno (1975) where the idea of the proof is to use hermitian metrics of a particular type. The proof was later taken up and improved, for example, in Barth et al. (2015) in 2015.

c) We deduce from the Enriques–Kodaira classification of surfaces (Griffiths and Harris 1978) that a non-algebraic surface whose first Betti number is even is either a K3 surface or an elliptic surface (for the latter case, which does not concern us, see Miyaoka (1974)).

6

Abelian Varieties

6.1. Definitions and examples

Consider a complex torus $\mathbb{T}^n = \mathbb{C}^n/\Lambda$ of dimension n, where $\Lambda \simeq H_1(\mathbb{T}^n, \mathbb{Z})$ is a lattice isomorphic to $\mathbb{Z}^{2n}$. Recall that the torus $\mathbb{T}^n$ is a holomorphic Kähler manifold and a commutative group.

DEFINITION 6.1.– *An Abelian variety is a complex torus that has a holomorphic embedding in a projective space.*

It is a complete abstract variety on which an algebraic group structure is defined. Such a variety is integral, projective and smooth; moreover, its group law is commutative.

EXAMPLE 6.1.– *A first important example of Abelian varieties is formed by elliptic curves. These are Abelian varieties of dimension 1 and are used in many fields: both in algebraic geometry and in complex analysis with elliptic functions and integrals, in number theory and unexpected maps in cryptography and then in the ultimate theorem completing the proof of Fermat conjecture, for example.*

An Abelian variety has been defined as a complex torus that has a holomorphic embedding in a projective space. But how can we explain that if $\mathcal{A}$ is an Abelian variety over $\mathbb{C}$, then the set of its complex points is a complex torus? Let us first recall that the set of complex points of $\mathcal{A}$ is by definition a complex, connected and compact Lie group since it is projective. The exponential map that sends the tangent space of this Lie group at the origin (Lie algebra) to this group is surjective. The kernel of this map is a discrete group (a lattice). Therefore, such a Lie group is isomorphic to the quotient of this tangent space by this lattice, and consequently, it is a complex torus. The question arises as to whether the converse is true, i.e. is every complex torus the set of complex points of a complex Abelian variety? As we will see in this chapter, the answer is positive in dimension 1; it is simply the theory of elliptic curves. On the contrary, the answer is false for complex tori of complex dimensions greater than or equal to 2 because the situation is completely different in this case.

DEFINITION 6.2.– *A closed analytic subset of the projective space $\mathbb{P}^N(\mathbb{C})$ is a locally closed subset defined by holomorphic equations.*

Every compact analytic subset is algebraic, i.e. defined by the simultaneous vanishing of a finite number of homogeneous polynomials P_j,

$$\bigcap_j \{Z \in \mathbb{P}^N(\mathbb{C}) : P_j(Z) = 0\}.$$

A remarkable result, which can be stated in various equivalent forms, is the following Chow theorem (Griffiths and Harris 1978):

THEOREM 6.1.– *Every closed analytic subvariety of $\mathbb{P}^N(\mathbb{C})$ is a projective variety. In other words, every closed complex submanifold of $\mathbb{P}^N(\mathbb{C})$ is algebraic, or again, every complex manifold admitting a holomorphic embedding in $\mathbb{P}^N(\mathbb{C})$ is algebraic.*

This theorem tells us that any complex analytic variety in a projective space is algebraic. In other words, it states that such a variety is defined by homogeneous polynomial equations and can therefore be studied by either analytic or algebraic methods. According to this theorem, it is equivalent to say that an Abelian variety is an algebraic complex torus, i.e. defined by the simultaneous cancellation of a finite number of homogeneous polynomials. Chow's theorem is a special case of the Serre comparison theorem (Géometrie Algébrique et Géométrie Analytique (GAGA) principle), which identifies the cohomology groups of a complex projective algebraic variety with the groups defined by analytic geometry. The GAGA principle asserts that a large number of analytic objects (e.g. a compact analytic subset, a holomorphic vector bundle and a coherent sheaf) defined from projective varieties are algebraic.

EXAMPLE 6.2.– *In dimension 1, every complex torus is an Abelian variety. This torus is isomorphic to a smooth cubic of $\mathbb{P}^2(\mathbb{C})$, and it is polarized. In this case, the embedding is carried out in a projective space of dimension 2. We obtain the models $\mathbb{C}/$lattice as planar projective curves, and it is easier in this case to work with the Weierstrass function $\wp$ and its derivative $\wp'$ (see Appendix, Chapter 10, for further information on these functions). Let us briefly recall that the function $\wp$ is defined by:*

$$\wp(z) = \frac{1}{z^2} + \sum_{\omega \in \Lambda \setminus \{0\}} \left(\frac{1}{(z-\omega)^2} - \frac{1}{\omega^2} \right),$$

where $\Lambda = \mathbb{Z}\omega_1 \oplus \mathbb{Z}\omega_2$ is the lattice generated by two complex numbers ω_1 and ω_2 different from 0 such that: $Im\left(\frac{\omega_2}{\omega_1}\right) > 0$. The above series converges normally on any *compact set not meeting the lattice* Λ*, i.e. on any compact set of* $\mathbb{C}\backslash\Lambda$*. We show that* $\wp(z)$ *is an elliptic function with periods* ω_1 *and* ω_2*; it is even and has double poles at the points* $\omega \in \Lambda$*, whose residue is zero. Furthermore, the derivative* $\wp'(z)$ *is a doubly periodic function, and it is odd. The function* $\wp(z)$ *is a solution in* Λ *of the differential equation:*

$$(\wp'(z))^2 = 4\,(\wp(z))^3 - g_2\wp(z) - g_3,$$

where g_2 *and* g_3 *are the Weierstrass invariants of the function* $\wp(z)$*, and they are defined by the Eisenstein series:*

$$g_2 = 60 \sum_{\omega\in\Lambda\backslash\{0\}} \frac{1}{\omega^4}, \qquad g_3 = 140 \sum_{\omega\in\Lambda\backslash\{0\}} \frac{1}{\omega^6}.$$

If $y^2 = 4x^3 - ax - b$ *is a cubic without singularities (the discriminant of the polynomial* $4x^3 - ax - b$ *is non-zero), then there exists a lattice* Λ *such that:* $a = g_2$ *and* $b = g_3$ *where* g_2 *and* g_3 *are the Weierstrass invariants. To characterize the field of elliptic functions* $\mathcal{E}_\Lambda$*, we form the quotient ring* $\mathbb{C}[x,y]$ *by the principal ideal corresponding to the equation:* $y^2 = 4x^3 - g_2x - g_3$*, i.e.*

$$\mathcal{E}_\Lambda = \mathbb{C}(\wp, \wp') \simeq \mathbb{C}[x,y]/\left(y^2 - 4x^3 + g_2x + g_3\right),$$

where $\wp = x$ *and* $\wp'$ *is identified with the image of* y *in the quotient. In this way, Weierstrass functions can be used to uniformize (to make uniform) an elliptic curve, and thus to realize it as a quotient of* $\mathbb{C}$ *by a certain lattice. Furthermore, the map:*

$$\mathbb{C}/\Lambda \longrightarrow \mathbb{P}^2(\mathbb{C}), \quad z \longmapsto [1 : \wp(z) : \wp'(z)], z \neq 0, \quad 0 \longmapsto [0 : 0 : 1],$$

is an isomorphism between the complex torus $\mathbb{C}/\Lambda$ *and the elliptic curve* $\mathcal{E}$ *with affine equation:* $y^2 = 4x^3 - g_2x - g_3$*. Indeed, it is enough to set* $x = \wp(z)$*,* $y = \wp'(z)$ *and use the differential equation above which defines a projective embedding in* $\mathbb{P}^2(\mathbb{C})$ *and an algebraic variety structure. This proves in particular that for any complex torus of dimension* 1*, there exists a holomorphic embedding of this torus in a projective space. Therefore, the Abelian varieties of dimension* 1 *are the elliptic curves. A similar reasoning applies to Riemann surfaces regardless of their genus. Note, however, that this is unfortunately not true for higher-dimensional tori because the situation is completely different for complex tori of complex dimensions greater than or equal to* 2 *(not all complex tori are Abelian varieties). In fact, in higher dimensions, it is more subtle; very few complex tori admit an algebraic structure where this imposes conditions called the Riemann bilinear relations. Abelian varieties are characterized by the existence of a Riemann form which is also positive definite.*

EXAMPLE 6.3.– *Another important example of an Abelian variety is provided by the Jacobian variety of a non-singular algebraic curve* C*, which we denote by Jac*(C)*. This can be defined in several ways: we denote by* $S^g(C)$ *the set of positive divisors on the* *curve $\mathcal{C}$ (of genus g), and we define on this set an equivalence relation: $\mathcal{D}_1 \equiv \mathcal{D}_2$ if and only if there exists a meromorphic function f such that: $(f) = \mathcal{D}_1 - \mathcal{D}_2$, i.e. if and only if:*

$$\int_{\mathcal{D}_1}^{\mathcal{D}_2} \omega = \int_{\gamma} \omega, \quad \forall \omega \in \Omega^1_{\mathcal{C}},$$

with γ a closed path on $\mathcal{C}$ and $\Omega^1_{\mathcal{C}}$ the sheaf of holomorphic 1-differential forms on $\mathcal{C}$. The Jacobian variety $Jac(\mathcal{C})$ of $\mathcal{C}$ is defined by:

$$Jac(\mathcal{C}) = S^g(\mathcal{C})/\equiv .$$

The Jacobian variety is also identified with the quotient of the group of divisors of degree 0 by the subgroup of principal divisors. This variety therefore appears as a subgroup of Pic($\mathcal{C}$), in this case the subgroup of Pic($\mathcal{C}$) formed by the classes of divisors of degree 0, which we also denote by Pic^0($\mathcal{C}$). We have:

$$Jac(\mathcal{C}) = Pic^0(\mathcal{C}) = \{lines\ bundle\ of\ degree\ 0\}.$$

Using the exact exponential sequence of sheaves,

$$0 \longrightarrow \mathbb{Z} \xrightarrow{i} \mathcal{O}_{\mathcal{C}} \xrightarrow{\exp} \mathcal{O}^*_{\mathcal{C}} \longrightarrow 0,$$

as well as duality, it is shown that the variety of $\mathcal{C}$ is a complex torus of dimension g. More specifically, we have:

$$\begin{aligned} Jac(\mathcal{C}) = H^1(\mathcal{C}, \mathcal{O}_{\mathcal{C}})/H^1(\mathcal{C}, \mathbb{Z}) &\simeq H^1(\mathcal{C}, \Omega^1_{\mathcal{C}})/H^1(\mathcal{C}, \mathbb{Z}), \\ &\simeq H^0(\mathcal{C}, \Omega^1_{\mathcal{C}})^*/H_1(\mathcal{C}, \mathbb{Z}), \\ &\simeq \mathbb{C}^g/\mathbb{Z}^{2g}. \end{aligned}$$

EXAMPLE 6.4.– *For Abelian varieties of dimension 2 (Abelian surfaces), we have the following relations. Let $\mathcal{D}$ be a divisor, possibly consisting of one or more algebraic curves, on an Abelian surface $\mathcal{A}$. Then, the geometric genus $g(\mathcal{D})$ of a divisor $\mathcal{D}$ on an Abelian surface $\mathcal{A}$ is given by the adjunction formula:*

$$g(\mathcal{D}) = 1 + \frac{1}{2}\deg K_{\mathcal{D}} = 1 + \frac{1}{2}\deg (K_{\mathcal{A}} + \mathcal{D}) \mid_{\mathcal{D}} = 1 + \frac{1}{2}(K_{\mathcal{A}}.\mathcal{D} + \mathcal{D}.\mathcal{D}),$$

where $K_{\mathcal{A}}$ is the canonical divisor on $\mathcal{A}$, i.e. the locus of zeros of a holomorphic 2-form and $\mathcal{D}.\mathcal{D}$ denotes the number of intersection points of $\mathcal{D}$ with $a + \mathcal{D}$, where $a + \mathcal{D}$ is a small translation by a of $\mathcal{D}$ on $\mathcal{A}$. The Euler–Poincaré characteristic of $\mathcal{D}$ is given by:

$$\chi(\mathcal{D}) = \dim H^0(\mathcal{A}, \mathcal{O}_{\mathcal{A}}(\mathcal{D})) - \dim H^1(\mathcal{A}, \mathcal{O}_{\mathcal{A}}(\mathcal{D})),$$

and taking into account the Riemann–Roch theorem for line bundles on a surface, we have:

$$\chi(\mathcal{D}) = p_a(\mathcal{A}) + 1 + \frac{1}{2}(\mathcal{D}.\mathcal{D} - \mathcal{D}.K_{\mathcal{A}}),$$

where $p_a(\mathcal{A})$ *is the arithmetic genus of* $\mathcal{A}$. *Using the Kodaira–Serre duality and Kodaira vanishing theorem, we obtain:*

$$\begin{aligned}\chi(\mathcal{D}) &= \dim \mathcal{L}(\mathcal{D}) - \dim H^1\left(\mathcal{A}, \Omega^2\left(\mathcal{D} \otimes \mathcal{O}_{\mathcal{A}}(-\mathcal{D})\right)\right),\\ &= \dim \mathcal{L}(\mathcal{D}) - \dim H^1(M, \Omega^2(\mathcal{D} \otimes K_M^*)) \text{ (Kodaira–Serre duality)},\\ &= \dim \mathcal{L}(\mathcal{D}) \text{ (Kodaira vanishing theorem)},\end{aligned}$$

whenever $\mathcal{D} \otimes K_M^*$ *defines a positive line bundle. However, for Abelian surfaces,* K_M *is trivial and* $p_a(M) = -1$; *therefore, combining the above relations,*

$$\chi(\mathcal{D}) = \dim \mathcal{L}(\mathcal{D}) = \frac{\mathcal{D}.\mathcal{D}}{2} = g(\mathcal{D}) - 1 = \delta_1\delta_2 = N + 1,$$

where the embedding of $\mathcal{A}$ *is done in a complex projective space* $\mathbb{P}^N(\mathbb{C})$.

Consider the exact sheaf sequence:

$$0 \longrightarrow \mathcal{O}_{\mathcal{C}} \longrightarrow \pi_* \mathcal{O}_{\widetilde{\mathcal{C}}} \longrightarrow Z_{\mathcal{C}} \longrightarrow 0,$$

where $\mathcal{C}$ is an analytic hypersurface on an irreducible Abelian surface, $\pi : \widetilde{\mathcal{C}} \to \mathcal{C}$ is its normalization (desingularization), and $Z_{\mathcal{C}}$ is the quotient sheaf $\pi_* \mathcal{O}_{\widetilde{\mathcal{C}}} \backslash \mathcal{O}_{\mathcal{C}}$ on $\mathcal{C}$. The exactness of the sheaf sequence (which is the normalization sequence of π) shows that the Euler characteristic:

$$\mathcal{X}(\mathcal{O}) = \dim H^0(\mathcal{O}) - \dim H^1(\mathcal{O}),$$

satisfies the relation:

$$\mathcal{X}(\mathcal{O}_{\mathcal{C}}) - \mathcal{X}(\pi_* \mathcal{O}_{\widetilde{\mathcal{C}}}) + \mathcal{X}(Z_{\mathcal{C}}) = 0.$$

Now

$$\mathcal{X}(\mathcal{O}_{\mathcal{C}}) = 1 - g(C), \qquad \mathcal{X}(\mathcal{O}_{\widetilde{\mathcal{C}}}) = n - \sum_{j=1}^{n} g(\widetilde{\mathcal{C}}_j),$$

where n denotes the number of connected components of $\widetilde{\mathcal{C}}$ and $\mathcal{C}_j$ (respectively $\widetilde{\mathcal{C}}_j$); $j = 1, ..., n$ denote the irreducible components (respectively normalizations) of $\mathcal{C}_j$. Substituting these expressions into the last equation, we obtain:

$$g(C) = \sum_{j=1}^{n} g(\widetilde{\mathcal{C}}_j) + \mathcal{X}(Z_{\mathcal{C}}) + 1 - n.$$

Note that $\mathcal{X}(Z_C)$ only accounts for the singular points p of C; $\mathcal{X}(Z_{C_p})$ is the dimension of the set of holomorphic functions on the different branches around p taken separately, modulo the holomorphic functions on the curve C near that singular point. Consider the case of a planar singularity (for which $\mathcal{X}(Z_C) = 2$, as well), i.e. the tangents to the branches lie in a plane. If $F_j(x, y) = 0$ denote the jth branch of C running through p with local parameter s_j, then:

$$\mathcal{X}(Z_C) = \dim \frac{\Pi_{j=1}^n \mathbb{C}[[s_j]]}{\mathbb{C}[[x, y]]/\Pi_{j=1}^n F_j(x, y)}.$$

For further information, see, for example, Adler et al. (2004, Chapter 5) and also Iitaka (2011, Chapter 9).

Finally, recall that a Kähler variety is a variety with a Kähler metric, i.e. a Hermitian metric whose associated differential 2-form of type $(1, 1)$ is closed. The complex torus $\mathbb{C}^2/lattice$ with the Euclidean metric $\sum dz_i \otimes d\overline{z}_i$ is a Kähler variety and any compact complex variety that can be embedded in projective space is also a Kähler variety. And as we explained at the end of section 5.1, a compact complex Kähler variety having as many independent meromorphic functions as its dimension is a projective variety.

6.2. Riemann conditions, polarization and Riemann form

Let $\mathbb{T}^n = \mathbb{C}^n/\Lambda$ be a complex torus where Λ is a lattice of rank $2n$ and $(e_1, ..., e_n)$ a (canonical) basis of $\mathbb{C}^n$. Let $\lambda_1, ..., \lambda_{2n}$ be a set of bases for the lattice Λ, and also a set of real bases for $\mathbb{C}^n$ and we will denote by $(x_1, ..., x_{2n})$ the corresponding dual basis, i.e. coordinates relative to the basis $(\lambda_1, ..., \lambda_{2n})$ in such a way that:

$$\int_{\lambda_j} dx_k = \delta_{jk}.$$

With a little abuse of notation, we will also denote by λ_j the corresponding 1-cycles on $\mathbb{T}^n$.

DEFINITION 6.3.– *The period matrix (for $\mathbb{T}^n$),*

$$\Omega = (\omega_{kj})_{n \times 2n} = \begin{pmatrix} \omega_{11} & \cdots & \omega_{1,2n} \\ \vdots & & \vdots \\ \omega_{n1} & \cdots & \omega_{n,2n} \end{pmatrix} \in M_{n \times 2n}(\mathbb{C}),$$

is defined via

$$\lambda_j = \sum_{k=1}^n \omega_{kj} e_k.$$

THEOREM 6.2.– *The complex torus $\mathbb{C}^n/\Lambda$ is an Abelian variety if and only if there exists an entire matrix Q of order $2n$, antisymmetric and satisfying the Riemann bilinear relations:*

$$\Omega.Q^{-1}.\Omega^{\top} = 0, \qquad -i\Omega.Q^{-1}.\overline{\Omega}^{\top} > 0, \tag{6.1}$$

($\overline{\Omega}$ denotes the complex conjugate matrix), or equivalently if and only if there exists a basis of the lattice Λ and a basis of the space $\mathbb{C}^n$ such that the period matrix Ω is written in the form:

$$\Omega = (\Delta_\delta, Z), \qquad \Delta_\delta = diag\ (\delta_1, ..., \delta_n),$$

where $\delta_\alpha \in \mathbb{N}^*$, $\delta_\alpha|\delta_{\alpha+1}$, $\alpha = 1, ..., n-1$, and Z is a matrix satisfying the following conditions: $Z^\intercal = Z$ and Im$Z > 0$.

PROOF.– Let $z_1, ..., z_n \in \mathbb{C}$. Taking into account the above notations, we have:

$$dz_k = \sum_{j=1}^{n} \omega_{kj} dx_j, \qquad d\overline{z}_k = \sum_{j=1}^{n} \overline{\omega}_{kj} dx_j. \tag{6.2}$$

Let $Q = (q_{kj})$ be an antisymmetric integer matrix (intersection matrix) of order $2n$. Let $\omega : \Lambda \times \Lambda \longrightarrow \mathbb{Z}$ be a Hodge form. Let:

$$\omega = \frac{1}{2} \sum q_{kj} dx_k \wedge dx_j.$$

Taking into account [6.2], we express ω in another form, and we immediately see that it is Hodge if and only if the Riemann bilinear relations are satisfied. According to the Kodaira embedding theorem, the torus $\mathbb{C}^n/\Lambda$ admits an embedding in the complex projective space. Conversely, if $\mathbb{C}^n/\Lambda$ is a projective variety, then there exists a Kähler form which represents an entire cohomology class and one can easily obtain a form which satisfies the previous relations. Moreover, one can choose a new basis of Λ on $\mathbb{Z}$ of $2n$ column vectors $\lambda_1, ..., \lambda_{2n}$ such that:

$$Q = \begin{pmatrix} 0 & \Delta_\delta \\ -\Delta_\delta & 0 \end{pmatrix}, \qquad \Delta_\delta = \text{diag}\ (\delta_1, ..., \delta_n),$$

where $\delta_1, \ldots, \delta_n \in \mathbb{N}^*$, $\delta_\alpha \mid \delta_{\alpha+1}$, $\alpha = 1, ..., n-1$, are the elementary divisors. These do not depend on the choice of $(\lambda_1, ..., \lambda_n)$. The $(1,1)$-form ω can therefore be expressed as:

$$\omega = \sum_{\alpha=1}^{n} \delta_\alpha dx_\alpha \wedge dx_{n+\alpha}.$$

We can choose a basis $(e_1, ..., e_n)$ of $\mathbb{C}^n$ by setting $e_\alpha = \dfrac{\lambda_\alpha}{\delta_\alpha}$, $\alpha = 1, ..., n$, and the period matrix Ω is written as $\Omega = (\Delta_\delta, Z)$, where Z is a matrix satisfying $Z^\intercal = Z$ and Im$Z > 0$. ■

Let $\Pi = (\pi)_{2n \times n}$ be a $2n \times n$ matrix. The expressions dz_k and $d\overline{z}_k$ in [6.2] can be written in the following condensed form:

$$dz = \Omega dx, \qquad d\overline{z} = \overline{\Omega} dx,$$

where $dx = \Pi dz + \overline{\Pi} dx$ or explicitly,

$$dx_j = \sum_{k=1}^{n} \pi_{jk} dz_k + \sum_{k=1}^{n} \overline{\pi}_{jk} d\overline{z}_k.$$

The matrices Ω and Π are related by:

$$\begin{pmatrix} \Omega \\ \overline{\Omega} \end{pmatrix} . (\Pi \quad \overline{\Pi}) = \mathbb{I}.$$

Using these notations, the relations [6.1] are equivalent to the following:

$$\Pi^\top . Q . \Pi = 0, \qquad -i\Pi^\top . Q . \overline{\Omega} > 0.$$

Another way to formulate the previous theorem in terms of the Riemann form (see section 4.4) is as follows:

THEOREM 6.3.– *A complex torus E/Λ, ($E \cong \mathbb{C}^n$, is an Abelian variety if and only if the lattice Λ admits a Riemann form. In other words, a complex torus E/Λ is an Abelian variety if and only if there exists on E a positive definite Hermitian form $h : E \times E \longrightarrow \mathbb{C}$ with imaginary part Im h that is an integer on the lattice Λ.*

An interesting result concerning Kähler varieties was obtained by Moishezon. Let us first recall that according to a theorem of Siegel (1983), any non-singular compact complex analytic variety of dimension n admits at most n algebraically independent meromorphic functions. The so-called Moishezon varieties (i.e. compact complex analytic varieties that become projective after a finite number of smooth center blow-ups) have the maximum number of algebraically independent meromorphic functions. More precisely, we have the following fundamental result obtained by Moishezon (1967): A compact complex analytic variety of dimension n is Moishezon if and only if it admits n algebraically independent meromorphic functions. Equivalently, Moishezon's theorem states that a compact Kähler variety of dimension n is projective if and only if it has n algebraically independent meromorphic functions.

Furthermore, using the Kähler form terminology (definition 5.6), we can define a polarization on a complex torus as being an entire Kähler form on that torus. Similarly, an Abelian variety is a complex torus on which exists a polarization and a polarized Abelian variety is a pair $(\mathcal{A}, \omega)$, where $\mathcal{A}$ is an Abelian variety and ω is a polarization on $\mathcal{A}$.

6.3. Isogenies and reducible Abelian varieties

DEFINITION 6.4.– *Let $\mathcal{A}_1$ and $\mathcal{A}_2$ be two Abelian varieties. We say that a morphism $f : \mathcal{A}_1 \longrightarrow \mathcal{A}_2$ is an isogeny if f is surjective and has a finite kernel, and if this is the case, we say that the varieties $\mathcal{A}_1$ and $\mathcal{A}_2$ are isogenic. We call the degree of the isogeny the cardinality of its kernel.*

If the dimensions of the two Abelian varieties $\mathcal{A}_1$ and $\mathcal{A}_2$ are identical, then for f to be an isogeny, it suffices that it be surjective or have a finite kernel (see proposition 6.1).

EXAMPLE 6.5.– *Let $\mathcal{A}$ be an Abelian variety and $k_{\mathcal{A}} \in End(\mathcal{A})$ be the multiplication morphism by k in $\mathcal{A}$, i.e.*

$$k_{\mathcal{A}} : \mathcal{A} \longrightarrow \mathcal{A}, \quad a \longmapsto ka = a + \cdots + a, \quad k \in \mathbb{N}.$$

(The notation $[k]$ is often used instead of $k_{\mathcal{A}}$.) This map determines an isogeny. Moreover, its degree (the cardinality of the kernel) is equal to k^{2m} where $m = \dim \mathcal{A}$ (the kernel $\ker k_{\mathcal{A}}$ is a $\mathbb{Z}$-module of cardinality k^{2m}, and it is called the group of k-torsion points of $\mathcal{A}$). Indeed, multiplication by $k \in \mathbb{N}^*$ is surjective (as an abstract group, the Abelian variety $\mathcal{A}$ is divisible), and its kernel is isomorphic to the group $(\mathbb{Z}/k\mathbb{Z})^{2m}$ which is finite:

$$\ker k_{\mathcal{A}} = \frac{1}{k}\Lambda/\Lambda \simeq \Lambda/k\Lambda \simeq (\mathbb{Z}/k\mathbb{Z})^{2m}.$$

It suffices to construct an isomorphism of real vector spaces $\mathbb{C}^m \longrightarrow \mathbb{R}^{2m}$ by choosing a basis of the $\mathbb{Z}$-free module $\Lambda \subset \mathbb{C}^m$ and by sending each basis element to a standard basis element of $\mathbb{R}^{2m}$. This induces an isomorphism of quotient groups:

$$\mathcal{A} \simeq \mathbb{C}^m/\Lambda \longrightarrow (\mathbb{R}/\mathbb{Z})^{2m}.$$

This isogeny is therefore of degree k^{2m}. Thus, the group $End(\mathcal{A})$ of endomorphisms of $\mathcal{A}$ is torsion-free and can be embedded in the ring $End(\mathcal{A}) \otimes \mathbb{Q}$.

EXAMPLE 6.6.– *Any complex torus isogenic to an Abelian variety is an Abelian variety. Any subtorus of an Abelian variety is an Abelian variety. Any quotient torus of an Abelian variety is an Abelian variety.*

PROPOSITION 6.1.– *Let $f : \mathcal{A}_1 \longrightarrow \mathcal{A}_2$ be a morphism of Abelian varieties $\mathcal{A}_1$ and $\mathcal{A}_2$. Then, the following conditions are equivalent:*

(i) f is an isogeny.

(ii) $\dim \mathcal{A}_1 = \dim \mathcal{A}_2$ and f is surjective.

(iii) $\dim \mathcal{A}_1 = \dim \mathcal{A}_2$ and $\ker f$ is finite.

PROOF.– Recall that the dimension of the fiber $f^{-1}(y)$, $y \in \mathcal{A}_2$ is expressed as a function of the dimensions of $\mathcal{A}_1$ and $\mathcal{A}_2$ using the formula:

$$\dim f^{-1}(y) = \dim \mathcal{A}_1 - \dim \mathcal{A}_2, \quad \forall y \in \mathcal{A}_2,$$

(according to the well-known fiber dimension theorem in algebraic geometry (Narasimhan 1995)).

– Let us first show that $(i) \Longrightarrow (ii)$. f is surjective and

$$\dim \mathcal{A}_1 - \dim \mathcal{A}_2 = \dim \ker f = 0.$$

– For the implication $(ii) \Longrightarrow (iii)$, we have $\dim \mathcal{A}_1 = \dim \mathcal{A}_2$ and since f is surjective, then $\dim \ker f = 0$, and therefore, $\ker f$ is finite.

– Finally, let us show that: $(iii) \Longrightarrow (i)$. Note that $f(\mathcal{A}_1)$ is irreducible because the continuous image of an irreducible set is still irreducible. Since,

$$\dim f(\mathcal{A}_1) = \dim \mathcal{A}_1 = \dim \mathcal{A}_2,$$

we deduce that: $f(\mathcal{A}_1) = \mathcal{A}_2$, which completes the proof. ■

From the above, the Hodge form ω can be seen as the curvature form of a positive line bundle $\mathcal{L}$ with $\omega = \left[\frac{i}{2\pi}\Theta_{\mathcal{L}}\right]$, $\mathcal{L} = [\mathcal{D}]$, where $\mathcal{D}$ is an ample divisor on an Abelian variety $\mathcal{A}$. Not every positive divisor on an Abelian variety is necessarily ample. We also show that if the Abelian variety is irreducible, then the line bundle $\mathcal{L} = [\mathcal{D}]$ associated with every positive divisor $\mathcal{D}$ is ample. We recall that an Abelian variety is irreducible when it does not contain any Abelian subvariety. Otherwise, it is said to be reducible.

The Poincaré reducibility theorem (see also exercise 6.9) states that if $\mathcal{A}$ is an Abelian variety containing an Abelian subvariety $\mathcal{B}$, then we can find another Abelian subvariety $\mathcal{C}$ of $\mathcal{A}$ such that $\mathcal{B} \cap \mathcal{C}$ is a finite subgroup of $\mathcal{A}$ and a surjective homomorphism $f : \mathcal{B} \oplus \mathcal{C} \longrightarrow \mathcal{A}$, such that: $\ker f = \mathcal{B} \cap \mathcal{C}$. In other words, if an Abelian variety contains an Abelian subvariety, then it is isogenous to the product of this subvariety by another subvariety.

We will demonstrate later (see the Lefschetz theorem (theorem 7.2) on theta functions) that if $\mathcal{L}$ is an ample line bundle on an Abelian variety, then $\mathcal{L}^k \equiv \mathcal{L}^{\otimes k}$ is very ample for $k \geq 3$.

6.4. Dual Abelian varieties

Let $\mathcal{A}$ be an Abelian variety and $Pic(\mathcal{A})$ be the Picard group of $\mathcal{A}$ (recall that it is the group of divisors of $\mathcal{A}$ up to linear equivalence, the group of invertible sheaves up to isomorphism, or in other words, the group of classes of isomorphisms of line bundles on $\mathcal{A}$). If for all $a \in \mathcal{A}$,

$$\tau_a : \mathcal{A} \longrightarrow \mathcal{A}, \quad z \longmapsto z + a$$

is the translation operator by a on $\mathcal{A}$, then for any line bundle $\mathcal{L}$ on $\mathcal{A}$, the line bundle $\mathcal{L}^{-1} \otimes \tau_a^* \mathcal{L}$ is algebraically trivial and the map:

$$\varphi_{\mathcal{L}} : \mathcal{A} \longrightarrow Pic(\mathcal{A}), \quad a \longmapsto \mathcal{L}^{-1} \otimes \tau_a^* \mathcal{L}$$

is a group morphism by virtue of the square theorem (exercise 6.7) because $\mathcal{L} \otimes \tau_a^* \mathcal{L} \otimes \tau_b^* \mathcal{L} \otimes \tau_{a+b}^* \mathcal{L}$ is trivial. By the same square theorem, we also have $\varphi_{\tau_a^*} \mathcal{L} = \varphi_{\mathcal{L}}$. Furthermore, we have $\varphi_{\mathcal{L}_1 \otimes \mathcal{L}_2} = \varphi_{\mathcal{L}_1} + \varphi_{\mathcal{L}_2}$, where $+$ is the group law on $\mathrm{Hom}(\mathcal{A}, \mathrm{Pic}(\mathcal{A}))$ induced by $\otimes$.

Let us also recall that:

$$Pic^0(\mathcal{A}) = \{\mathcal{L} \in Pic(\mathcal{A}) : \tau_a^* \mathcal{L} \simeq \mathcal{L}, a \in \mathcal{A}\},$$

i.e. the subgroup of $Pic(\mathcal{A})$ of translation-invariant divisor classes, or again, it is the subgroup of $Pic(\mathcal{A})$ consisting of the classes of isomorphisms of bundles algebraically equivalent to the trivial bundle. In other words,

$$Pic^0(\mathcal{A}) = Div_a(\mathcal{A})/Princ(\mathcal{A}),$$

where $Div_a(\mathcal{A})$ is the subgroup of divisors whose associated bundle has a zero Riemann form and $Princ(\mathcal{A})$ denotes the group of principal divisors. In other words, $Pic^0(\mathcal{A})$ consists of the bundles $\mathcal{L}$ such that $\varphi_{\mathcal{L}} = 0$, i.e. in terms of the first Chern class, the kernel:

$$Pic^0(\mathcal{A}) = \{\mathcal{L} \in Pic(\mathcal{A}) : c_1(\mathcal{L}) = 0\},$$

is formed by the classes of isomorphisms of topologically trivial bundles. The quotient denoted:

$$NS(\mathcal{A}) = Pic(\mathcal{A})/Pic^0(\mathcal{A}),$$

is called the Néron–Severi group. This injects itself into $H^2(\mathcal{A}, \mathbb{Z})$ and is of the finite type.

In the special case where $\mathcal{A}$ is a curve, then $Pic^0(\mathcal{A})$ is the group of divisors of degree 0. In the general case, this is no longer true, and these are groups of divisors algebraically equivalent to zero. Consider the long sequence in cohomology:

$$\cdots \to H^1(\mathcal{A}, \mathcal{O}_{\mathcal{A}}) \to H^1(\mathcal{A}, \mathcal{O}_{\mathcal{A}}^*) \xrightarrow{c_1} H^2(\mathcal{A}, \mathbb{Z}) \to H^2(\mathcal{A}, \mathcal{O}_{\mathcal{A}}) \to \cdots,$$

where c_1 is the Chern class. We know that to any divisor $\mathcal{D}$ on $\mathcal{A}$ with $c_1(\mathcal{D}) = [\omega]$ corresponds a positive line bundle $\mathcal{L} \in H^1(\mathcal{A}, \mathcal{O}_{\mathcal{A}}^*)$, such that: $c_1(\mathcal{L}) = [\omega] \in H^2(\mathcal{A}, \mathbb{Z})$. Therefore, to an Abelian variety $\mathcal{A}$ is associated a dual Abelian variety of $\mathcal{A}$, which we denote by $\mathcal{A}^*$ (which is also denoted by $\mathcal{A}^\vee$, or again by $Pic^0(\mathcal{A})$). More precisely, we have

DEFINITION 6.5.– *The dual Abelian variety $\mathcal{A}^*$ of $\mathcal{A}$ is defined by:*[1]

$$\mathcal{A}^* \equiv Pic^0(\mathcal{A}) = H^1(\mathcal{A}, \mathcal{O}_\mathcal{A})/H^1(\mathcal{A}, \mathbb{Z}) \simeq H^1(\mathcal{A}, \mathcal{O}^*_\mathcal{A}),$$

i.e. the group of holomorphic line bundles on $\mathcal{A}$ whose Chern class is zero.

In the definition of $Pic^0(\mathcal{A})$, the fact that $\tau_a^*\mathcal{L} \simeq \mathcal{L}$, $a \in \mathcal{A}$, means that $\varphi_\mathcal{L}$ is trivial which implies that $[\mathcal{L}] \in Pic^0(\mathcal{A})$ and reciprocally.

In general, $\mathcal{A}$ and $\mathcal{A}^*$ are not isomorphic. However, we can show that $\mathcal{A}$ and $\mathcal{A}^*$ are isogenous. If:

$$(\Delta_\delta \quad Z), \quad Z^\top = Z, \quad \mathrm{Im} Z > 0, \quad \delta_\alpha \in \mathbb{N}^*, \quad \delta_\alpha | \delta_{\alpha+1},$$

is the period matrix of the Abelian variety $\mathcal{A}$, then the period matrix of the dual Abelian variety $\mathcal{A}^*$ is:

$$\left(\delta_n \Delta_\delta^{-1} \quad \delta_n \Delta_\delta^{-1} Z \Delta_\delta^{-1}\right), \quad \text{where} \quad \Delta_\delta = diag(\delta_1, ..., \delta_n).$$

Note that in the case of a Jacobian variety $Jac(\mathcal{C})$ and its dual variety $Jac^*(\mathcal{C})$ where $\mathcal{C}$ is a compact Riemann surface, these two varieties are not only isogenous but isomorphic.

We have already defined a polarization on an Abelian variety $\mathcal{A}$. It is an algebraic equivalence class of ample line bundles $\mathcal{L}$ over $\mathcal{A}$. In other words, a polarization of $\mathcal{A}$ is an isogeny $\varphi : \mathcal{A} \longrightarrow \mathcal{A}^*$, which can be written in the form:

$$\varphi_\mathcal{L} : \mathcal{A} \longrightarrow \mathcal{A}^*, \quad a \longmapsto \mathcal{L}^{-1} \otimes \tau_a^*\mathcal{L},$$

where τ_a is the translation by a. In this case, the line bundles $\mathcal{L}'$ such that: $\varphi = \varphi_{\mathcal{L}'}$ are none other than the translates $\tau_a^*\mathcal{L}$, $a \in \mathcal{A}$. The classes of these bundles in $Pic(\mathcal{A})$ form a translate of $\mathcal{A}^*$ and the positive integer $\dim \mathcal{L} = \dim \tau_a^*\mathcal{L}$, is called the degree of the polarization φ and in addition $H^j(\mathcal{A}, \mathcal{L}) = 0$, for all $j \geq 1$. The varieties $\mathcal{A}$ and $\mathcal{A}^*$ have the same dimension and:

$$\deg \mathcal{L} = (\dim \mathcal{A})! \dim H^0(\mathcal{A}, \mathcal{L}) = (\dim \mathcal{A}^*)! \dim H^0(\mathcal{A}, \mathcal{L}).$$

If the polarization $\varphi = \varphi_\mathcal{L}$ is an isomorphism or of degree 1 or $\dim \mathcal{L} = 1$, then we obtain a principal polarization.

---

1 Here, we prefer to use the notation $\mathcal{A}^*$ but to avoid any confusion, the reader can use one of the notations common in the literature: $\mathcal{A}^\vee$ or $\widehat{\mathcal{A}}$, etc. Same note for the case of the dual $Jac^*(\mathcal{C})$ of a Jacobian variety where $\mathcal{C}$ is a compact Riemann surface.

6.5. Prym varieties

This section is devoted to the explicit study of Prym varieties. These appear as subvarieties of certain Jacobian varieties and belong to a general class of Abelian varieties. They form an important class of principally polarized Abelian varieties, in fact more general than Jacobian varieties but nevertheless geometrically accessible. The theory of Prym varieties plays a crucial role in several current research projects (Schottky problem, integrable systems, etc.), their geometry is very rich and one of the advantages of having an explicit description of Prym varieties is the possibility of applying them to the modern theory of algebraically integrable dynamical systems. We describe them here explicitly.

Let $\varphi : C \longrightarrow C_0$ be a double covering with $2n$ branch points, where C and C_0 are the two smooth complex algebraic curves (or compact Riemann surfaces). Let $\sigma : C \longrightarrow C$ be the involution exchanging the sheets of a ramified double covering along C_0 and such that φ identifies C_0 with the quotient C/σ. We denote by g the genus of C and by g_0 the genus of C_0. The genus g of C is expressed in terms of the genus g_0 of C_0 using the Riemann–Hurwitz formula: $g = 2g_0 + n - 1$.

We have seen (example 6.3) that the Jacobian variety of the curve C, is defined by $Jac(C) = S^g/\equiv$ and that $S^g/\equiv$ is homeomorphic to the complex torus $\mathbb{C}^g/\Lambda$ of dimension g, where Λ is the lattice defined by the period matrix. In other words, $Jac(C) = Pic^0(C) = \{\text{line bundles of degree zero}\} \simeq \mathbb{C}^g/\mathbb{Z}^{2g}$. We have an application (Abel) given by:

$$S^g(C) \longrightarrow \mathbb{C}^g/L_\Omega,\ \sum_{k=1}^{g} s_k \longmapsto \sum_{k=1}^{g} \int^{s_k(t)} (\omega_1, \dots, \omega_g) = t(k_1, \dots, k_g),$$

where $(\omega_1, \dots, \omega_g)$ is a basis of holomorphic differentials on C, L_Ω is the lattice defined by the matrix of the periods Ω, and $s_1, \dots, s_g$ are the functions defined on an open Zariski set by algebraic relations. In the particular case where C is a genus g hyperelliptic curve, i.e. $C = \overline{\{(w,z) : w^2 = P_n(z)\}}$, with $P_n(z)$ being a polynomial of degree $2g+1$ or $2g+2$, we have:

$$\omega_1 = \frac{dz}{\sqrt{P_n(z)}},\ \omega_2 = \frac{zdz}{\sqrt{P_n(z)}}, \dots, \omega_g = \frac{z^{g-1}dz}{\sqrt{P_n(z)}},$$

$$\sum_{k=1}^{g} \frac{ds_k}{\sqrt{P_{n+1}(s_k)}} = k_1 dt, \dots, \sum_{k=1}^{g} \frac{s_k^{g-1} ds_k}{\sqrt{P_{n+1}(s_k)}} = k_g dt.$$

Note also that in genus 2, principally polarized is equivalent to hyperelliptic, which shows that any Abelian surface is isogenous to a hyperelliptic Jacobian.

Let $(a_1, \ldots, a_{g_0}, \ldots, a_g, b_1, \ldots, b_{g_0}, \ldots, b_g)$, be a canonical homology basis of $H_1(C, \mathbb{Z})$ such that for the involution σ, we have:

$$\sigma(a_1) = a_{g_0+n}, \ldots, \sigma(a_{g_0}) = a_g,$$

$$\sigma(a_{g_0+1}) = -a_{g_0+1}, \ldots, \sigma(a_{g_0+n-1}) = -a_{g_0+n-1},$$

$$\sigma(b_1) = b_{g_0+n}, \ldots, \sigma(b_{g_0}) = b_g,$$

$$\sigma(b_{g_0+1}) = -b_{g_0+1}, \ldots, \sigma(b_{g_0+n-1}) = -b_{g_0+n-1}.$$

Note that $\varphi(a_{g_0+1}), \ldots, \varphi(a_{g_0+n-1}), \varphi(b_{g_0+1}), \ldots, \varphi(b_{g_0+n-1})$ are homologous to zero on C_0. Let $(\omega_1, \ldots, \omega_g)$ be a basis of holomorphic differentials on the curve C where $\omega_{g_0+n}, \ldots, \omega_g$, are holomorphic differentials on C_0 and the pullback of ω_j is,

$$\sigma^*(\omega_j) = \begin{cases} -\omega_j, \ 1 \leq j \leq g_0 + n - 1, \\ \omega_j, \ g_0 + n \leq j \leq g, \end{cases}$$

In the following, we use the abusive notation $\omega_j(\gamma_k)$ to designate the periods $\int_{\gamma_k} \omega_j$ of ω_j following the cycles γ_k on C.

THEOREM 6.4.– *Let $\varphi : C \longrightarrow C_0$ be a double covering where C and C_0 are the non-singular algebraic curves with Jacobians $Jac(C)$ and $Jac(C_0)$, respectively. Let $\sigma : C \longrightarrow C$ be the involution exchanging sheets of C over $C_0 = C/\sigma$. This involution extends by linearity to a map (which will again be denoted by σ), $\sigma : Jac(C) \longrightarrow Jac(C)$ and up some points of order two, $Jac(C)$ splits into an even part, i.e. $Jac(C)$, and an odd part (called the Prym variety) denoted $Prym(C/C_0)$ and defined by:*

$$Prym(C/C_0) = \left(H^0\left(C, \Omega_C^1\right)^-\right)^* / H_1(C, \mathbb{Z})^-,$$

where Ω_C^1 is the sheaf of holomorphic 1-forms on C and $-$ denotes the -1 eigenspace for a vector space on which σ acts. To be precise, we have:

$$Jac(C) = Prym(C/C_0) \oplus Jac(C_0),$$

with

$$\dim Jac(C_0) = \text{genus } g_0 \text{ of } C_0, \quad \dim Jac(C) = \text{genus } g \text{ of } C = 2g_0 + n - 1,$$

$$\dim Prym(C/C_0) = g - g_0 = g_0 + n + 1,$$

with $2n$ branch points. Moreover, if:

$$\Omega = \begin{pmatrix} A & B & C & D & E & F \\ G & H & I & J & K & L \end{pmatrix},$$

is the period matrix of $Jac(\mathcal{C})$, where $A, \ldots, L$ denote the following matrices (where k is the number of rows and j is the number of columns):

$$A=(\omega_j(a_k))_{\substack{1\leq k\leq g_0 \\ 1\leq j\leq g_0+n-1}}, B=(\omega_j(a_k))_{\substack{g_0+1\leq k\leq g_0+n-1 \\ 1\leq j\leq g_0+n-1}}, C=(\omega_j(a_k))_{\substack{g_0+n\leq k\leq g \\ 1\leq j\leq g_0+n-1}},$$

$$D=(\omega_j(b_k))_{\substack{1\leq k\leq g_0 \\ 1\leq j\leq g_0+n-1}}, E=(\omega_j(b_k))_{\substack{g_0+1\leq k\leq g_0+n-1 \\ 1\leq j\leq g_0+n-1}}, F=(\omega_j(b_k))_{\substack{g_0+n\leq k\leq g \\ 1\leq j\leq g_0+n-1}},$$

$$G=(\omega_j(a_k))_{\substack{1\leq k\leq g_0 \\ g_0+n1\leq j\leq g}}, H=(\omega_j(a_k))_{\substack{g_0+1\leq k\leq g_0+n-1 \\ g_0+n\leq j\leq g}}, I=(\omega_j(a_k))_{\substack{g_0+n\leq k\leq g \\ g_0+n\leq j\leq g}},$$

$$J=(\omega_j(b_k))_{\substack{1\leq k\leq g_0 \\ g_0+n1\leq j\leq g}}, K=(\omega_j(b_k))_{\substack{g_0+1\leq k\leq g_0+n-1 \\ g_0+n\leq j\leq g}}, L=(\omega_j(b_k))_{\substack{g_0+n\leq k\leq g \\ g_0+n\leq j\leq g}},$$

then the period matrices of $Jac(\mathcal{C}_0)$, $Prym(\mathcal{C}/\mathcal{C}_0)$, and the dual $Prym^*(\mathcal{C}/\mathcal{C}_0)$ (which is also denoted by $Prym^\vee(\mathcal{C}/\mathcal{C}_0)$) of $Prym(\mathcal{C}/\mathcal{C}_0)$ are given, respectively, by $\Delta = (G \quad H), \Gamma = (2A \quad B \quad 2D \quad E)$ and $\Gamma^* = (A \quad B \quad D \quad E)$.

PROOF.– We have,

$$\begin{aligned}\omega_j(a_{g_0+1}) = -\omega_j(\sigma(a_{g_0+1})) &= -\int_{a_{g_0+1}} \sigma^*(\omega_j), \\ &= \begin{cases} \omega_j(a_{g_0+1}), & 1 \leq j \leq g_0+n-1, \\ -\omega_j(a_{g_0+1}), & g_0+n \leq j \leq g, \end{cases}\end{aligned}$$

$$\vdots$$

$$\omega_j(a_{g_0+n-1}) = \begin{cases} \omega_j(a_{g_0+n-1}), & 1 \leq j \leq g_0+n-1, \\ -\omega_j(a_{g_0+n-1}), & g_0+n \leq j \leq g, \end{cases}$$

$$\begin{aligned}\omega_j(b_{g_0+1}) = -\omega_j(\sigma(b_{g_0+1})) &= -\int_{b_{g_0+1}} \sigma^*(\omega_j), \\ &= \begin{cases} \omega_j(b_{g_0+1}), & 1 \leq j \leq g_0+n-1, \\ -\omega_j(b_{g_0+1}), & g_0+n \leq j \leq g, \end{cases}\end{aligned}$$

$$\vdots$$

$$\omega_j(a_{g_0+n-1}) = \begin{cases} \omega_j(a_{g_0+n-1}), \ 1 \le j \le g_0+n-1, \\ -\omega_j(a_{g_0+n-1}), \ g_0+n \le j \le g, \end{cases}$$

$$\omega_j(a_{g_0+n}) = \omega_j(\sigma\,(a_1)) = \int_{a_1} \sigma^*\,(\omega_j)\,,$$
$$= \begin{cases} -\omega_j(a_1), \ 1 \le j \le g_0+n-1, \\ \omega_j(a_1), \ g_0+n \le j \le g, \end{cases}$$

$$\vdots$$

$$\omega_j(a_{g_0+n}) = \begin{cases} -\omega_j(a_{g_0}), \ 1 \le j \le g_0+n-1, \\ \omega_j(a_{g_0}), \ g_0+n \le j \le g, \end{cases}$$

$$\omega_j(b_{g_0+n}) = \omega_j(\sigma\,(b_1)) = \int_{b_1} \sigma^*\,(\omega_j)\,,$$
$$= \begin{cases} -\omega_j(b_1), \ 1 \le j \le g_0+n-1, \\ \omega_j(b_1), \ g_0+n \le j \le g, \end{cases}$$

$$\vdots$$

$$\omega_j(b_{g_0+n}) = \begin{cases} -\omega_j(b_{g_0}), \ 1 \le j \le g_0+n-1, \\ \omega_j(b_{g_0}), \ g_0+n \le j \le g, \end{cases}$$

Then, $C = -A$, $F = -D$, $H = O$, $I = G$, $K = O$, $L = J$, where O is the null matrix and therefore,

$$\Omega = \begin{pmatrix} A\ B\ -A\ D\ E\ -D \\ G\ O\ \ G\ \ J\ O\ \ J \end{pmatrix} \equiv \left(C_1\ C_2\ C_3\ C_4\ C_5\ C_6\right).$$

By an elementary column operation, we obtain the following matrices:

$$\Omega_1 = \left(C_1\ C_2\ C_1 + C_3\ C_4\ C_5\ C_4 + C_6\right),$$
$$= \begin{pmatrix} A\ B\ \ O\ \ D\ E\ \ O \\ G\ O\ 2G\ J\ O\ 2J \end{pmatrix},$$

$$\begin{aligned}
\Omega_2 &= \left(C_1\ C_2\ C_1 - C_3\ C_4\ C_5\ C_4 - C_6\right), \\
&= \begin{pmatrix} A\ B\ 2A\ D\ E\ 2D \\ G\ O\ \ O\ \ J\ O\ \ O \end{pmatrix}, \\
\Omega_3 &= \left(C_1 - C_3\ C_2\ C_4 - C_6\ C_5\ C_1 + C_3\ C_4 + C_6\right), \\
&= \begin{pmatrix} 2A\ B\ 2D\ E\ \ O\ \ O \\ O\ \ O\ \ O\ \ O\ 2G\ 2J \end{pmatrix}, \\
&= \begin{pmatrix} \Gamma\ O \\ O\ 2\Delta \end{pmatrix},
\end{aligned}$$

where

$$\begin{aligned}
\Delta &= \left(G\ J\right), \\
&= \begin{pmatrix} \omega_{g_0+n}(a_1) \ldots \omega_{g_0+n}(a_{g_0})\ \omega_{g_0+n}(b_1) \ldots \omega_{g_0+n}(b_{g_0}) \\ \vdots \qquad\qquad \vdots \qquad\qquad \vdots \qquad\qquad \vdots \\ \omega_g(a_1) \ \ldots\ \omega_g(a_{g_0}) \quad \omega_g(b_1) \ \ldots\ \omega_g(b_{g_0}) \end{pmatrix},
\end{aligned} \qquad [6.3]$$

and

$$\begin{aligned}
\Gamma &= \left(2A\ B\ 2D\ E\right), \\
&= \begin{pmatrix}
2\omega_1(a_1) & \ldots & 2\omega_{g_0+n-1}(a_1) \\
\vdots & & \vdots \\
2\omega_1(a_{g_0}) & \ldots & 2\omega_{g_0+n-1}(a_{g_0}) \\
\omega_1(a_{g_0+1}) & \ldots & \omega_{g_0+n-1}(a_{g_0+1}) \\
\vdots & & \vdots \\
\omega_1(a_{g_0+n-1}) & \ldots & \omega_{g_0+n-1}(a_{g_0+n-1}) \\
2\omega_1(b_1) & \ldots & 2\omega_{g_0+n-1}(b_1) \\
\vdots & & \vdots \\
2\omega_1(b_{g_0}) & \ldots & 2\omega_{g_0+n-1}(b_{g_0}) \\
\omega_1(b_{g_0+1}) & \ldots & \omega_{g_0+n-1}(b_{g_0+1}) \\
\vdots & & \vdots \\
\omega_1(b_{g_0+n-1}) & \ldots & \omega_{g_0+n-1}(b_{g_0+n-1})
\end{pmatrix}^{\top},
\end{aligned} \qquad [6.4]$$

where $\top$ denotes the transpose of the matrix. Let

$$L_\Omega = \left\{ \sum_{i=1}^{g} m_i \int_{a_i} \begin{pmatrix} \omega_1 \\ \vdots \\ \omega_g \end{pmatrix} + n_i \int_{b_i} \begin{pmatrix} \omega_1 \\ \vdots \\ \omega_g \end{pmatrix} : m_i, n_i \in \mathbb{Z} \right\},$$

the lattice generated by $\mathbb{Z}^g$ and the columns of the period matrix Ω. Let us denote also by L_{Ω_1}, L_{Ω_2}, L_{Ω_3} and L_Δ the period lattices associated respectively with $\Omega_1, \Omega_2, \Omega_3$ and Δ. Since $L_\Omega = L_{\Omega_1} = L_{\Omega_2} = L_{\Omega_3}$, it follows that the maps:

$$\mathbb{C}^{g_0+n-1}/L_\Gamma : \begin{pmatrix} t_1 \\ \vdots \\ t_{g_0+n-1} \end{pmatrix} \text{mod. } L_\Gamma \hookrightarrow \mathbb{C}^g/L_\Omega : \begin{pmatrix} t_1 \\ \vdots \\ t_{g_0+n-1} \\ 0 \\ \vdots \\ 0 \end{pmatrix} \text{mod. } L_\Omega,$$

$$\mathbb{C}^{g_0}/2L_\Delta : \begin{pmatrix} t_1 \\ \vdots \\ t_{g_0} \end{pmatrix} \text{mod. } 2L_\Delta \hookrightarrow \mathbb{C}^g/L_\Omega : \begin{pmatrix} 0 \\ \vdots \\ 0 \\ t_1 \\ \vdots \\ t_{g_0} \end{pmatrix} \text{mod. } L_\Omega,$$

are injective. Therefore, the tori $\mathbb{C}^{g_0+n-1}/L_\Gamma$ and $\mathbb{C}^{g_0}/2L_\Delta$ can be embedded into $\mathbb{C}^g/L_\Omega$ and the map,

$$\mathbb{C}^g/L_{\Omega_3} = \mathbb{C}^{g_0+n-1}/L_\Gamma \oplus \mathbb{C}^{g_0}/2L_\Delta \longrightarrow \mathbb{C}^g/L_\Omega,$$

$$\begin{pmatrix} t_1 \\ \vdots \\ t_g \end{pmatrix} \text{mod. } L_{\Omega_3} \longmapsto \begin{pmatrix} t_1 \\ \vdots \\ t_g \end{pmatrix} \text{mod. } L_\Omega,$$

shows that the Jacobian variety $Jac(\mathbb{C}_0)$ intersects the torus $\mathbb{C}^{g_0+n-1}/L_\Gamma$ in 2^{2g_0} points. We have then the following diagram:

$$\begin{array}{lllll}
\mathcal{C} & KerN_\varphi & & & \\
\downarrow\sigma & \downarrow & & & \\
\mathcal{C} \overset{u}{\longrightarrow} & Jac(\mathcal{C}) & \overset{u^*}{\longleftarrow} & Jac^*(\mathcal{C}) & \\
\downarrow\varphi & \downarrow N_\varphi & & \uparrow N_\varphi^* & \\
\mathcal{C}_0 \overset{u_0}{\longrightarrow} & Jac(\mathcal{C}_0) & \overset{u_0^*}{\longleftarrow} & Jac^*(\mathcal{C}_0) & \\
& \downarrow\varphi^* & & & \\
& Jac(\mathcal{C}) & \subset & H^1(\mathcal{C}, \mathcal{O}_{\mathcal{C}}^*) \simeq & H^1(\mathcal{C}_0, (\varphi_*\mathcal{O}_{\mathcal{C}})^*) \\
& \downarrow\sigma & & & \downarrow N_\varphi \\
& Jac(\mathcal{C}) & \overset{N_\varphi}{\longrightarrow} & Jac(\mathcal{C}_0) \hookrightarrow & H^1(\mathcal{C}_0, \mathcal{O}_{\mathcal{C}_0}^*)
\end{array}$$

where

$$
\begin{aligned}
u &: z \mapsto \text{divisor class}(z-p),\ p \in \mathcal{C}, \text{fixed},\\
u_0 &: z_0 \mapsto \text{divisor class}(z_0-p_0),\ p_0 \in \mathcal{C}_0,\ \text{fixed with } p_0=\varphi(p),\\
N_\varphi &: Jac(\mathcal{C}) \to Jac(\mathcal{C}_0),\ \sum m_i p_i \mapsto \sum m_i \varphi(p_i)\\
&\quad \text{is the norm mapping.}\\
Jac^*(\mathcal{C}) &= \text{dual of } Jac(\mathcal{C})\,,\\
Jac^*(\mathcal{C}_0) &= \text{dual of } Jac(\mathcal{C}_0).
\end{aligned}
$$

The norm map N_φ is surjective. The maps u and u_0 determine isomorphisms and up to isomorphism, the curve $\mathcal{C}$ (respectively $\mathcal{C}_0$) depends on $3g-3$ (respectively $3g_0-3$) complex parameters, while the Jacobian variety $Jac(\mathcal{C})$ (respectively $Jac(\mathcal{C}_0)$) depends on $\frac{1}{2}g(g+1)$ (respectively $\frac{1}{2}g_0\,(g_0+1)$) complex parameters. Moreover, the dual $Jac^*\,(\mathcal{C})$ of $Jac(\mathcal{C})$ is isomorphic to $Jac(\mathcal{C})$. The Prym variety denoted $Prym(\mathcal{C}/\mathcal{C}_0)$ is defined by:

$$
Prym(\mathcal{C}/\mathcal{C}_0) = \left(H^0\left(\mathcal{C},\Omega^1_{\mathcal{C}}\right)^-\right)^* / H_1\left(\mathcal{C},\mathbb{Z}\right)^-,
$$

where $\Omega^1_{\mathcal{C}}$ is the sheaf of 1-holomorphic differential forms on the curve $\mathcal{C}$, the exponent $-$ denotes the fixed part under $-\sigma$ and $*$ the dual. In other words, $Prym(\mathcal{C}/\mathcal{C}_0)$ is the connected component at the origin of the fixed part under $-\sigma$. Let $D \in Prym(\mathcal{C}/\mathcal{C}_0)$, i.e.

$$
\begin{pmatrix} 0\\ \vdots\\ 0\\ 2\int_0^D \omega_{g_0+n}\\ \vdots\\ 2\int_0^D \omega_g \end{pmatrix} \in L_\Omega \Longleftrightarrow \begin{pmatrix} \int_0^D \omega_{g_0+n}\\ \vdots\\ \int_0^D \omega_g \end{pmatrix} \in L_\Delta,
$$

$$
\Longleftrightarrow \begin{pmatrix} \int_0^D \omega_1\\ \vdots\\ \int_0^D \omega_{g_0+n-1}\\ \int_0^D \omega_{g_0+n}\\ \vdots\\ \int_0^D \omega_g \end{pmatrix} = \begin{pmatrix} t_1\\ \vdots\\ t_{g_0+n-1}\\ 0\\ \vdots\\ 0 \end{pmatrix}.
$$

Consequently, the Abel–Jacobi map:

$$Jac(\mathcal{C}) \longrightarrow \mathcal{C}^g/L_\Omega,\; D \longmapsto \begin{pmatrix} \int_0^D \omega_1 \\ \vdots \\ \int_0^D \omega_{g_0+n-1} \\ \int_0^D \omega_{g_0+n} \\ \vdots \\ \int_0^D \omega_g \end{pmatrix},$$

maps $Prym(\mathcal{C}/\mathcal{C}_0))$ bi-analytically onto $\mathcal{C}^{g_0+n-1}/L_\Gamma$. By Chow's theorem, $Prym(\mathcal{C}/\mathcal{C}_0)$ is thus a sub-Abelian variety of $Jac(\mathcal{C})$. More precisely, the kernel of N_φ has two connected components and the Prym variety $Prym(\mathcal{C}/\mathcal{C}_0)$ associated with the covering φ is the neutral component of $KerN_\varphi$, we have:

$$Prym(\mathcal{C}/\mathcal{C}_0) = (KerN_\varphi)^0 = Ker\left(1_{Jac(\mathcal{C})} + \sigma\right)^0 = Im\left(1_{Jac(\mathcal{C})} - \sigma\right) \subset Jac(\mathcal{C}).$$

In other words, the involution σ induces an involution:

$$\sigma : Jac(\mathcal{C}) \longrightarrow Jac(\mathcal{C}), \quad \text{class of } D \longmapsto \text{class of } \sigma D,$$

and up to some points of order two, $Jac(\mathcal{C})$ splits into an even part $Jac(\mathcal{C}_0)$ and an odd part $Prym(\mathcal{C}/\mathcal{C}_0)$:

$$Jac(\mathcal{C}) = Prym(\mathcal{C}/\mathcal{C}_0) \oplus Jac(\mathcal{C}_0),$$

with $\dim Jac(\mathcal{C}_0) = g_0$, $\dim Jac(\mathcal{C}) = g = 2g_0 + n - 1$ and $\dim Prym(\mathcal{C}/\mathcal{C}_0) = g - g_0 = g_0 + n - 1$. From the above, Δ [6.3] is the period matrix of $Jac(\mathcal{C}_0)$, while Γ [6.4] is the period matrix of $Prym(\mathcal{C}/\mathcal{C}_0)$. To determine the period matrix of the dual $Prym^*(\mathcal{C}/\mathcal{C}_0)$, let us write the period matrix Γ of $Prym(\mathcal{C}/\mathcal{C}_0)$ in the form $\Gamma = (U \quad V)$, where:

$$U =$$

$$\begin{pmatrix} 2\omega_1(a_1) & \dots & 2\omega_1(a_{g_0}) & \omega_1(a_{g_0+1}) & \dots & \omega_1(a_{g_0+n-1}) \\ \vdots & & \vdots & \vdots & & \vdots \\ 2\omega_{g_0+n-1}(a_1) & \dots & 2\omega_{g_0+n-1}(a_{g_0}) & \omega_{g_0+n-1}(a_{g_0+1}) & \dots & \omega_{g_0+n-1}(a_{g_0+n-1}) \end{pmatrix}$$

and

$$V =$$

$$\begin{pmatrix} 2\omega_1(b_1) & \dots & 2\omega_1(b_{g_0}) & \omega_1(b_{g_0+1}) & \dots & \omega_1(b_{g_0+n-1}) \\ \vdots & & \vdots & \vdots & & \vdots \\ 2\omega_{g_0+n-1}(b_1) & \dots & 2\omega_{g_0+n-1}(b_{g_0}) & \omega_{g_0+n-1}(b_{g_0+1}) & \dots & \omega_{g_0+n-1}(b_{g_0+n-1}) \end{pmatrix}$$

Let us write

$$e_1 = \begin{pmatrix} 2\omega_1(a_1) \\ \vdots \\ 2\omega_{g_0+n-1}(a_1) \end{pmatrix}, \ldots, e_{g_0} = \begin{pmatrix} 2\omega_1(a_{g_0}) \\ \vdots \\ 2\omega_{g_0+n-1}(a_{g_0}) \end{pmatrix},$$

$$e_{g_0+1} = \begin{pmatrix} \omega_1(a_{g_0+1}) \\ \vdots \\ \omega_{g_0+n-1}(a_{g_0+1}) \end{pmatrix}, \ldots, e_{g_0+n-1} = \begin{pmatrix} \omega_1(a_{g_0+n-1}) \\ \vdots \\ \omega_{g_0+n-1}(a_{g_0+n-1}) \end{pmatrix}.$$

We can choose a new base $(\lambda_1, \ldots, \lambda_{g_0+n-1})$ with:

$$\lambda_j = \frac{e_j}{\delta_j}, \qquad \delta_j = \begin{cases} 1 & \text{for } 1 \leq j \leq g_0, \\ 2 & \text{for } g_0 + 1 \leq j \leq g_0 + n - 1, \end{cases}$$

such that the period matrix Γ is expressed in the form $(\Delta_\delta \quad Z)$, where $\Delta_\delta = diag(\delta_1, \ldots, \delta_n)$ and $Z = \Delta_\delta U^{-1} V$, is a symmetric matrix, with a positive definite imaginary part. Then:

$$\Gamma^* = (\delta_{g_0+n-1}\Delta_\delta^{-1}, \delta_{g_0+n-1}\Delta_\delta^{-1} Z \Delta_\delta^{-1}) = (\delta_{g_0+n-1}\Delta_\delta^{-1}, \delta_{g_0+n-1}U^{-1}V\Delta_\delta^{-1}),$$

$$\Gamma^* = \left(\delta_{g_0+n-1}\Delta_\delta^{-1}, \delta_{g_0+n-1}\Delta_\delta^{-1}(U^*)^{-1}V^*\right) = (U^*, V^*) = (A \quad B \quad D \quad E),$$

is the period matrix of the dual Abelian variety $Prym^*(\mathcal{C}/\mathcal{C}_0)$. This completes the proof of the theorem. ■

REMARK 6.1.– *The principal polarization of the variety* $Jac(\mathcal{C})$ *induces on the variety* $Prym(\mathcal{C}/\mathcal{C}_0)$ *twice the principal polarization. Let* $\mathcal{L}$ *be a line bundle on an Abelian variety* $\mathcal{A}$*, in particular the variety* $Prym(\mathcal{C}/\mathcal{C}_0)$*. We assume that* $\mathcal{L}$ *defines a polarization of the type* $(\delta_1, ..., \delta_n)$ *on* $\mathcal{A}$*. Recall that the exact exponential sequence:*

$$0 \longrightarrow \mathbb{Z} \longrightarrow \mathcal{O}_\mathcal{A} \overset{\exp}{\longrightarrow} \mathcal{O}_\mathcal{A}^* \longrightarrow 0,$$

defines a morphism c_1 *(first Chern class), which associates with any line bundle a cohomology class in* $H^2(\mathcal{A}, \mathbb{Z})$*. Then, the Chern class of the bundle* $\mathcal{L}^k$ *is related to that of* $\mathcal{L}$ *by the relation* $c_1(\mathcal{L}^k) = kc_1(\mathcal{L})$*. Hence,* $\mathcal{L}^k$ *defines a polarization of the type* $(k\delta_1, ..., k\delta_n)$ *and by the Lefschetz theorem (theorem 7.2, demonstrated in detail in Chapter 7), we have:*

$$\dim H^0(\mathcal{A}, \mathcal{O}(\mathcal{L}^k)) = k^n \prod_{\alpha=1}^{n} \delta_\alpha.$$

6.6. Projectively normal embedding

Let $\mathcal{D}$ be a very ample divisor, and consider the following map:

$$\mathcal{L}(\mathcal{D})^{\otimes k} \longrightarrow \mathcal{L}(kD).$$

DEFINITION 6.6.– *If this map is surjective, then we say that such a divisor is projectively normal, i.e. any function in* $\mathcal{L}(kD)$ *can be expressed as a linear combination of the* $k^{\text{èth}}$ *product of functions in* $\mathcal{L}(\mathcal{D})$.

According to a criterion of Koizumi (1976), Mumford (1974b), and Mumford (1975b), we have the following proposition:

PROPOSITION 6.2.– *The embedding in question will be projectively normal if and only if there exists an ample divisor* $\mathcal{D}_0$ *such that:* $\mathcal{D} = k\mathcal{D}_0$, $k \geq 3$.

DEFINITION 6.7.– *An embedding of an Abelian variety* $\mathcal{A}$ *into* $\mathbb{P}^N(\mathbb{C})$ *is projectively normal if the image of* $\mathcal{A}$ *into* $\mathbb{P}^N(\mathbb{C})$ *is a projectively normal variety; i.e. if its ring of homogeneous coordinates is integrally closed.*

Let $\mathcal{L}$ be an ample line bundle on an Abelian variety $\mathcal{A}$. The above definition is equivalent to the following proposition (which can therefore be taken as a definition).

PROPOSITION 6.3.– *The embedding* $\mathcal{A} \longrightarrow \mathbb{P}^N(\mathbb{C})$ *is projectively normal if for all* $k \geq 1$ *the restriction:*

$$H^0\left(\mathbb{P}^N(\mathbb{C}), \mathcal{O}_{\mathbb{P}^N(\mathbb{C})}\right) \longrightarrow H^0\left(\mathcal{A}, \mathcal{L}^{\otimes k}\right),$$

is surjective.

Furthermore, we show (Birkenhake and Lange 2004) that if $\mathcal{L}$ is an ample line bundle on an Abelian variety $\mathcal{A}$ of dimension g and of type $(\delta_1, ..., \delta_g)$, then the map $\psi_{\mathcal{L}^k} : \mathcal{A} \longrightarrow \mathbb{P}^N(\mathbb{C})$, associated with the bundle $\mathcal{L}^k \equiv \mathcal{L}^{\otimes k}$, determines a projectively normal embedding for all $k \geq 3$ and also for $k = 2$ when the points of $K(\mathcal{L}^2))$ are not base points of $\mathcal{L}$ where $K(\mathcal{L})$ is the fixed group of $\mathcal{L}$, i.e. the group:

$$K(\mathcal{L}) = \{z \in \mathcal{A} : \mathcal{L} \simeq \tau_z^* \mathcal{L}\},$$

with $\tau_z^* : \mathcal{A} \longrightarrow \mathcal{A}, x \longmapsto x + z$, the translation by z. Note that:

$$K(\mathcal{L}) = \ker f \simeq (\mathbb{Z}/\delta_1\mathbb{Z} \times ... \times \mathbb{Z}/\delta_g\mathbb{Z})^2,$$

is the kernel of the isogeny:

$$f : \mathcal{A} \longrightarrow \mathcal{A}^* = Pic^0(\mathcal{A}), \quad z \longmapsto \tau_z^* \mathcal{L} \otimes \mathcal{L}^{-1}.$$

In the case where $\mathcal{A}$ is an Abelian surface, the rational map $\psi_{\mathcal{L}} : \mathcal{A} \longrightarrow \mathbb{P}^{\delta_1\delta_2-1}(\mathbb{C})$ is a projectively normal embedding if the following conditions are satisfied:

(i) When $\delta_1 \geq 3$, $\psi_{\mathcal{L}}$ is a projectively normal embedding (Lang 1995; Birkenhake and Lange 2004).

(ii) When $\delta_1 = 2$, $\psi_{\mathcal{L}}$ is a projectively normal embedding if and only if the points of $K(\mathcal{L}^2)$ are not base points of $\mathcal{L}$ (Ramanan 1980; Birkenhake and Lange 2004).

(iii) When $\mathcal{L}$ is primitive, i.e. $(\delta_1, \delta_2) = (1, \delta_2)$ and $\delta_2 \geq 7$, $\psi_{\mathcal{L}}$ is a projectively normal embedding if and only if $\mathcal{L}$ is very ample (Lazarsfeld et al. 1990).

(iv) When $\mathcal{L}$ is primitive, $\delta_2 \geq 7$ and $\mathcal{A}$ is generic, $\psi_{\mathcal{L}}$ is a projectively normal embedding (Iyer 1999).

A necessary condition for $\psi_{\mathcal{L}}$ to be a projectively normal embedding is that $\mathcal{L}$ is very ample. Note that when $\delta_1\delta_2 \leq 4$, $\mathcal{L}$ is not even very ample. Similarly, if $\delta_1\delta_2 \leq 7$, then:

$$\dim \mathrm{Sym}^2 H^2(\mathcal{A}, \mathcal{L}) < \dim \mathrm{Sym}\, H^0(\mathcal{A}, \mathcal{L}^2),$$

and thus, the embedding in question can never be projectively normal.

6.7. Number of even and odd sections of a line bundle

When studying certain Abelian surfaces, it is sometimes essential to know the number of even and odd sections of a line bundle. We will explicitly provide two formulas that allow us to calculate the number of these sections.

Let $\mathcal{L}$ be a line bundle on an Abelian surface $\mathcal{A}$ of type (δ_1, δ_2). Assume that $\mathcal{L}$ is symmetric, i.e. $\sigma^*\mathcal{L} \simeq \mathcal{L}$, where σ denotes the "$-$" involution on $\mathcal{A}$. The involution σ appears in $\mathcal{L}$ as an involution $\widetilde{\sigma}$ acting in two different ways depending on the sign and therefore for each section (theta function) $s \in H^0(\mathcal{A}, \mathcal{L})$, we have $\widetilde{\sigma}s = \pm s$. Recall that a section $s \in H^0(\mathcal{A}, \mathcal{L})$ is said to be even if $\widetilde{\sigma}s = +s$ and odd if $\widetilde{\sigma}s = -s$. Under $\widetilde{\sigma}$, the vector space $H^0(\mathcal{A}, \mathcal{L})$ is decomposed into two subspaces:

$$H^0(\mathcal{A}, \mathcal{L}) = H^0(\mathcal{A}, \mathcal{L})^{even} \oplus H^0(\mathcal{A}, \mathcal{L})^{odd},$$

where $H^0(\mathcal{A}, \mathcal{L})^{even}$ contains all even sections and $\mathcal{H}^0(\mathcal{A}, \mathcal{L})^{odd}$ all odd sections. Using the inversion formula [30, p.331], we show that:

$$\dim H^0(\mathcal{A}, \mathcal{L})^{even} = \begin{cases} \frac{\delta_1\delta_2}{2} + 2\left(1 + [\frac{\delta_2}{2}] - \frac{\delta_2}{2}\right) & \text{for } \delta_1 \text{ even} \\ \frac{\delta_1\delta_2}{2} + \left(1 + [\frac{\delta_2}{2}] - \frac{\delta_2}{2}\right) & \text{for } \delta_1 \text{ uneven} \end{cases}$$

$$\dim H^0(\mathcal{A},\mathcal{L})^{uneven} = \begin{cases} \frac{\delta_1\delta_2}{2} - 2\left(1+[\frac{\delta_2}{2}]-\frac{\delta_2}{2}\right) \text{ for } \delta_1 \text{ even} \\ \frac{\delta_1\delta_2}{2} - \left(1+[\frac{\delta_2}{2}]-\frac{\delta_2}{2}\right) \text{ for } \delta_1 \text{ uneven} \end{cases}$$

or in condensed form,

$$\dim H^0(\mathcal{A},\mathcal{L})^{even} = \frac{\delta_1\delta_2}{2} + 2^{-1+\sharp \text{ even } \delta_k}, \quad k=1,2$$

$$\dim H^0(\mathcal{A},\mathcal{L})^{uneven} = \frac{\delta_1\delta_2}{2} - 2^{-1+\sharp \text{ even } \delta_k}, \quad k=1,2$$

6.8. Exercises

EXERCISE 6.1.– *Show that $\Omega \in M_{n\times 2n}(\mathbb{C})$ is a period matrix (definition 6.3) (of a complex torus) if and only if the matrix $\begin{pmatrix}\Omega\\ \overline{\Omega}\end{pmatrix} \in M_{2n}(\mathbb{C})$ is invertible.*

Solution: Recall that a period matrix of a complex torus $\mathbb{C}^n/\Lambda$ is a matrix whose columns form a basis for a lattice Λ in $\mathbb{C}^n$, where the torus is defined as $\mathbb{C}^n$ modulo this lattice. let $A \equiv \begin{pmatrix}\Omega\\ \overline{\Omega}\end{pmatrix}$. Suppose that the columns of the matrix Ω are linearly dependent over $\mathbb{R}$. Hence, $\Omega x = 0$ vanishes where $x \in \mathbb{R}^{2n}$, $x \neq 0$, and we obtain that $Ax = 0$ which implies $A = 0$. Reciprocally, suppose that the matrix A is not invertible. In this case, $A(x+iy) = 0$ where $x, y \in \mathbb{R}^{2n}$, $xy \neq 0$, and the same is true of the period matrix Ω at $x \pm iy$; $\Omega(x \pm iy) = 0$. We deduce that $\Omega x = \Omega y = 0$. This shows that the columns of the matrix Ω are linearly dependent over $\mathbb{R}$, which is absurd since the columns of Ω must be independent.

EXERCISE 6.2.– *Show that if $\mathcal{C}$ is a smooth projective curve of genus 3, then for it to be a double covering of an elliptic curve $\mathcal{E}$, it is necessary and sufficient that $\mathcal{C}$ admits an embedding in an Abelian surface $\mathcal{A}$. Deduce that if $\mathcal{C} \hookrightarrow \mathcal{A}$ is the embedding corresponding to the above covering in the Abelian surface $\mathcal{A}$, then the two Abelian surfaces $\mathcal{A}$ and $Prym\,(\mathcal{C}/\mathcal{E})$ admit natural polarizations of type $(1,2)$ identifying one with the dual of the other.*

Hint: See Barth (1985).

EXERCISE 6.3.– *Show that any Abelian surface is isogenous to a hyperelliptic Jacobian.*

Solution: In genus 2, principally polarized is equivalent to hyperelliptic.

EXERCISE 6.4.– *Let $\mathcal{C}$ be a smooth hyperelliptic projective curve of genus g. Assume that $\varphi : \Gamma \longrightarrow \mathcal{C}$ is a cyclic étale covering of degree n, associated with a fixed element* *belonging to $Jac(C)$ of order n (recall that φ is étale at a if φ is differentiable at a and if its differential $d\varphi_a$ is an isomorphism). Show that if n is not divisible by 4, then the variety $Prym(\Gamma/C)$ is isomorphic to a product of Jacobian varieties.*

Hint: See Scheidemann (2023).

EXERCISE 6.5.– *Let $\mathcal{A}_1$ and $\mathcal{A}_2$ be the two Abelian varieties and $f : \mathcal{A}_1 \longrightarrow \mathcal{A}_2$ be an isogeny of degree $k \in \mathbb{N}^*$. Using the notations of definition 6.4 and example 6.5, show that there exists a unique isogeny $g : \mathcal{A}_2 \longrightarrow \mathcal{A}_1$ such that: $fog = k_{\mathcal{A}_2}$ and $gof = k_{\mathcal{A}_1}$. (The isogeny g is called the dual isogeny of f).*

Solution: The map f is surjective and its degree (i.e. the cardinality of the kernel) is equal to $k^{2\dim \mathcal{A}_1}$ (see example 6.5), which implies that: $\ker f \subset \ker k_{\mathcal{A}_1}$. It follows that there is a unique f such that: $gof = k_{\mathcal{A}_1}$. In addition, we have $fogof(x) = f(k_{\mathcal{A}_1}(x)) = k_{\mathcal{A}_2}(f(x))$, and we deduce that we also have, $fog = k_{\mathcal{A}_2}$.

EXERCISE 6.6.– *Let $\mathcal{A}$ be an Abelian variety and $\mathcal{A}_1$ be an Abelian subvariety of $\mathcal{A}$. Show (Poincaré reducibility theorem) that there exists an Abelian subvariety $\mathcal{A}_2$ of $\mathcal{A}$ such that $\mathcal{A} = \mathcal{A}_1 + \mathcal{A}_2$ with $\mathcal{A}_1 \cap \mathcal{A}_2$ finite. In other words, $\mathcal{A}$ is isogenous to $\mathcal{A}_1 \times \mathcal{A}_2$.*

Hint: The idea is to choose the polarizations on $\mathcal{A}$ and $\mathcal{A}/\mathcal{A}_1$ in order to identify them with their duals. The dual to the quotient map $\mathcal{A} \longrightarrow \mathcal{A}/\mathcal{A}_1$ is a map $\mathcal{A}/\mathcal{A}_1 \longrightarrow \mathcal{A}$, and we denote its image $\mathcal{A}_2$ (see Mumford 1975b).

EXERCISE 6.7.– *Let $\mathcal{L}$ be an ample line bundle over an Abelian variety $\mathcal{A}$. Show that for $\mathcal{L}^k \equiv \mathcal{L}^{\otimes k}$, $k \geq 3$, the map $\psi_{\mathcal{L}^k} : \mathcal{A} \longrightarrow \mathbb{P}^N(\mathbb{C})$, determines a projectively normal embedding.*

Hint: See Birkenhake and Lange (2004, p. 187).

EXERCISE 6.8.– *An Abelian variety is said to be simple if it contains no Abelian subvariety distinct from zero and itself. Show that there is no non-trivial homomorphism between two non-isogenic simple Abelian varieties.*

Hint: Show that any non-constant holomorphic map between simple Abelian varieties is an isogeny. For a simple Abelian variety $\mathcal{A}$, the ring $\mathrm{End}^0(\mathcal{A})$ is a division ring (every non-zero element is invertible). A homomorphism between two simple Abelian varieties is either an isogeny or zero.

EXERCISE 6.9.– *Show that any Abelian variety $\mathcal{A}$ is isogenous to a product of simple pairwise not isogenous Abelian varieties $\mathcal{A}_j^{n_j}$ of the form (Poincaré complete reducibility theorem):*

$$\mathcal{A} = \mathcal{A}_1^{n_1} \times \cdots \times \mathcal{A}_r^{n_r}.$$

Solution: We will use inductive reasoning on the dimension k of $\mathcal{A}$. For $k = 1$, it is obvious; any Abelian variety of dimension 1 is simple. Let us assume that the result is true for lower dimensions. If $\mathcal{A}$ is simple, it is finished; otherwise, according to the Exercise 6.6, the variety $\mathcal{A}$ is isogenous to a product $\mathcal{A}_1 \times \mathcal{A}_2$, and it is enough to apply the induction hypothesis to the two factors to obtain the result.

EXERCISE 6.10.– *Let $\mathcal{D}$ be an effective divisor on a complex torus $\mathbb{T} = \mathbb{C}^n/\Lambda$. Show that there exists an entire theta function with respect to the lattice Λ that represents this divisor (Poincaré theorem).*

Hint: See Hindry and Silverman (2000, p. 98), Lang (1995, p. 157) or Vanhaecke (2001, p. 35).

EXERCISE 6.11.– *(a) Let $\mathcal{A}$ be an Abelian variety, $\mathcal{B}$ be any variety and $f, g, h : \mathcal{B} \longrightarrow \mathcal{A}$ be maps. Show that for all line bundles $\mathcal{L}$ on A, we have:*

$$(f+g+h)^*\mathcal{L} \cong (f+g)^*\mathcal{L} \otimes (f+h)^*\mathcal{L} \otimes (g+h)^*\mathcal{L} \otimes f^*\mathcal{L}^{-1} \otimes g^*\mathcal{L}^{-1} \otimes h^*\mathcal{L}^{-1}.$$

(b) Let $\mathcal{L}$ be a line bundle on an Abelian variety $\mathcal{A}$. Show that for all $a, b \in \mathcal{A}$, the tensor product of the pullback of $\mathcal{L}$ by translation by $a + b$ with $\mathcal{L}$ is isomorphic to the tensor product of the pullback of $\mathcal{L}$ by translation by a and the pullback of $\mathcal{L}$ by translation by b. In other words, show that we have,

$$\tau_{a+b}^*\mathcal{L} \otimes \mathcal{L} \simeq \tau_a^*\mathcal{L} \otimes \tau_b^*\mathcal{L} \quad \textit{(theorem of the square)},$$

where $\tau_a : \mathcal{A} \longrightarrow \mathcal{A}$, $z \longmapsto z + a$, is the translation operator by a on $\mathcal{A}$.

Hint: *(a)* See Mumford (1975b, p. 56). *(b)* Apply the f, g, h in question *(a)* with $\mathcal{A} = \mathcal{B}$, $f = a$, $g = b$ (constant maps), and $h = identity$.

EXERCISE 6.12.– *Let $\mathcal{A}$ be an Abelian variety and $\mathcal{L} \in Pic(\mathcal{A})$. Show that the image of $\varphi_{\mathcal{L}} : \mathcal{A} \longrightarrow Pic(\mathcal{A})$ is contained in $Pic^0(\mathcal{A})$ (see section 6.4 for notations)*

Hint: Use theorem of the square (previous exercise).

7

Theta Functions and Complex Projective Tori

7.1. Meromorphic functions and theta functions

The aim of this section is to introduce theta functions, which provide privileged coordinates for the projective embedding associated with a very large line bundle. They play a major role in much current research work and are powerful tools for studying some problems. Theta functions are special functions that are used in several areas, such as quadratic forms, moduli spaces, when studying integrable systems and in the study of Abelian varieties. The purpose of this section is to provide a short and quick exposition of some important aspects of meromorphic theta functions for compact Riemann surfaces, both for their own interests and because this leads to clarifying a certain number of results obtained in this chapter. For example, Lesfari (2023) can be consulted for the proof of the results as well as for other additional information.

DEFINITION 7.1.– *The function defined using the following (Fourier) series is called the theta function:*

$$\theta(z|B) = \sum_{m\in\mathbb{Z}^g} e^{\pi i\langle Bm,m\rangle+2\pi i\langle z,m\rangle}, \quad z \in \mathbb{C}^g \qquad [7.1]$$

where $B = (b_{jk})_{1\leq j,k\leq g}$ *is a square matrix of order* g*, which is symmetric and whose imaginary part is positive definite (Riemann matrix),*

$$\langle Bm,m\rangle = \sum_{j,k=1}^{g} b_{jk}m_jm_k, \qquad \langle z,m\rangle = \sum_{j=1}^{g} z_jm_j.$$

When theta functions are used later in the study of meromorphic functions or in other problems on a Riemann surface or complex algebraic curve C of genus g, then B will play the role of the period matrix of C with respect to a chosen cycle basis.

The convergence of this series for all $z \in \mathbb{C}^g$ results from the fact that Im $B > 0$. We show that this series converges absolutely and uniformly on compact sets, and thus, the function $\theta(z|B)$ is holomorphic over $\mathbb{C}^g$. More precisely, taking into account the estimate $\text{Re}(\langle Bm, m\rangle) \leq -b\langle m, m\rangle$, where $-b < 0$ is the largest eigenvalue of the matrix Re(B), we show that this series converges absolutely and uniformly on compact sets and that, moreover, the function $\theta(z|B)$ is holomorphic on $\mathbb{C}^g$. Several properties concerning the sum of this series (continuity, differentiation, integration, etc.) are valid, and we can, for example, interchange the summation with integrals, limits, derivatives, etc. When the matrix B is fixed, we will put in the sequel $\theta(z) \equiv \theta(z|B)$. Let $(e_1, ..., e_g)$ be a basis of $\mathbb{C}^g$ with $(e_j)_k = \delta_{jk}$ and let $f_j = (b_{1j}...b_{gj})^\top$ be the columns of the matrix B or in condensed form $f_j = Be_j, j = 1, ..., g$.

PROPOSITION 7.1.– *The function θ satisfies the functional equations:*

$$\theta(z + e_j) = \theta(z), \qquad \theta(z + f_j) = e^{-\pi i b_{jj} - 2\pi i z_j}.\theta(z).$$

For any $m, n \in \mathbb{Z}^g$, we have:

$$\theta(z + n + Bm) = e^{-\pi i\langle Bm,m\rangle - 2\pi i\langle m,z\rangle}\theta(z). \qquad [7.2]$$

Any vector of the form $n + Bm$ is a period of the Riemann theta function, and it constitutes the period lattice.

Thus, the vectors $e_1, ..., e_g$ are the periods of function $\theta(z)$. The vectors $f_1, ..., f_g$ are called the quasi-periods. The function θ is quasi-periodic and is well defined on the Jacobian variety of $\mathcal{C}$. Consider a generalization of the theta function [7.1] called the theta function with characteristics $\alpha, \beta \in \mathbb{R}^g$,

DEFINITION 7.2.– *We call the theta function with characteristics $\begin{bmatrix}\alpha\\ \beta\end{bmatrix}$ the function defined by:*

$$\theta\begin{bmatrix}\alpha\\ \beta\end{bmatrix}(z|B) = \sum_{m\in\mathbb{Z}^g} e^{\pi i\langle B(m+\alpha), m+\alpha\rangle + 2\pi i\langle z+\beta, m+\alpha\rangle}, \qquad [7.3]$$

$$= e^{\pi i\langle B\alpha,\alpha\rangle + 2\pi i\langle z+\beta,\alpha\rangle}.\theta(z + \beta + B\alpha). \qquad [7.4]$$

where $\alpha, \beta \in \mathbb{R}^g$.

To simplify the formulas, we simply note: $\theta\begin{bmatrix}\alpha\\ \beta\end{bmatrix}(z) \equiv \theta\begin{bmatrix}\alpha\\ \beta\end{bmatrix}(z|B)$, when the matrix B is fixed. In particular, $\theta\begin{bmatrix}0\\ 0\end{bmatrix}(z) = \theta(z)$. From equation [7.2], we have also $\theta\begin{bmatrix}m\\ n\end{bmatrix}(z) = \theta(z)$, $m, n \in \mathbb{Z}^g$. Consequently, it is sufficient to consider the functions $\theta\begin{bmatrix}\alpha\\ \beta\end{bmatrix}(z)$ where $\alpha = (\alpha_1, ..., \alpha_g)$, $\beta = (\beta_1, ..., \beta_g) \in \mathbb{R}^g$ are such that: $0 < \alpha_j, \beta_j < 1, j = 1, ..., g$.

PROPOSITION 7.2.– *The periodicity property of the theta functions with characteristics is given by the following relation:*

$$\theta\begin{bmatrix}\alpha\\ \beta\end{bmatrix}(z)(z+n+Bm) = e^{-\pi i\langle Bm,m\rangle-2\pi i\langle z,m\rangle+2\pi i(\langle\alpha,n\rangle-\langle\beta,m\rangle)}.\theta\begin{bmatrix}\alpha\\ \beta\end{bmatrix}(z).$$

If $\alpha_1, ..., \alpha_g$ and $\beta_1, ..., \beta_g$ are 0 or $\frac{1}{2}$, we will say that the set $\begin{bmatrix}\alpha\\ \beta\end{bmatrix}$ is a half period. In addition, a half period $\begin{bmatrix}\alpha\\ \beta\end{bmatrix}$ is said to be even if $4\langle\alpha,\beta\rangle \equiv 0$ (mod. 2) and odd if not.

PROPOSITION 7.3.– *The function $\theta\begin{bmatrix}\alpha\\ \beta\end{bmatrix}(z)$ is even if the half-period $\begin{bmatrix}\alpha\\ \beta\end{bmatrix}$ is even and odd if the half-period $\begin{bmatrix}\alpha\\ \beta\end{bmatrix}$ is odd. In addition, we have $\theta(z) = \theta(-z)$.*

EXAMPLE 7.1.– *The number of even half-periods is equal to $2^{g-1}(2^g + 1)$ and that of odd half-periods to $2^{g-1}(2^g - 1)$.*

We will explain how to express meromorphic functions on a Riemann surface or complex algebraic curve C of genus $g \geq 1$ in terms of theta functions. The matrix B will play the role of the period matrix of C with respect to a chosen cycle basis. Let us first consider the case of Riemann surfaces of genus $g = 1$ (i.e. the case of elliptic curves). Recall that an elliptic function is a doubly periodic meromorphic function. The matrix B in this case reduces to a number denoted b with $\operatorname{Im} b \geq 0$. The numbers 1 and b generate a parallelogram of the periods denoted Ω. The four theta functions corresponding to the half-periods $\begin{bmatrix}1/2\\ 1/2\end{bmatrix}$, $\begin{bmatrix}1/2\\ 0\end{bmatrix}$, $\begin{bmatrix}0\\ 0\end{bmatrix}$, $\begin{bmatrix}0\\ 1/2\end{bmatrix}$ are:

$$i\theta_1(z) \equiv \theta\begin{bmatrix}1/2\\ 1/2\end{bmatrix}(z), \qquad \theta_2(z) \equiv \theta\begin{bmatrix}1/2\\ 0\end{bmatrix}(z),$$

$$\theta_3(z) \equiv \theta\begin{bmatrix}0\\ 0\end{bmatrix}(z) = \theta(z), \qquad \theta_4(z) \equiv \theta\begin{bmatrix}0\\ 1/2\end{bmatrix}(z).$$

These functions are holomorphic on $\mathbb{C}$. Moreover, we deduce from proposition 7.3 that $\theta_1(z)$ is odd and that $\theta_2(z)$, $\theta_3(z)$, $\theta_4(z)$ are even. To determine the zeros of the functions θ_j, it is sufficient, from proposition 7.1, to look for them in the parallelogram of the periods Ω. Since $\theta_1(z)$ is odd, then $\theta_1(0) = 0$ and the other zeros of $\theta_j(z)$ are obtained via proposition 7.1. In particular, we see that $\theta_3(\frac{1}{2} + \frac{b}{2}) = 0$. We claim that this zero of $\theta_3(z)$ in Ω is unique.

PROPOSITION 7.4.– *The function $\theta(z)$ has in the parallelogram of the periods Ω (generated by 1 and b) only one zero at the point $z = \frac{1}{2}(1 + b)$.*

By putting $z = x \in \mathbb{R}$, $b = it$, $t \in \mathbb{R}_+$, we will see $\theta(x|it)$ as the fundamental periodic solution to the heat equation (Mumford 1983). Equation [7.1] is written:

$$\theta(x|it) = \sum_{m\in\mathbb{Z}} e^{-\pi m^2 t + 2\pi imx} = 1 + 2\sum_{m=1}^{\infty} e^{-\pi m^2 t} \cos 2\pi mx.$$

Thus, θ is a real-valued function of two real variables. This function is periodic with respect to x, i.e. $\theta(x + 1|it) = \theta(x|it)$. Since:

$$\frac{\partial\theta(x|it)}{\partial t} = 2\sum_{m=1}^{\infty}(-\pi m^2)e^{-\pi m^2 t} \cos 2\pi mx,$$

$$\frac{\partial^2\theta(x|it)}{\partial x^2} = 2\sum_{m=1}^{\infty}(-4\pi^2 m^2)e^{-\pi m^2 t} \cos 2\pi mx,$$

then

$$4\pi\frac{\partial\theta(x|it)}{\partial t} = \frac{\partial^2\theta(x|it)}{\partial x^2}.$$

This suggests that we characterize the theta function $\theta(x|it)$ as the unique solution to the heat equation with a certain periodic initial data when $t = 0$. To examine the limiting behavior of f, we integrate it against a test periodic function $f(x) = \sum_n a_n e^{2\pi inx}$. Then:

$$\begin{aligned}
\lim_{t\to 0}\int_0^1 \theta(x|it)f(x)dx &= \int_0^1 \sum_{m,n} a_n e^{-\pi m^2 t + 2\pi i(m+n)x} dx,\\
&= \lim_{t\to 0}\sum_{m,n} a_n e^{-\pi m^2 t}\int_0^1 e^{2\pi i(m+n)x} dx,\\
&= \lim_{t\to 0}\sum_m a_m e^{-\pi m^2 t},\\
&= \sum_m a_m,\\
&= f(0).
\end{aligned}$$

Hence, $\theta(x|it)$ converges, as a distribution, to the sum of the delta functions at all integral points x$\in\mathbb{Z}$ as $t\to 0$. The uniqueness of this solution results from the fact that $\lim_{t\to 0}\theta(x|it)=\sum_{m=-\infty}^{\infty}\delta_m(x)$, where δ_m is the distribution of Dirac at m. We have mentioned previously the convergence of the theta function. Thus, $\theta(x|it)$ may be seen as the fundamental solution to the heat equation when the space variable x lies on a circle $\mathbb{R}/\mathbb{Z}$. Similarly, the function $\theta_1(z)$ satisfies a third-order differential equation. Indeed, it is enough to use the relation $\wp(z)=-\frac{\partial^2}{\partial z^2}\log\theta_1(z)+C$, where C is a constant and $\wp(z)$ is the Weierstrass function.

Let us recall a few classic identities verified by theta functions (see, for example, Ohbuchi 1988):

PROPOSITION 7.5.– *The theta function satisfies the addition formulas:*

$$\theta\begin{bmatrix}\alpha\\ \beta\end{bmatrix}(z_1+z_2)=\sum_{2\delta\in(\mathbb{Z}_2)^g}\widehat{\theta}\begin{bmatrix}\frac{\alpha+\beta}{2}+\delta\\ \\ \gamma+\varepsilon\end{bmatrix}(2z_1).\widehat{\theta}\begin{bmatrix}\frac{\alpha-\beta}{2}+\delta\\ \\ \gamma-\varepsilon\end{bmatrix}(2z_2),$$

where $\alpha,\beta,\gamma,\varepsilon\in\mathbb{R}^g$,

$$\theta\begin{bmatrix}\alpha\\ \beta\end{bmatrix}(z)\equiv\theta\begin{bmatrix}\alpha\\ \beta\end{bmatrix}(z|B),\qquad \widehat{\theta}\begin{bmatrix}\alpha\\ \beta\end{bmatrix}(z)\equiv\theta\begin{bmatrix}\alpha\\ \beta\end{bmatrix}(z|2B).$$

$$\theta\begin{bmatrix}m_1\\ n_1\end{bmatrix}(z_1)\,\theta\begin{bmatrix}m_2\\ n_2\end{bmatrix}(z_2).\theta\begin{bmatrix}m_3\\ n_3\end{bmatrix}(z_3).\theta\begin{bmatrix}m_4\\ n_4\end{bmatrix}(z_4)$$

$$=\frac{1}{2^g}\sum_{2(a_1,a_2)\in(\mathbb{Z}_2)^{2g}}e^{-4\pi i\langle m_1,a_2\rangle}.\theta\begin{bmatrix}k_1+a_1\\ l_1+a_2\end{bmatrix}(w_1)...\theta\begin{bmatrix}k_4+a_1\\ l_4+a_2\end{bmatrix}(w_4),$$

where

$$(z_1,...,z_4)=\frac{1}{2}(w_1,...,w_4)\begin{pmatrix}\mathrm{I}&\mathrm{I}&\mathrm{I}&\mathrm{I}\\ \mathrm{I}&\mathrm{I}&-\mathrm{I}&-\mathrm{I}\\ \mathrm{I}&-\mathrm{I}&\mathrm{I}&-\mathrm{I}\\ \mathrm{I}&-\mathrm{I}&-\mathrm{I}&\mathrm{I}\end{pmatrix}.$$

Here, $\begin{pmatrix}m_1\\ n_1\end{pmatrix},...,\begin{pmatrix}m_4\\ n_4\end{pmatrix},\begin{pmatrix}k_1\\ l_1\end{pmatrix},...,\begin{pmatrix}k_4\\ l_4\end{pmatrix}$ *are the arbitrary vectors of order* $2g$ *with:*

$$\left(\begin{pmatrix}m_1\\ n_1\end{pmatrix},...,\begin{pmatrix}m_4\\ n_4\end{pmatrix}\right)=\frac{1}{2}\left(\begin{pmatrix}k_1\\ l_1\end{pmatrix},...,\begin{pmatrix}k_4\\ l_4\end{pmatrix}\right)\begin{pmatrix}\mathrm{I}&\mathrm{I}&\mathrm{I}&\mathrm{I}\\ \mathrm{I}&\mathrm{I}&-\mathrm{I}&-\mathrm{I}\\ \mathrm{I}&-\mathrm{I}&\mathrm{I}&-\mathrm{I}\\ \mathrm{I}&-\mathrm{I}&-\mathrm{I}&\mathrm{I}\end{pmatrix},$$

and I *denotes the unit matrix of order* g *or* $2g$.

In particular, we have the formulas:

$$\left(\theta\begin{bmatrix}0\\0\end{bmatrix}(z)\right)^2\cdot\left(\theta\begin{bmatrix}0\\0\end{bmatrix}(0)\right)^2=$$

$$\left(\theta\begin{bmatrix}0\\1/2\end{bmatrix}(z)\right)^2\cdot\left(\theta\begin{bmatrix}0\\1/2\end{bmatrix}(0)\right)^2+\left(\theta\begin{bmatrix}1/2\\0\end{bmatrix}(z)\right)^2\cdot\left(\theta\begin{bmatrix}1/2\\0\end{bmatrix}(0)\right)^2$$

and

$$\left(\theta\begin{bmatrix}1/2\\1/2\end{bmatrix}(z)\right)^2\cdot\left(\theta\begin{bmatrix}0\\0\end{bmatrix}(0)\right)^2=$$

$$\left(\theta\begin{bmatrix}0\\1/2\end{bmatrix}(z)\right)^2\cdot\left(\theta\begin{bmatrix}1/2\\0\end{bmatrix}(0)\right)^2-\left(\theta\begin{bmatrix}1/2\\0\end{bmatrix}(z)\right)^2\cdot\left(\theta\begin{bmatrix}0\\1/2\end{bmatrix}(0)\right)^2$$

as well as the Jacobi identity obtained by posing $z=0$,

$$\left(\theta\begin{bmatrix}0\\0\end{bmatrix}(0)\right)^4=\left(\theta\begin{bmatrix}0\\1/2\end{bmatrix}(0)\right)^4+\left(\theta\begin{bmatrix}1/2\\0\end{bmatrix}(0)\right)^4.$$

We will see how to express meromorphic functions on the torus $\mathbb{C}/\Lambda$, in terms of the theta function. Several approaches are possible:

Method 1: We know that any rational fraction, which therefore is a meromorphic function on $\mathbb{P}^1(\mathbb{C})$, can be written in the form:

$$f(z)=\prod_{j=1}^{m}\frac{z-p_j}{z-q_j}.$$

By analogy, let $p_1,\ldots,p_m,q_1,\ldots,q_m$ be points of the Riemann surface C and $f(z)$ be a function with zeros at the points $p_1,\ldots,p_m$ and poles at the points $q_1,\ldots,q_m$. Recall that in general, according to Abel's theorem (see Appendix 9 or any other work on Riemann surfaces), there exists a meromorphic function f such that:

$$(f)=\sum_{j=1}^{m}q_j-\sum_{j=1}^{m}p_j,$$

if and only if there exists a closed path γ such that for any differential form ω on C,

$$\sum_{j=1}^{m}\int_{p_j}^{q_j}\omega=\int_\gamma\omega.$$

We assume that one of these conditions is satisfied. Moreover, we know (see Appendix, Chapter 9) that the dimension of the space of holomorphic differentials on a Riemann surface of genus g is equal to g. Since $\mathcal{C}$ has genus 1, then there exists a single holomorphic differential ω on $\mathcal{C}$. Still according to Abel's theorem, the existence of the function $f(z)$ imposes the condition $\sum_{j=1}^{m} p_j = \sum_{j=1}^{m} q_j$. Note that for $m = 1$; $p_1 = q_1$, the only valid case is $f(z) = \text{constant}$.

In the case where $m \geq 2$, the function $f(z)$ is expressed as a function of θ using the formula:

$$f(z) = c \prod_{j=1}^{m} \frac{\theta\left(z - p_j - \frac{1}{2}(1+b)\right)}{\theta\left(z - q_j - \frac{1}{2}(1+b)\right)},$$

where c is a constant. Note that $f(z+1) = f(z)$. From relation [7.1] and the fact that $\sum_{j=1}^{m} p_j = \sum_{j=1}^{m} q_j$, we also have $f(z+b) = f(z)$. Therefore, f is doubly periodic. The function f is meromorphic with zeros at $q_j + \frac{1+b}{2}$ and poles at $p_j + \frac{1+b}{2}$.

Method 2: The function $\log \theta(z)$ is equal to the sum of a doubly periodic function, with periods 1 and b, and a linear function. Therefore, the function $\frac{d^2}{dz^2} \log \theta(z)$ is doubly periodic and meromorphic on $\mathcal{C}$, with a double pole at $z = \frac{1+b}{2}$. This function coincides with the Weierstrass function $\wp(z)$:

$$\wp(z) = -\frac{d^2}{dz^2} \log \theta(z) + c, \qquad [7.5]$$

where c is a constant chosen such that the Laurent series expansion of $\wp(z)$ at $z = 0$ has no constant term. Using relation [7.1] as well as the differential equation: $\left(\wp'(z)\right)^2 = 4\left(\wp(z)\right)^3 - g_2\wp(z) - g_3$ (see Appendix, Chapter 9), we show that the function $\theta(z)$ satisfies a third-order differential equation.

Method 3: Meromorphic functions with simple poles on the projective line $\mathbb{P}^1(\mathbb{C})$ can be written in the form:

$$f(z) = \sum_j \frac{\lambda_j}{z - p_j} + c,$$

where $\lambda_j \in \mathbb{C}$ and c is a constant. By analogy, we consider on $\mathcal{C}$ the function:

$$f(z) = \sum_j \lambda_j \frac{d}{dz} \log \theta(z - p_j) + c,$$

where $p_j \in \mathcal{C}$, $\lambda_j \in \mathbb{C}$ such that: $\sum_j \lambda_j = 0$ and c is a constant. This function is doubly periodic and meromorphic with simple poles at the points $p_j + \frac{1+b}{2}$ and residues λ_j at these same points.

We have seen how meromorphic functions on the torus $\mathbb{C}/\Lambda$ can be expressed in terms of the theta function. Moreover, for $g = 1$, we know that: $\mathcal{C} \simeq \mathbb{C}/\Lambda \simeq \text{Jac}(\mathcal{C})$. Therefore, the construction that was done previously on the torus $\mathbb{C}/\Lambda$ or what amounts to the same thing (since $\mathcal{C}$ is an elliptic curve) on the Jacobian variety $\text{Jac}(\mathcal{C})$ is also valid on the Riemann surface $\mathcal{C}$. For example, consider the case of a function with poles at $p_1, ..., p_m$ and zeros at $q_1, ..., q_m$ on the Riemann surface $\mathcal{C}$. According to Abel's theorem, we have $\sum_{j=1}^m \varphi(p_j) = \sum_{j=1}^m \varphi(q_j)$, and we can, using Method 1 described above, express the function $f(p)$ in terms of the theta function using the formula:

$$f(p) = c\prod_{j=1}^{m} \frac{\theta\left(\varphi(p) - \varphi(q_j) - \frac{1}{2}(1+b)\right)}{\theta\left(\varphi(p) - \varphi(p_j) - \frac{1}{2}(1+b)\right)}.$$

Let us now study the case where the Riemann surface $\mathcal{C}$ is of genus $g > 1$. When studying Abel's theorem as well as the well-known Jacobi inversion problem (see Appendix, Chapter 9), it is equivalent to say that this consists of determining g points $p_1, ..., p_g$ on $\mathcal{C}$ such that:

$$\sum_{k=1}^{g} \int_{p_0}^{p_k} \omega_j \equiv z_j \pmod{\Lambda}, \quad j = 1, ..., g$$

where $(z_1, ..., z_g) \in \text{Jac}(\mathcal{C})$, $(\omega_1, ..., \omega_g)$ is a basis of holomorphic differentials on $\mathcal{C}$, p_0 is a basis point on $\mathcal{C}$, and Λ is a lattice generated by the column vectors of the period matrix. In other words, the problem is to determine the divisor $\mathcal{D} = \sum_{j=1}^g p_j$ in terms of $z = (z_1, ..., z_g) \in \text{Jac}(\mathcal{C})$ such that if:

$$\varphi : \text{Div}(\mathcal{C}) \longrightarrow \text{Jac}(\mathcal{C}), \quad \mathcal{C} \longmapsto \left(\int_0^{\mathcal{C}} \omega_1, ..., \int_0^{\mathcal{C}} \omega_g\right),$$

is the Abel–Jacobi map, where $(\omega_1, ..., \omega_g)$ is a normalized basis of the space of holomorphic differential forms on $\mathcal{C}$, then the equation $\varphi(\mathcal{D}) = z$ is satisfied. We will study the Jacobi inversion problem using theta functions.

Let $(a_1, \ldots, a_g, b_1, \ldots, b_g)$ be a basis of cycles (closed contours) in the homology group $H_1(\mathcal{C}, \mathbb{Z}) \simeq \mathbb{Z}^{2g}$ such that the intersection products of cycles two by two are written: $(a_j, a_j) = (b_j, b_j) = 0$, $(a_j, b_k) = \delta_{jk}$, $1 \leq j, k \leq g$. We will say that $(a_1, \ldots, a_g, b_1, \ldots, b_g)$ is a symplectic basis of the homology group $H_1(\mathcal{C}, \mathbb{Z})$. We denote by $\mathcal{C}^*$ the normal representation of the Riemann surface $\mathcal{C}$ of genus g, i.e. a polygon with $4g$ sides identified two by two according to the symbol $a_1 b_1 a_1^{-1} b_1^{-1} ... a_g b_g a_g^{-1} b_g^{-1}$ and which can be defined from a triangulation of the surface $\mathcal{C}$. The $4g$ sides are named in the order they appear on the oriented edge of the polygon, a_1^{-1} being identified with a_1 after reversing its orientation, etc. (see Figure 7.1 for $g = 2$).

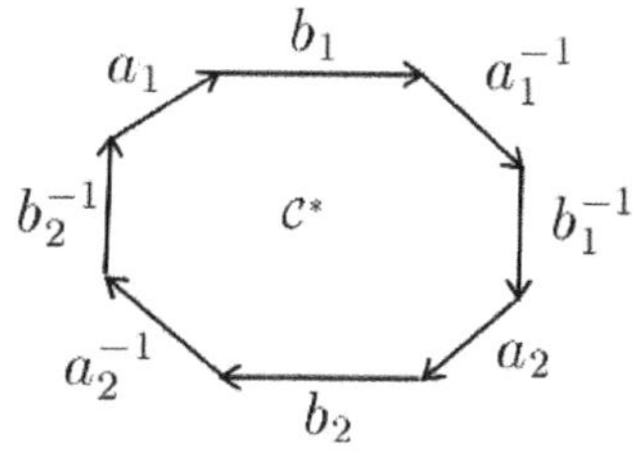

Figure 7.1. *Normal representation (genus two)*

PROPOSITION 7.6.– *If the function defined by:*

$$\zeta(p) = \theta(\varphi(p) - c), \quad c \in \mathbb{C}^g,$$

is not identically zero, then it admits g zeros (counted with their order of multiplicity) on the normal representation C^* of C. Furthermore, if $p_1, ..., p_g$ denote the zeros of this function, then we have on the Jacobian variety $Jac(C)$ the formula:

$$\sum_{k=1}^{g} \varphi(p_k) \equiv c - \delta \quad (\textit{mod. periods}),$$

where $\delta \in \mathbb{C}^g$ is the vector (called the Riemann constant vector) defined by:

$$\delta_j = \frac{1}{2}(1 + b_{jj}) - \sum_{k \neq j} \left(\int_{a_k} \omega_k(p) \int_{p_0}^{p} \omega_j \right), \quad j = 1, ..., g. \qquad [7.6]$$

In general, the vector δ depends on p_0 except in the special case $g = 1$ where $\delta = \frac{1+b}{2}$. We show that $2\delta = -\varphi(\mathcal{K})$, where $\mathcal{K}$ is the canonical divisor. Therefore, by skillfully choosing the point p_0, we can express $\mathcal{K}$ in a very simple way. For example, consider the case where C is a hyperelliptic curve of genus g with affine equation:

$$w^2 = \prod_{j=1}^{2g+2} (\xi - \xi_j),$$

where all ξ_j are distinct. Let $(a_1, \ldots, a_g, b_1, \ldots, b_g)$ be a symplectic basis of the homology group $H_1(C, \mathbb{Z})$), and let:

$$\sigma : C \longrightarrow C, \quad (w, \xi) \longmapsto (-w, \xi),$$

be the hyperelliptic involution (which consists of exchanging the two sheets of the curve C) with $\sigma(a_j) = -a_j$ and $\sigma(b_j) = -b_j$. Note that:

$$\int_{a_j} \omega_k = -\int_{\sigma(a_j)} \omega_k = -\int_{a_j} \sigma^* \omega_k.$$

So, by choosing $p_0 = \xi_1$, we obtain:

$$\delta_j = \frac{1}{2}(1+b_{jj}) + \sum_{k\neq j}\int_{a_k}\omega_k\left(\int_{\xi_1}^{\xi_{2k+1}}\omega_j + \int_{\xi_{2k+1}}^{p}\omega_j\right), \quad j = 1,...,g,$$

$$= \frac{1}{2}(1+b_{jj}) + \sum_{k\neq j}\int_{\xi_1}^{\xi_{2k+1}}\omega_j\int_{a_k}\omega_k$$

$$+\sum_{k\neq j}\int_{\xi_{2k+1}}^{\xi_{2k+2}}\left(\left(\int_{\xi_{2k+1}}^{p}\omega_j\right)\omega_k(p)\left(\int_{\xi_{2k+1}}^{\sigma p}\omega_j\right)\omega_k(\sigma p)\right).$$

Taking into account the fact that $\omega_k(\sigma p) = -\omega_k(p)$ and modulo a linear combination $n + Bm$ (a lattice generated by the column vectors of the period matrix), we finally obtain for this specific case the formula:

$$\delta_j = \sum_{k=1}^{g} b_{jk} + \frac{j}{2}, \quad j = 1,...,g.$$

The zeros of a theta function on $\mathbb{C}^g$ form a submanifold of Jac(C) of dimension $g-1$, denoted $\Theta = \{z : \theta(z) = 0\}$, called the theta divisor. It is invariant under a finite number of translations and can be singular. Equation [7.2] implies that Θ is well defined on the Jacobian manifold Jac(C). Since $\theta(-z) = \theta(z)$, we deduce that Θ is symmetric: $-\Theta = \Theta$.

The Riemann fundamental theorem (Mumford 1983) is stated as follows:

THEOREM 7.1 (Riemann).– *The function* $\zeta(p) = \theta(\varphi(p) - c)$, $c \in \mathbb{C}^g$, *is either identically zero, or admits exactly g zeros* $q_1, ..., q_g$ *on* C *such that:*

$$\sum_{j=1}^{g}\varphi(q_j) = c + \delta,$$

where δ *is defined by equation [7.6].*

REMARK 7.1.– *Let* C *be a compact Riemann surface of genus* $g > 0$. *We denote by* $Sym^d C$ *the set of all positive divisors* $\mathcal{D} = \sum_{j=1}^{d} p_j$, *of degree d on* C. *We say that* $Sym^d C$ *is the* d^{th} *symmetric product of* C. *We show that* $Sym^d C$ *can be equipped with a complex manifold structure. Let* $C^d = C \times ... \times C$ *be the direct product of* C*; it is a complex manifold. Consider the symmetric group* Σ_d *of permutations of* $\{1, ..., d\}$. *Then, for all* $\sigma \in \Sigma_d$, *we define the map* $\sigma : C^d \longrightarrow C^d$, *by setting* $\sigma(p_1, ..., p_d) = (p_{\sigma_1}, ..., p_{\sigma_d})$, *where* $\sigma_1, ..., \sigma_d$ *denote the elementary symmetric functions. This map is biholomorphic, hence* $\Sigma_d \subset Aut(C^d)$, *i.e.* Σ_d *is a subgroup of the automorphism* *group of C^d. Note that $Sym^d C$ inherits from C^d a topological space structure. The quotient space $Sym^d C = C^d/\Sigma_d$ is a compact separated space. The projection π : $C^d \longrightarrow Sym^d C$ provides $Sym^d C$ with a complex manifold structure. Indeed, let $p_j \in C$, $\mathcal{D} = \sum p_j \in Sym^d C$, $p_j \neq p_k$. Around each p_j, we choose a local coordinate system (U_j, z_j) in C. We assume that for $p_j \neq p_k$, we have $U_j \cap U_k \neq 0$ and that for $p_j = p_k$, we have $z_j = z_k$ in $U_j = U_k$. The map*

$$\sum q_j \longmapsto (\sigma_1(z_j(q_j)), ..., \sigma_d(z_j(q_j))),$$

determines, according to the fundamental theorem of algebra, a local chart on $\pi(U_1 \times ... \times U_d) \subset Sym^d C$. Apart from the branch points of the surface, the map π is a covering, and we can take $(z_1(p_1), ..., z_d(p_d))$ as coordinates around $\mathcal{D} \in Sym^d C$. Therefore, around a point d.p, the set $(z_1 + \cdots + z_d, ..., z_1...z_d)$, forms a local coordinate system.

The result mentioned in the previous theorem can be explained by noting that when we embed the Riemann surface C into its Jacobian manifold Jac(C) via the Abel map φ, then either its image is entirely included in the theta divisor, or it meets it at exactly g points. In fact, if $\zeta(p)$ is not identically zero on C, then its zeros coincide with the points $p_1, ..., p_g$ and determine the solution to the inverse Jacobi problem $\varphi(\mathcal{D}) = z$ for the vector $z = c - \delta$. Note that the special divisors of the form:

$$\mathcal{D} = p_1 + \cdots + p_N, \quad N = \deg \mathcal{D} \geq g,$$

coincide with the critical points of the Abel–Jacobi map restricted to the space $\text{Sym}^N C$,

$$\text{Sym}^N C \longrightarrow \text{Jac}(C), \quad \mathcal{D} \longmapsto \left(\int_0^{\mathcal{D}} \omega_1, ..., \int_0^{\mathcal{D}} \omega_N \right),$$

or what amounts to the same thing, $\varphi(p_1, ..., p_N) = \varphi(p_1) + \cdots + \varphi(p_N)$. These critical points are the points $p_1, ..., p_N$ where the rank of the differential of this map is less than g. According to the Riemann fundamental theorem, the function $\zeta(p) = \theta(\varphi(p) - c)$, is identically zero if and only if $c \equiv \varphi(q_1) + \cdots\cdots + \varphi(q_g) + \delta$, where $q_1 + \cdots + q_g$ is a special divisor.

PROPOSITION 7.7.– *Let $z = (z_1, ..., z_g) \in \mathbb{C}^g$ be a vector such that the function $\zeta(p) = \theta(\varphi(p) - z - \delta)$, is not identically zero on C. Then, the function $\zeta(p)$ has exactly g zeros $p_1, ..., p_g$ on C which determine the solution to the Jacobi inversion problem $\varphi(\mathcal{D}) = z$, where $\mathcal{D} = \sum_{j=1}^{g} p_j$. In other words, we have:*

$$\varphi_j(p_1) + \cdots + \varphi_j(p_g) = \sum_{k=1}^{g} \int_{p_0}^{p_k} \omega_j \equiv z_j, \quad j = 1, ..., g.$$

Furthermore, the divisor $\mathcal{D}$ is non-special, and the points $p_1, ..., p_g$ are uniquely determined from the above system.

We have the following characterization of the divisor theta:

PROPOSITION 7.8.– $\theta(c) = 0$, *if and only if there exist* $p_1, ..., p_{g-1} \in C$ *such that:*

$$c \equiv \varphi(p_1) + \cdots + \varphi(p_{g-1}) + \delta = \sum_{j=1}^{g-1} \int_{p_0}^{p_j} \omega + \delta$$

with base point p_0 *and* δ *is the Riemann constant vector.*

PROPOSITION 7.9.– *Let* $\mathcal{D}_1$ *be a non-special divisor of degree* g, $\mathcal{D}_2$ *be a positive divisor of degree* n, $(\omega_1, ..., \omega_g)$ *be a basis of holomorphic differentials on* C, $\varphi(p) = \left(\int_{p_0}^{p} \omega_1, ..., \int_{p_0}^{p} \omega_g\right)$ *be the Abel map with base point* p_0, η *be a normalized differential of third kind*[1] *on* C *having poles on* $\mathcal{D}_2$ *and residues* -1, $u = (u_1, ..., u_g)$ *be the vector of periods with* $u_k = \int_{b_k} \eta$ *and, finally,* δ *be the vector defined using the Riemann constants by equation [7.6]. If* ψ *is a meromorphic function on* C *having* $g + n$ *poles on* $\mathcal{D}_1 + \mathcal{D}_2$*, then this function is expressed in terms of theta function using the formula:*

$$\psi(p) = a \frac{\theta(\varphi(p) - \varphi(\mathcal{D}) + u - \delta)}{\theta(\varphi(p) - \varphi(\mathcal{D}_1) - \delta)} e^{\int_{p_0}^{p} \eta},$$

where a *is a constant, and* δ *is the Riemann constant vector.*

On the Riemann surface C of genus g, singular functions possessing g poles and essential singularities play a crucial role in the study of integrable systems, in particular the Korteweg–de Vries equation (K-dV):

$$\frac{\partial u}{\partial t} - 6u\frac{\partial u}{\partial x} + \frac{\partial^3 u}{\partial x^3} = 0,$$

the Kadomtsev–Petviashvili equation (KP):

$$\frac{\partial^2 u}{\partial y^2} - \frac{\partial}{\partial x}\left(4\frac{\partial u}{\partial t} - 12u\frac{\partial u}{\partial x} - \frac{\partial^3 u}{\partial x^3}\right) = 0,$$

the nonlinear Schrödinger equation:

$$i\frac{\partial \psi}{\partial t} + \frac{\partial^2 \psi}{\partial x^2} + \psi|^2\psi = 0,$$

the Boussinesq equation:

$$\frac{\partial^2 u}{\partial t^2} - \frac{\partial^2 u}{\partial x^2} + \frac{\partial^4 u}{\partial x^4} + \frac{\partial^2 u^2}{\partial x^2} = 0,$$

1 A meromorphic (Abelian) differential on C is said to be of the third kind if it has only simple poles (and its residue is non-zero at at least one point of C).

and the Camassa–Holm equation:

$$\frac{\partial u}{\partial t} - \frac{\partial^3 u}{\partial t \partial x^2} + 3u\frac{\partial u}{\partial x} = 2\frac{\partial u}{\partial x}\frac{\partial^2 u}{\partial x^2} + u\frac{\partial^3 u}{\partial x^3},$$

whose exact solutions are solitons, i.e. self-reinforcing solitary waves that maintain their shape as they propagate at a constant rate. We will see by analogy from the previous theorem how to express these functions (known as Baker–Akhiezer functions) in terms of theta functions and at the same time prove their existence. Let $Q_1, ..., Q_n$ be the points on a Riemann surface $\mathcal{C}$ of genus g and z_j be the local parameters such that: $z_j(Q_j) = \infty$. We associate with each point Q_j an arbitrary polynomial denoted by $q_j(z_j)$. Let $\mathcal{D} = P_1 + \cdots + P_g$, be a positive divisor on $\mathcal{C}$ and $\psi(P)$ a function (called the Baker–Akhiezer function) satisfying the following conditions: (i) $\psi(P)$ is meromorphic on $\mathcal{C} \setminus \{Q_1, ..., Q_n\}$ and admits poles only at the points $p_1, ..., p_n$ of the divisor $\mathcal{D}$ and (ii) the function $\psi(P)e^{-q_j(z_j(P))}$ is analytic in the neighborhood of Q_j, $j = 1, ..., n$. The condition (ii) can be replaced by this condition: the function ψ admits an essential singularity of the form $\psi(P) \sim ce^{q_j(z_j(P))}$ at the points Q_j, $j = 1, ..., n$, where c is a constant. These functions $\psi(P)$ form a vector space that we note $L \equiv L(\mathcal{D}; Q_1, ..., Q_n, q_1, ..., q_n)$.

PROPOSITION 7.10.– *Let $\mathcal{D} = p_1 + \cdots + p_g$ be a non-special divisor of degree g. Then, the space L is of dimension 1 and its basis is described using:*

$$\psi_1(P) = \frac{\theta\left(\varphi(P) - \varphi(\mathcal{D}) + V - \delta\right)}{\theta\left(\varphi(p) - \varphi(\mathcal{D}) - \delta\right)} e^{\int_{p_0}^{P} \eta}, \qquad [7.7]$$

where η is a normalized differential of the second kind having poles at the points $Q_1, ..., Q_n$, the main parts coincide with the polynomials $q_j(z_j)$, $j = 1, ..., n$, $V = (V_1, ..., V_g)$ with $V_k = \int_{b_k} \eta$, $k = 1, ..., g$, $\varphi(P) = \left(\int_{p_0}^{p} \omega_1, ..., \int_{p_0}^{p} \omega_g\right)$, and the Abel map with base point p_0, δ is defined by Riemann constants [7.6]. The integration contour in integrals $\int_{p_0}^{p} \eta$ and $\int_{p_0}^{p} \omega_j$, $j = 1, ..., g$ is the same.

REMARK 7.2.– *Let $\mathbb{T} = E/\Lambda$ be a complex torus where Λ is a lattice of the complex vector space E. By the Appell–Humbert theorem (theorem 4.1), any line bundle over the torus $\mathbb{T}$ is of the form $\mathcal{L}(h, \alpha)$ where h is a Riemann form (definition 4.5 or exercise 4.2) on E and α a semi-character (definition 4.1) for the form h. The sections of $\mathcal{L}(h, \alpha)$ are holomorphic functions on the space E. Moreover, these functions satisfy a functional equation of the form:*

$$\theta(z + \lambda) = \alpha(\lambda)e^{\pi h(z,\lambda) + \frac{\pi}{2}h(\lambda,\lambda)}.$$

These functions are none other than the theta functions for $\mathcal{L}(h, \alpha)$.

We will now cite a few examples of how theta functions can be applied.

EXAMPLE 7.2.– *Consider the sine-Gordon equation:*

$$\frac{\partial^2 u}{\partial t^2} - \frac{\partial^2 u}{\partial x^2} + \sin u = 0,$$

where $u = u(x,t)$. Its solution can be written in the form (Belokolos et al. 1994):

$$u(x,t) = 2i \ln \frac{\theta \begin{bmatrix} \alpha \\ \beta \end{bmatrix} (Ux + Vt + W|B)}{\theta \begin{bmatrix} 0 \\ 0 \end{bmatrix} (Ux + Vt + W|B)} + C + 2\pi m,$$

where $U, V, W \in \mathbb{C}^g, C \in \mathbb{R}, m \in \mathbb{Z}$.

EXAMPLE 7.3.– *We cite yet another example which concerns the Landau–Lifshitz equation (Belokolos et al. 1994):*

$$\frac{\partial S}{\partial t} = S \times \frac{\partial^2 S}{\partial x^2} + S \times JS,$$

where $S = (S_1, S_2, S_3)$, $S_1^2 + S_2^2 + S_3^2 = 1$ and $J = diag(J_1, J_2, J_3)$. This equation describes the effects of a magnetic field on ferromagnetic materials. The real solutions (with magnetic anisotropy of the axis of easy magnetization type) are given by:

$$S_1 = \frac{\theta(\omega + d + m)\theta(\omega + d + m + r) - \theta(\omega + d)\theta(\omega + d + r)}{\theta(\omega + d)\theta(\omega + d + m + r) - \theta(\omega + d + r)\theta(\omega + d + m)},$$

$$S_2 = -i\frac{\theta(\omega + d + m)\theta(\omega + d + m + r) + \theta(\omega + d)\theta(\omega + d + r)}{\theta(\omega + d)\theta(\omega + d + m + r) - \theta(\omega + d + r)\theta(\omega + d + m)},$$

$$S_3 = \frac{\theta(\omega + d)\theta(\omega + d + m + r) + \theta(\omega + d + r)\theta(\omega + d + m)}{\theta(\omega + d)\theta(\omega + d + m + r) - \theta(\omega + d + r)\theta(\omega + d + m)}.$$

Here, the theta function is linked to a hyperelliptic curve of genus g,

$$w^2 = (\lambda^2 - a^2) \prod_{j=1}^{2g} (\lambda - e_j),$$

whose cycle $\sum a = a_1 + \cdots + a_g$ encircles the cut $[-a, a]$ and $r = \int_{\infty^+}^{\infty^-} du$ where the integration path crosses the cycle $\sum a$, the vector $d \in \mathbb{C}^g$ is such that $Im\, d = -\frac{1}{2} Im\, r$, $m = (m_1, ..., m_g)$, $\omega = \frac{1}{2\pi}(Vx + Wt)$.

In Chapter 8, we will see other important examples where theta functions are used in the study of integrable systems associated with compact Riemann surfaces.

7.2. Lefschetz theorem

We will give a direct proof of the projectivity criterion of complex tori using theta functions. This is the subject of the Lefschetz theorem, which states that if a line bundle $\mathcal{L}$ over an Abelian variety $\mathcal{A}$ is positive, then $H^0\left(\mathcal{A}, \mathcal{O}(\mathcal{L}^k)\right)$ has no base point for $k \geq 2$ and provides a projective embedding for $k \geq 3$.

THEOREM 7.2.– *Let $\mathbb{T}^n = \mathbb{C}^n/\Lambda$ be a complex torus where Λ is a lattice of $\mathbb{C}^n$, $\mathcal{L}$ is a positive line bundle on $\mathbb{T}^n$ and $\delta_1, ..., \delta_n$ are the elementary divisors of the polarization $c_1(\mathcal{L})$. Then, these divisors are related by the formula*

$$\dim H^0\left(\mathbb{T}^n, \mathcal{O}(\mathcal{L}^k)\right) = \prod_{\alpha=1}^{n} \delta_\alpha.$$

Furthermore, $H^0\left(\mathbb{T}^n, \mathcal{O}(\mathcal{L}^k)\right)$ has no base point for $k \geq 2$ and provides a projective embedding for $k \geq 3$.

PROOF.– Let $(\lambda_1, ..., \lambda_{2n})$ be a basis of Λ and $(x_1, ..., x_{2n})$ be the corresponding dual basis. We have already shown (see proof of proposition 4.5) that we have:

$$c_1(\mathcal{L}) = \sum_{\alpha=1}^{n} \delta_\alpha dx_\alpha \wedge dx_{n+\alpha},$$

and that, in addition, we can choose a basis $(e_1, ..., e_n)$ of $\mathbb{C}^n$ by setting $e_\alpha = \delta_\alpha^{-1}\lambda_\alpha$, $\alpha = 1, ..., n$, in such a way that the period matrix Ω is written:

$$\Omega = (\Delta_\delta, Z), \quad \Delta_\delta = \text{diag}\,(\delta_1, ..., \delta_n), \quad \delta_\alpha \in \mathbb{N}^*, \quad Z^\top = Z, \quad \text{Im}\, Z > 0.$$

By proposition 4.6, the line bundle $\mathcal{L}$ is a translation of the bundle $\mathcal{L}_0$ of multiples $e_{\lambda_\alpha} \equiv 1$, $e_{\lambda_{n+\alpha}}(z) = e^{-2\pi i z_\alpha}$. Let:

$$\mathcal{L} = \tau_\mu^* \mathcal{L}_0, \qquad \mu = \frac{1}{2}\sum Z_{\alpha\alpha}.e_\alpha \in \mathbb{T}^n,$$

and $\tau_\mu : \mathbb{T}^n \longrightarrow \mathbb{T}^n$, be a translation. The multiples of $\mathcal{L}$ are:

$$e_{\lambda_\alpha} \equiv 1, \qquad e_{\lambda_{n+\alpha}}(z) = e^{-2\pi i z_\alpha - \pi i Z_{\alpha\alpha}},$$

and hence, the sections of $\mathcal{L}$ are given (via the pullback of $\mathbb{C}^n$) by the functions θ with:

$$\theta(z + \lambda_\alpha) = \theta(z), \qquad [7.8]$$

$$\theta(z + \lambda_{n+\alpha}) = e^{-2\pi i z_\alpha - \pi i Z_{\alpha\alpha}}.\theta(z). \qquad [7.9]$$

According to the first periodicity condition, the function θ has a Fourier series expansion of the form:

$$\theta(z) = \sum_{l \in \mathbb{Z}^n} a_l.e^{2\pi i \left\langle l, \Delta_\delta^{-1} z\right\rangle}.$$

So,

$$\theta(z+\lambda_{n+\alpha}) = \sum_{l\in\mathbb{Z}^n} a_l.e^{2\pi i\langle l,\Delta_\delta^{-1}(z+\lambda_{n+\alpha})\rangle},$$
$$= \sum_{l\in\mathbb{Z}^n} a_l.e^{2\pi i\langle l,\Delta_\delta^{-1}\lambda_{n+\alpha}\rangle}.e^{2\pi i\langle l,\Delta_\delta^{-1}z\rangle}.$$

Also, the second functional equation [7.9] is written:

$$\theta(z+\lambda_{n+\alpha}) = e^{-2\pi i z_\alpha - \pi i Z_{\alpha\alpha}} \sum_{l\in\mathbb{Z}^n} a_l e^{2\pi i\langle l,\Delta_\delta^{-1}z\rangle},$$
$$= \sum_{l\in\mathbb{Z}^n} a_{l+\delta_\alpha e_\alpha}.e^{-\pi i Z_{\alpha\alpha}}.e^{2\pi i\langle l,\Delta_\delta^{-1}z\rangle}.$$

Identifying the coefficients of the basic functionals, we find the recurrence relation:

$$a_{l+\delta_\alpha e_\alpha} = e^{2\pi i\langle l,\Delta_\delta^{-1}\rangle + \pi i Z_{\alpha\alpha}}.a_l.$$

Knowing the a_l, $0 \le l_\alpha \le \delta_\alpha$, the other coefficients are deduced inductively, the function θ is completely determined, and therefore,

$$\dim H^0\left(\mathbb{T}^n, \mathcal{O}(\mathcal{L}^k)\right) \le \prod_{\alpha=1}^n \delta_\alpha.$$

We will show that this dimension is exactly $\prod_{\alpha=1}^n \delta_\alpha$. Let $\mathcal{L} \longrightarrow \mathbb{T}^n$ be mainly polarized and normalized as before. We have:

$$H^0\left(\mathbb{T}^n, \mathcal{O}(\mathcal{L}^k)\right)$$
$$=$$
$$\left\{\theta : \theta(z+e_\alpha) = \theta(z), \theta(z+\lambda_{z+\lambda_{n+\alpha}}) = e^{-2\pi i k\left(z_\alpha + \frac{Z_{\alpha\alpha}}{2}\right)}.\theta(z)\right\}.$$

If θ is the Riemann theta function on $\mathcal{L}$, then:

$$\Theta_\mu(z) = \theta(z+\mu)\theta(z-\mu) \in H^0\left(\mathbb{T}^n, \mathcal{O}(\mathcal{L}^2)\right), \quad \forall \mu \in \mathbb{T}^n.$$

For any z, we can choose μ such that $\Theta_\mu(z) \neq 0$, or precisely so that $\theta(z \pm \mu) \neq 0$, i.e. so that the $z \pm \mu$ do not belong to the variety $V \neq \emptyset$ of zeros of θ, or which amounts to the same thing as $\mu \notin V - z$ and $\mu \notin z - V$. Since the function θ is holomorphic and not identically zero, the variety V has no interior point, and the same is true of the closed sets $V - z$ and $z - V$. Therefore, the union of the latter has no interior point, and it suffices to choose μ from their complement, which is a dense open set. It follows that the space $H^0\left(\mathbb{T}^n, \mathcal{O}(\mathcal{L}^2)\right)$ has no base point, the map $\psi_{\mathcal{L}^2} : \mathbb{T}^n \longrightarrow \mathbb{P}^N(\mathbb{C})$ is a morphism. Also, for any basis $(\theta_0, ..., \theta_N)$ of the space $H^0\left(\mathbb{T}^n, \mathcal{O}(\mathcal{L}^3)\right)$, the θ_α have no common zero and the map $z \longmapsto (\theta_\alpha(z))$ defines a map $\psi_{\mathcal{L}^3} : \mathbb{T}^n \longrightarrow \mathbb{P}^N(\mathbb{C})$. Let us show that this is an embedding. We first start by showing that the map is of maximum rank, i.e.

$$r \equiv \text{rang} \begin{pmatrix} \theta_0(z) & \theta_1(z) & \dots & \theta_N(z) \\ \frac{\partial \theta_0}{\partial z_1}(z) & \frac{\partial \theta_1}{\partial z_1}(z) & \dots & \frac{\partial \theta_N}{\partial z_1}(z) \\ \vdots & \vdots & \ddots & \vdots \\ \frac{\partial \theta_0}{\partial z_n}(z) & \frac{\partial \theta_1}{\partial z_n}(z) & \dots & \frac{\partial \theta_N}{\partial z_n}(z) \end{pmatrix} = n + 1,$$

on $\mathbb{C}^n$ or that the vectors $\overrightarrow{u} = (\theta_0, ..., \theta_N)(z)$ and $\overrightarrow{v}_\alpha = \left(\frac{\partial \theta_0}{\partial z_\alpha}, ..., \frac{\partial \theta_N}{\partial z_\alpha}\right)(z)$, $\alpha = 1, ..., n$, are $\mathbb{C}$-linearly independent and therefore that the map $\psi_{\mathcal{L}^3}$ is an immersion. Assume that $r < n + 1$ in $z_0 \in \mathbb{C}^n$, fixed. In other words, a linear relationship between the vectors $\overrightarrow{u}$ and $\overline{v}_\alpha$ is written:

$$a_0 \overrightarrow{u} = \sum_{\alpha=1}^{n} a_\alpha \overrightarrow{v}_\alpha, \qquad [7.10]$$

or even

$$a_0 \theta_\beta(z_0) = \sum_{\alpha=1}^{n} a_\alpha \frac{\partial \theta_\beta}{\partial z_\alpha}(z_0), \quad \beta = 1, ..., n \qquad [7.11]$$

Hence, for all $\Theta \in H^0\left(\mathbb{T}^n, \mathcal{O}(\mathcal{L}^3)\right)$, we have also:

$$a_0 \Theta(z_0) = \sum_{\alpha=1}^{n} a_\alpha \frac{\partial \Theta}{\partial z_\alpha}(z_0).$$

Let

$$\Theta(z, \mu, \nu) = \theta(z + \mu)\theta(z + \nu)\theta(z - \mu - \nu) \in H^0\left(\mathbb{T}^n, \mathcal{O}(\mathcal{L}^3)\right), \qquad [7.12]$$

be the product of the translated Riemann theta function. Hence, for all μ, ν, we have:

$$a_0 \Theta(z_0, \mu, \nu) = \sum_{\alpha=1}^{n} a_\alpha \frac{\partial \Theta}{\partial z_\alpha}(z_0, \mu, \nu), \quad z_0 \text{ fixed.}$$

Let us define the meromorphic function:

$$\varphi(z) = \sum a_\alpha \frac{\partial \log \theta(z)}{\partial z_\alpha},$$

from which, for z_0 fixed and all μ, ν,

$$\begin{aligned}
&\varphi(z_0+\mu)+\varphi(z_0+\nu)+\varphi(z_0-\mu-\nu)\\
&=\sum a_\alpha \frac{\partial \log\theta(z_0+\mu)}{\partial z_\alpha}+\sum a_\alpha \frac{\partial \log\theta(z_0+\nu)}{\partial z_\alpha}\\
&\qquad+\sum a_\alpha \frac{\partial \log\theta(z_0-\mu-\nu)}{\partial z_\alpha},\\
&=\sum a_\alpha \frac{\partial \log\Theta}{\partial z_\alpha}(z_0,\mu,\nu),\\
&=\frac{1}{\Theta(z_0,\mu,\nu)}\sum a_\alpha \frac{\partial\Theta}{\partial z_\alpha}(z_0,\mu,\nu),\\
&=\frac{1}{\Theta(z_0,\mu,\nu)}.a_0\Theta(z_0,\mu,\nu)=a_0.
\end{aligned}$$

The meromorphic function φ is such that $\varphi(z_0+\mu)+\varphi(z_0+\nu)+\varphi(z_0-\mu-\nu)$ is independent of μ and ν. For any μ, we can find ν such that $z+\nu$ and $z-\mu-\nu$ do not belong to the set of poles of φ, i.e. such that:

$$\varphi(z_0+\nu)\neq\infty, \qquad \varphi(z_0-\mu-\nu)\neq\infty.$$

Therefore, $\varphi(z_0+\mu)$ is a holomorphic function for all μ. On the contrary, using relations [7.8] and [7.9], we immediately obtain:

$$\varphi(z+e_\alpha)=\varphi(z), \qquad \varphi(z+\lambda_{n+\alpha})=\varphi(z)-2\pi i a_\alpha. \qquad [7.13]$$

The holomorphic functions $\dfrac{\partial\varphi}{\partial z_\alpha}$ are Λ-periodic, and are therefore bounded on $\mathbb{C}^n$ and are constants by virtue of Liouville theorem. Thus, φ is linear:

$$\varphi(z)=\sum_\alpha b_\alpha z_\alpha + C.$$

The function φ is holomorphic and linear, and since $\varphi(z+e_\alpha)=\varphi(z)$, we deduce that $b_\alpha=0$ for all α and:

$$\varphi(z+\lambda_{n+\alpha})=\varphi(z)=\text{constant}.$$

From [7.13], we get that $a_\alpha=0$ for all α and consequently the relation [7.11] is trivial, thus showing the linear independence between θ and its derivatives and thus proving that $\psi_{\mathcal{L}^3}$ is an immersion. We will now show that the map:

$$\psi_{\mathcal{L}^3}:\mathbb{T}^n\longrightarrow\mathbb{P}^N(\mathbb{C}), \quad z \bmod \lambda\longmapsto[\theta_0(z):\ldots:\theta_N(z)],$$

is injective. Let $z_1, z_2 \in \mathbb{C}^n$ with:

$$[\theta_0(z_1) : ... : \theta_N(z_1)] = [\theta_0(z_2) : ... : \theta_N(z_2)].$$

So, there exists a $\rho \in \mathbb{C}$, $\rho \neq 0$ such that:

$$\theta_\alpha(z_1) = \rho\theta_\alpha(z_2), \quad \alpha = 1, ..., N,$$

or again:

$$\Theta(z_1) = \rho\Theta(z_2), \quad \forall\Theta \in H^0\left(\mathbb{T}^n, \mathcal{O}(\mathcal{L}^3)\right),$$

and it suffices to show that z_1 and z_2 represent the same point on $\mathbb{T}^n$. From relation [7.12], we derive:

$$\frac{\Theta(z_1, \mu, \nu)}{\Theta(z_2, \mu, \nu)} = \frac{\theta(z_1 + \mu)\theta(z_1 + \nu)\theta(z_1 - \mu - \nu)}{\theta(z_2 + \mu)\theta(z_2 + \nu)\theta(z_2 - \mu - \nu)} = \rho. \quad [7.14]$$

For any $\mu \in \mathbb{C}^n$, we can choose ν such that the $z_1+\nu$, $z_2+\nu$, $z_1-\mu-\nu$, $z_2-\mu-\nu$, do not belong to the zero variety of θ, i.e. so that the $\theta(z_1+\nu)$, $\theta(z_2+\nu)$, $\theta(z_1-\mu-\nu)$, $\theta(z_2 - \mu - \nu)$, are without zero. By equation [7.14], the function $\mu \longmapsto \dfrac{\theta(z_1 + \mu)}{\theta(z_2 + \mu)}$ is holomorphic and without zero. We deduce that we have $\dfrac{\theta(z_1 + \mu)}{\theta(z_2 + \mu)} = e^{\varphi(\mu)}$, i.e.

$$\varphi(\mu) = \log \frac{\theta(z_1 + \mu)}{\theta(z_2 + \mu)}.$$

From the periodicity of θ, we derive the relations:

$$\varphi(\mu + e_\alpha) = \varphi(\mu) + 2\pi n_\alpha, \quad n_\alpha \in \mathbb{Z}, \quad [7.15]$$

$$\varphi(\mu + \lambda_{n+\alpha}) = \varphi(\mu) - 2\pi i(z_1 - z_2)_\alpha + 2\pi i m_\alpha, \quad m_\alpha \in \mathbb{Z}. \quad [7.16]$$

Reasoning as before, we show that $\dfrac{\partial\varphi}{\partial\mu_\alpha}$ is constant and that:

$$\varphi(\mu) = 2\pi i \sum_\alpha a_\alpha\mu_\alpha + c.$$

Relation [7.15] shows that $a_\alpha = n_\alpha \in \mathbb{Z}$ and:

$$\varphi(\mu + \lambda_{n+\alpha}) - \varphi(\mu) = 2\pi i(z_1 - z_2)_\alpha + 2\pi i m_\alpha = 2\pi i \sum a_\alpha Z_{\alpha\beta}.$$

Since φ is linear, then:

$$\varphi(\mu + \lambda_\alpha) - \varphi(\mu) = 2\pi i(z_1 - z_2)_\alpha + 2\pi i m_\alpha,$$

and composing with relation [7.16], we obtain:

$$-(z_1 - z_2)_\alpha + m_\alpha = \sum a_\alpha Z_{\alpha\beta}, \qquad z_1 - z_2 = \sum m_\alpha e_\alpha - \sum a_\alpha \lambda_{n+\alpha},$$

which shows that $z_1 - z_2 \in \Lambda$. Therefore, $\psi_{\mathcal{L}^3} : \mathbb{T}^n \longrightarrow \mathbb{P}^N(\mathbb{C})$, is an embedding. Finally, for any positive line bundle $\mathcal{L} \longrightarrow \mathbb{T}^n$, we can find an Abelian variety $\mathbb{T}'^n$ with a line bundle $\mathcal{L}' \longrightarrow \mathbb{T}'^n$ of principal polarization so that the map $\pi' : \mathbb{T}^n \longrightarrow \mathbb{T}'^n$, is a covering with $\prod \delta_\alpha$ sheets and $\pi'^*(\mathcal{L}') = \mathcal{L}$. The reasoning we used previously for $\mathcal{L}'$ extends easily to $\mathcal{L}$. Furthermore, the fact that $\psi_{\mathcal{L}^3}$ separates the points in $\pi^{-1}(p)$ for all $p \in \mathbb{T}'^n$ follows from the explicit construction of the theta functions, and the proof is complete. ■

DEFINITION 7.3.– *A polarization of the type* $(\delta_1, ..., \delta_n)$ *on an Abelian variety* $\mathcal{A}$ *is the cohomology class* $[\omega]$ *of a Hodge form on this variety. The polarization is said to be principal if* $\delta_\alpha = 1$ *for all* α*, and in this case, we say that* $\mathcal{A}$ *is principally polarized.*

EXAMPLE 7.4.– *The case of Riemann surfaces is particularly interesting. Let* C *be a Riemann surface of genus* g *and let* $Jac(C) = \mathbb{C}^g/\Lambda$ *be its Jacobian variety. Recall that here* Λ *is the lattice generated by the column vectors of the period matrix* $\Omega = (I, Z)$, $Z = Z^\top$, *Im* $Z > 0$, *of* C *(or what amounts to the same thing of* $Jac(C)$*). The Jacobian variety* $Jac(C)$ *is an Abelian variety of dimension* g *and the polarization on this variety is principal with:*

$$\omega = \sum_{\alpha=1}^{g} dx_\alpha \wedge dx_{n+\alpha}.$$

In other words, the Jacobian variety associated with a Riemann surface is a principally polarized Abelian variety.

We have already pointed out that the line bundle associated with a divisor on an irreducible Abelian variety is always ample. Furthermore, the Lefschetz theorem proved previously states that if a line bundle $\mathcal{L}$ is ample on an Abelian variety, then $\mathcal{L}^k$ is very ample for $k \geq 3$. Moreover, for the case of an irreducible Abelian surface $\mathcal{A}$ of type (δ_1, δ_2), a theorem (Ramanan 1980) states that if $\mathcal{D}$ is an ample divisor on $\mathcal{A}$, then $\mathcal{D}$ is very ample if one of the following conditions is satisfied: *(i)* $\delta_1 = 1, \delta_2 \geq 5$, *(ii)* $\delta_1 = 2, \delta_2 \geq 4$ and *(iii)* $\delta_1 \geq 1$.

In the next chapter, we discuss the concept of projectively normal embedding, a result of which the Lefschetz theorem is a consequence.

7.3. Exercises

EXERCISE 7.1.– *(a) Show that the function θ satisfies the following functional equations:*

$$\theta(z+e_j)=\theta(z), \qquad \theta(z+f_j)=e^{-\pi i b_{jj}-2\pi i z_j}.\theta(z).$$

(b) Show that for all $m,n\in\mathbb{Z}^g$, we have:

$$\theta(z+n+Bm)=e^{-\pi i\langle Bm,m\rangle-2\pi i\langle m,z\rangle}\theta(z),$$

and deduce that the function θ is quasi-periodic.

(c) Show that the function θ satisfies the following periodicity property:

$$\theta\begin{bmatrix}\alpha\\ \beta\end{bmatrix}(z)(z+n+Bm)=e^{-\pi i\langle Bm,m\rangle-2\pi i\langle z,m\rangle+2\pi i(\langle\alpha,n\rangle-\langle\beta,m\rangle)}\theta\begin{bmatrix}\alpha\\ \beta\end{bmatrix}(z).$$

(d) Show that the function $\theta\begin{bmatrix}\alpha\\ \beta\end{bmatrix}(z)$ is even (respectively odd) if the half-period $\begin{bmatrix}\alpha\\ \beta\end{bmatrix}$ is even (respectively odd).

Solution: *(a)* The first relation follows immediately from formula [7.1]. Concerning the second relation, we have:

$$\begin{aligned}\theta(z+f_j)&=\sum_{m\in\mathbb{Z}^g}e^{\pi i\langle Bm,m\rangle+2\pi i\langle m,z+f_j\rangle}\\ &=\sum_{m\in\mathbb{Z}^g}e^{\pi i\langle B(n-e_j),n-e_j\rangle+2\pi i\langle n-e_j,z+f_j\rangle},\end{aligned}$$

where $n\equiv m+e_j$. Hence,

$$\begin{aligned}&\theta(z+f_j)\\ &=\sum_{m\in\mathbb{Z}^g}e^{\pi i\langle Bn,n\rangle-2\pi i\langle Be_j,n\rangle+\pi i\langle Be_j,e_j\rangle+2\pi i\langle n,z\rangle+2\pi i\langle n,f_j\rangle-2\pi i\langle e_j,z\rangle-2\pi i\langle e_j,f_j\rangle},\end{aligned}$$

and finally,

$$\theta(z+f_j)=e^{-\pi i\langle Be_j,e_j\rangle-2\pi i\langle e_j,z\rangle}.\theta(z)=e^{-\pi i b_{jj}-2\pi i z_j}.\theta(z),$$

hence the result.

(b) Vectors of the form $n+Bm$ form a lattice of periods, and the result is immediately deduced from the previous question.

(c) It is enough to reason as in question *(a)*.

(d) By substituting $z \mapsto -z$, $m \mapsto -m-2$, in relation [7.3], we immediately obtain for the general term of the series, the expression:

$$e^{\pi i\langle B(-m-\alpha),-m-\alpha\rangle+2\pi i\langle -z+\beta,-m-\alpha\rangle} = e^{\pi i\langle B(m+\alpha),m+\alpha\rangle+2\pi i\langle z+\beta,m+\alpha\rangle}.e^{4\pi i\langle\alpha,\beta\rangle}.$$

The sign of $e^{4\pi i\langle\alpha,\beta\rangle}$ is determined by the parity of the number $4\langle\alpha,\beta\rangle$, hence the result.

EXERCISE 7.2.– *The function $\theta(z)$ has in the parallelogram of periods Ω (generated by 1 and b), a single zero at the point $z = \dfrac{1+b}{2}$.*

Solution: The numbers 1 and b generate a parallelogram of periods denoted Ω. The four theta functions corresponding to the half-periods $\begin{bmatrix}1/2\\1/2\end{bmatrix}$, $\begin{bmatrix}1/2\\0\end{bmatrix}$, $\begin{bmatrix}0\\0\end{bmatrix}$, $\begin{bmatrix}0\\1/2\end{bmatrix}$ are:

$$i\theta_1(z) \equiv \theta\begin{bmatrix}1/2\\1/2\end{bmatrix}(z), \qquad \theta_2(z) \equiv \theta\begin{bmatrix}1/2\\0\end{bmatrix}(z),$$

$$\theta_3(z) \equiv \theta\begin{bmatrix}0\\0\end{bmatrix}(z) = \theta(z), \qquad \theta_4(z) \equiv \theta\begin{bmatrix}0\\1/2\end{bmatrix}(z).$$

These functions are holomorphic on $\mathbb{C}$. Furthermore, we immediately deduce from exercise 7.1*(d)* that $\theta_1(z)$ is odd and that $\theta_2(z)$, $\theta_3(z)$ and $\theta_4(z)$ are even. To determine the zeros of the functions θ_j, it suffices according to proposition 7.2 to look for them in the parallelogram of periods Ω. Since $\theta_1(z)$ is odd, then $\theta_1(0) = 0$, and the other zeros of $\theta_j(z)$ are obtained via proposition 7.1. Let us take, for example, the case of $\theta_3(z)$. We have $\theta_3(\frac{1}{2}+\frac{b}{2}) = 0$, and furthermore, $z = \dfrac{1+b}{2}$ is the only zero of this function in Ω. Indeed, we have:

$$\begin{aligned}\frac{1}{2\pi i}\int_{\partial\Omega}\frac{d\theta_3(z)}{\theta_3(z)} &= \frac{1}{2\pi i}\int_{\partial\Omega} d\log\theta_3(z),\\ &= \frac{1}{2\pi i}\int_0^1 (d\log\theta_3(z) - d\log\theta_3(z+b))\\ &\quad+\frac{1}{2\pi i}\int_0^b (d\log\theta_3(z+1) - d\log\theta_3(z)).\end{aligned}$$

Now, according to proposition 7.1, we have:

$$\theta_3(z+1) = \theta_3(z), \qquad \theta_3(z+b) = e^{-\pi i b - 2\pi i z}\theta_3(z),$$

so,

$$\int_0^b (d\log\theta_3(z+1) - d\log\theta_3(z)) = 0,$$

and

$$\int_0^1 (d\log\theta_3(z) - d\log\theta_3(z+b)) = \int_0^1 (d\log\theta_3(z) - d\log e^{-\pi i b - 2\pi i z}\theta_3(z)),$$
$$= \int_0^1 2\pi i dz = 1.$$

Consequently,

$$\frac{1}{2\pi i}\int_{\partial\Omega}\frac{d\theta_3(z)}{\theta_3(z)} = \frac{1}{2\pi i}\int_{\partial\Omega} d\theta_3(z) = 1.$$

EXERCISE 7.3.– *Setting $z = x \in \mathbb{R}$, $b = it$, $t \in \mathbb{R}_+$, equation [7.1] is written:*

$$\theta(x|it) = \sum_{m\in\mathbb{Z}} e^{-\pi m^2 t + 2\pi i m x} = 1 + 2\sum_{m=1}^{\infty} e^{-\pi m^2 t}\cos 2\pi m x.$$

(a) Show that this function is periodic with respect to x.

(b) Show that this function satisfies the heat equation and that this solution is unique.

Solution: (a) We have:

$$\theta(x+1|it) = \theta(x|it).$$

(b) We obtain:

$$4\pi\frac{\partial\theta(x|it)}{\partial t} = \frac{\partial^2\theta(x|it)}{\partial x^2}.$$

The uniqueness of this solution results from the fact that:

$$\lim_{t\to 0}\theta(x|it) = \sum_{m=-\infty}^{\infty}\delta_m(x),$$

where δ_m is the Dirac distribution at point m.

EXERCISE 7.4.– *Determine a relationship between the Weierstrass function $\wp(z)$ and the theta function $\theta_1(z)$.*

Solution: We have:

$$\wp(z) = -\frac{\partial^2}{\partial z^2} \log \theta_1(z) + c,$$

where c is a constant (see, for example, Dubrovin 1981).

EXERCISE 7.5.– *Show that the function $\theta_1(z)$ satisfies a third-order differential equation.*

Solution: Simply use the relationship obtained in the previous exercise and the differential equation: $(\wp'(z))^2 = 4(\wp(z))^3 - g_2\wp(z) - g_3$, mentioned in the Appendix in Chapter 10 (theorem 10.2).

EXERCISE 7.6.– *The Weierstrass function $\sigma(z)$ relative to a Λ lattice is an odd function that satisfies the following equation:*

$$\sigma(z) = z \prod_{\omega \in \Lambda \setminus \{0\}} \left(1 - \frac{z}{\omega}\right) e^{\frac{z}{\omega} + \frac{1}{2}\left(\frac{z}{\omega}\right)^2}.$$

(See, for example, Lesfari (2023) for more information on this function.)

Solution: It suffices to use the Weierstrass function $\wp(z)$ and prove the relation:

$$\sigma(z + \omega_k) = -e^{\tau_k\left(z + \frac{\omega_k}{2}\right)}.\sigma(z), \quad (k = 1, 2),$$

where τ_k are the constants (see Lesfari 2023, proposition 2.20, p. 134).

EXERCISE 7.7.– *Let $\mathcal{D} = p_1 + \cdots + p_g$, be a non-special divisor on a Riemann surface C of genus g. Show that the function defined by $\theta(\varphi(p) - \varphi(\mathcal{D}) - \delta)$ admits exactly g zeros on the surface C at points $p = p_1, ..., p_g$ (φ being the Abel–Jacobi map and δ the Riemann constant vector [7.6]).*

Solution: This result follows immediately from proposition 7.7.

8

Algebraically Completely Integrable Systems

8.1. Preliminaries

Let us first recall some notions that can be found treated in detail in, for example, Lesfari (2022). Let M be a differentiable manifold of dimension m. Let $TM = \bigcup_{x\in M} T_xM$ be the bundle tangent to M (union of spaces tangent to M at all its points x). This bundle has a structure of a differentiable manifold of dimension $2m$ and allows us to immediately transfer to manifolds the theory of ordinary differential equations. A vector field (also called a tangent bundle section) on M is an application, noted by X, which at any point $x \in M$ associates a tangent vector $X_x \in T_xM$. In other words, it is an application: $X : M \longrightarrow TM$, such that if $\pi : TM \longrightarrow M$, is the natural projection, we have $\pi \circ X = id_M$. In a local coordinate system $(x_1, ..., x_m)$ in a neighborhood $U \subset M$, the vector field X is written in the form $X = \sum_{k=1}^{m} f_k(x)\frac{\partial}{\partial x_k}$, $x \in U$, where the functions $f_1, \ldots, f_m : U \longrightarrow \mathbb{R}$, are the components of X with respect to $(x_1, ..., x_m)$. A vector field X is differentiable if its components $f_k(x)$ are differentiable functions. This definition of differentiability does not obviously depend on the choice of the local coordinate system. To the vector field X corresponds a system of differential equations:

$$\frac{dx_1}{dt} = f_1(x_1, ..., x_m), ..., \frac{dx_m}{dt} = f_m(x_1, ..., x_m).$$

If $\sum_{k=1}^{m} f_k(x)\frac{\partial}{\partial x_k}$ is the local expression of X, then the integral curves (or trajectories) of X are the solutions $\gamma(t) = \{x_k(t)\}$ of the above system. We assume in the following that the vector field X is differentiable (of class $\mathcal{C}^\infty$) and with compact support (i.e. X is zero outside of a compact of M), which will be particularly the case if the manifold M is compact.

Given a point $x \in M$, we denote by $g_t^X(x)$ (or quite simply $g_t(x)$) the position of x after a displacement of a duration $t \in \mathbb{R}$. We therefore have an application $g_t^X : M \longrightarrow M$, $t \in \mathbb{R}$, which is a diffeomorphism (a one-to-one differentiable mapping with a differentiable inverse), by virtue of the theory of differential equations. More precisely, to the vector field X, we associate a one-parameter group of diffeomorphisms g_t^X on M, i.e. a differentiable application (of class $\mathcal{C}^\infty$): $M \times \mathbb{R} \longrightarrow M$, verifying a group law: for all $t \in \mathbb{R}$, $g_t^X : M \longrightarrow M$ is a diffeomorphism of M on M, and for all $t, s \in \mathbb{R}$, $g_{t+s}^X = g_t^X \circ g_s^X$. This last condition means that the correspondence $t \longmapsto g_t^X$, is a homomorphism of the additive group $\mathbb{R}$ in the group of diffeomorphisms from M to M. It implies that $g_{-t}^X = \left(g_t^X\right)^{-1}$, because $g_0^X = id_M$ is the identical transformation which leaves each point invariant.

The one-parameter group of diffeomorphism g_t^X on M is called flow. It admits the vector field X for velocity field

$$\frac{d}{dt} g_t^X(x) = X\left(g_t^X(x)\right),$$

with the initial condition: $g_0^X(x) = x$. Obviously, $\frac{d}{dt} g_t^X(x)\big|_{t=0} = X(x)$. So through these formulas $g_t^X(x)$ is the curve on the manifold that passes through x such that the tangent at each point is the vector $X\left(g_t^X(x)\right)$. The vector field X generates a unique one-parameter group of diffeomorphism of the manifold M.

Consider now a system of nonlinear differential equations of the form

$$X_H : \dot{x} = J\frac{\partial H}{\partial x} \equiv f(x), \; x \in \mathbb{R}^m, \qquad [8.1]$$

where H is a function of class $\mathcal{C}^\infty$ (Hamiltonian) and $J = J(x)$ is a skew-symmetric matrix with polynomial entries in x, for which the corresponding Poisson bracket $\{H_i, H_j\} = \langle \frac{\partial H_i}{\partial x}, J\frac{\partial H_j}{\partial x} \rangle$ satisfies the Jacobi identities. Recall that the Jacobi identity is described by the expression,

$$\{\{H, F\}, G\} + \{\{F, G\}, H\} + \{\{G, H\}, F\} = 0,$$

where H, F, G are $\mathcal{C}^\infty$ and

$$\{H, F\} = \left\langle \frac{\partial H}{\partial x}, J\frac{\partial F}{\partial x} \right\rangle = \sum_{i,j} J_{ij} \frac{\partial H}{\partial x_i}\frac{\partial F}{\partial x_j},$$

are the Poisson brackets. A Hamiltonian system is completely integrable or Liouville integrable if there exist as many independent integrals of motion in involution (i.e. $\{H_i, H_j\} = 0$) as the dimension of the configuration space. According to the Arnold–Liouville theorem (Arnold 1989), if for almost all $c_j \in \mathbb{R}$ the invariant manifolds:

$$\bigcap_{j=1}^{n+k} \{x : H_j(x) = c_j\} \subset \mathbb{R}^m,$$

are compact (and connected), then they are diffeomorphic to the real tori $\mathbb{R}^n/lattice$ and the flows $g_t^{X_j}(x)$ defined by the Hamiltonian vector fields X_j generated by $H_1, ..., H_n$ are linear (rectilinear motions) on these tori.

EXAMPLE 8.1.– *If we consider the case where* $m = 2n$, $J = \begin{pmatrix} \mathbb{O} & \mathbb{I} \\ -\mathbb{I} & \mathbb{O} \end{pmatrix}$ *with* $\mathbb{O}$ *(respectively* $\mathbb{I}$*) the zero (respectively unit) matrix of order* n*. The matrix* J *does not depend on the variables* x_j*, and we have*

$$\{H, F\} = \left\langle \frac{\partial H}{\partial x}, J\frac{\partial F}{\partial x} \right\rangle = \sum_{j=1}^{2n} \frac{\partial H}{\partial x_j} \sum_{l=1}^{2n} J_{jl} \frac{\partial F}{\partial x_l},$$

$$= \sum_{j=1}^{n} \left(\frac{\partial H}{\partial x_{n+j}} \frac{\partial F}{\partial x_j} - \frac{\partial H}{\partial x_j} \frac{\partial F}{\partial x_{n+j}} \right).$$

We thus find the definition of the Poisson bracket found in classical Hamiltonian formalism. Furthermore, equation [8.1] immediately transforms into a system of n *differential equations:*

$$\dot{q}_1 = \frac{\partial H}{\partial p_1}, ..., \dot{q}_n = \frac{\partial H}{\partial p_n}, \qquad \dot{p}_1 = -\frac{\partial H}{\partial q_1}, ..., \dot{p}_n = -\frac{\partial H}{\partial q_n},$$

where $q_1 = x_1, ..., q_n = x_n$, $p_1 = x_{n+1}, ..., p_n = x_{2n}$*. These are Hamilton's equations, also called canonical Hamilton equations; they show that it is sufficient to know the Hamiltonian function* H *to determine the equations of motion. They are often interpreted by considering that the variables* p_k *and* q_k *are the coordinates of a point moving in a* $2n$*-dimensional space, called phase space. The flow associated with the above system obviously leaves invariant each hypersurface of constant energy* $H = c$*. The above Hamilton equations can be written as:*

$$\dot{q}_j = \{H, q_j\} = \frac{\partial H}{\partial p_j}, \qquad \dot{p}_j = \{H, p_j\} = -\frac{\partial H}{\partial q_j},$$

where $1 \leq j \leq n$*. Note that the functions* $1, q_j, p_j$ $(1 \leq j \leq n)$ *satisfy the following commutation relations:*

$$\{q_j, q_l\} = \{p_j, p_l\} = \{q_j, 1\} = \{p_j, 1\} = 0, \quad \{p_j, q_l\} = \delta_{jl}, \quad 1 \leq j, l \leq n.$$

These functions constitute a basis of a real Lie algebra (Heisenberg algebra), of dimension $2n + 1$.

Recall that integrable Hamiltonian systems are nonlinear ordinary differential equations described by a Hamiltonian function and possessing sufficiently many independent constants of motion in involution. The regular compact-level manifolds defined by the intersection of the constants of motion are diffeomorphic to a real torus on which the motion is quasi-periodic as a consequence of the following purely differential geometric fact: a compact and connected n-dimensional manifold on which there exist n vector fields that commute and are independent at every point is diffeomorphic to an n-dimensional real torus, and each vector field will define a linear flow there. Algebraic integrability means that the system is completely integrable in the sense of the phase space being foliated by tori, which, in addition, are real parts of a complex algebraic tori (Abelian varieties).

To introduce the complete algebraic integrability of Hamiltonian systems, we will work with complex numbers (instead of real numbers). The concepts of integrability in the Liouville sense, involution, commutativity of vector fields, etc., can be defined as usual in the real case. However, difficulties arise: we know that there are no compact holomorphic subvarieties in the complex space $\mathbb{C}^m$ (by virtue of the maximum principle) so the complex tori that we can obtain in the Arnold–Liouville theorem are not compact. So, the problem of compactification of invariant manifolds arises. Moreover, the solutions of the system in question are not uniform. We will now consider the complex space $\mathbb{C}^m$, instead of $\mathbb{R}^m$.

According to Adler et al. (2004), the system [8.1] with polynomial right-hand side will be called algebraic complete integrable (a.c.i.) when:

(i) The system possesses $n+k$ independent polynomial invariants $H_1, ..., H_{n+k}$ of which k lead to zero vector fields $J\frac{\partial H_{n+i}}{\partial x}(x) = 0, 1 \leq i \leq k$, the n remaining ones are in involution (i.e. $\{H_i, H_j\} = 0$) and $m = 2n+k$. For most values of $c_i \in \mathbb{R}$, the invariant manifolds $\bigcap_{i=1}^{n+k} \{x \in \mathbb{R}^m : H_i = c_i\}$ are assumed compact and connected. Then, according to the Arnold–Liouville theorem, there exists a diffeomorphism $\bigcap_{i=1}^{n+k} \{x \in \mathbb{R}^m : H_i = c_i\} \longrightarrow \mathbb{R}^n/Lattice$, and the solutions of the system [8.1] are straight-line motions on these tori.

(ii) The invariant manifolds

$$M = \bigcap_{i=1}^{n+k} \{x \in \mathbb{C}^m, H_i = c_i\}$$

can be completed into complex algebraic tori, i.e. $M \cup \mathcal{D} = \mathbb{C}^n/Lattice$, where $\mathbb{C}^n/Lattice$ is a complex algebraic torus (i.e. Abelian variety) and $\mathcal{D}$ a divisor (called the Painlevé divisor). In the natural coordinates $(t_1, ..., t_n)$ of $\mathbb{C}^n/Lattice$ coming from $\mathbb{C}^n$, the functions $x_i = x_i(t_1, ..., t_n)$ are meromorphic and [8.1] defines straight-line motion on $\mathbb{C}^n/Lattice$.

Condition (i) means there is an algebraic map:

$$(x_1(t), ..., x_m(t)) \longmapsto (\mu_1(t), ..., \mu_n(t)),$$

making the following sums linear in t:

$$\sum_{i=1}^{n} \int_{\mu_i(0)}^{\mu_i(t)} \omega_j = d_j t \ , \ 1 \leq j \leq n, \ d_j \in \mathbb{C},$$

where $\omega_1, ..., \omega_n$ denote the holomorphic differentials on some algebraic curves.

If the flow is algebraically completely integrable, the differential equations [8.1] must admit Laurent series solutions depending on $m - 1$ free parameters. Indeed, since the flow evolves on the complex torus $\mathcal{A} \simeq \mathbb{C}^n/lattice$, the coordinates x_j remain finite on the affine (i.e. non-compact) part M of this torus, and furthermore, the coordinates $x_1, ..., x_m$ must tend to infinity along a divisor $\mathcal{D} \subset \mathbb{C}^n/lattice$. By the definition of complete algebraic integrability, the flow [8.1] is linear (rectilinear motion) on $\mathcal{A}$; each of these trajectories must, however, hit the submanifold $\mathcal{D}$ in at least one place. Conversely, through any point of $\mathcal{D}$, there exists a rectilinear motion and consequently a Laurent development around this point of intersection. Hence the differential equations [8.1] must admit Laurent series that depend on the $n - 1$ parameters defining $\mathcal{D}$ and the $n + k$ constants c_j defining $\mathcal{A}$, i.e. the series around the points of intersection depend on $(n + k) + (n - 1) = m - 1$ free parameters (m being the dimension of the phase space).

Adler et al. (2004) have shown that the existence of a coherent set of Laurent solutions:

$$x_i = \frac{1}{t^{k_i}} \left(x^{(0)} + x^{(1)} t + x^{(2)} t^2 + \cdots \right), \quad k_i \in \mathbb{Z}, \quad \text{some } k_i > 0, \qquad [8.2]$$

depending on $dim\ (phase\ space) - 1 = m - 1$ free parameters is necessary and sufficient for a Hamiltonian system with the right number of constants of motion to be a.c.i. So, if the Hamiltonian flow [8.1] is a.c.i., it means that the variables x_i are meromorphic on the torus $\mathbb{C}^n/Lattice$ and by compactness they must blow up along a codimension-one subvariety (a divisor) $\mathcal{D} \subset \mathbb{C}^n/Lattice$. By the a.c.i. definition, the flow [8.1] is a straight line motion in $\mathbb{C}^n/Lattice$, and thus, it must hit the divisor $\mathcal{D}$ in at least one place. Moreover, through every point of $\mathcal{D}$, there is a straight line motion and therefore a Laurent expansion around that point of intersection. It is interesting to note that these systems have not only Laurent solutions depending on $m - 1$ free parameters but also depending on $m - 2$ free parameters; these solutions are related to the canonical divisor of a curve on the torus.

Assume now Hamiltonian flows to be (weight-)homogeneous with a weight $\nu_i \in \mathbb{N}$, going with each variable x_i, i.e.

$$f_i \left(\alpha^{\nu_1} x_1, ..., \alpha^{\nu_m} x_m \right) = \alpha^{\nu_i + 1} f_i \left(x_1, ..., x_m \right), \ \forall \alpha \in \mathbb{C}.$$

Observe that then the constants of the motion H can be chosen to be (weight-)homogeneous: $H\left(\alpha^{\nu_1}x_1,...,\alpha^{\nu_m}x_m\right) = \alpha^k H\left(x_1,...,x_m\right)$, $k \in \mathbb{Z}$.

We must have $k_i = \nu_i$ and coefficients in the series must satisfy at the 0th step non-linear equations:

$$f_i\left(x_1^{(0)},...,x_m^{(0)}\right) + \nu_i x_i^{(0)} = 0,\ 1 \le i \le m, \qquad [8.3]$$

and at the k^{th}step, linear systems of equations:

$$(L - kI)x^{(k)} = \begin{cases} 0 \ \text{ for } k = 1 \\ \text{some polynomial in } x^{(1)},...,x^{(k-1)} \text{ for } k > 1, \end{cases} \qquad [8.4]$$

where

$$L = \text{Jacobian map of [8.3]} = \frac{\partial f}{\partial x} + gI\ |_{x=x^{(0)}}\ .$$

If $m - 1$ free parameters are to appear in the Laurent series, they must either come from the nonlinear equations [8.3] or from the eigenvalue problem [8.4], i.e. L must have at least $m - 1$ integer eigenvalues. These are much less conditions than expected, because of the fact that the homogeneity k of the constant H must be an eigenvalue of L. Moreover, all the formal series must be convergent, and the study of convergence will be carried out via the majorant method (exercise 8.1) and the concept of projective normality or normally generated line bundle. Next, we assume that the divisor $\mathcal{D}$ is very ample and, in addition, projectively normal (the line bundle $\mathcal{L} = [\mathcal{D}]$ is normally generated). Consider a point $p \in \mathcal{D}$, a chart U_j around p on the torus and a function y_j in $\mathcal{L}(\mathcal{D})$ having a pole of maximal order at p. Then, the vector $(1/y_j, y_1/y_j, \ldots, y_N/y_j)$ provides a good system of coordinates in U_j. Then, taking the derivative with regard to one of the flows:

$$\left(\frac{y_i}{y_j}\right)^{\cdot} = \frac{\dot{y}_i y_j - y_i \dot{y}_j}{y_j^2}, \quad 1 \le j \le N,$$

are finite on U_j as well. Therefore, since y_j^2 has a double pole along $\mathcal{D}$, the numerator must also have a double pole (at worst), i.e. the Wronskian of y_i and y_j (with respect to X_H [8.1]) $\dot{y}_i y_j - y_i \dot{y}_j \in \mathcal{L}(2\mathcal{D})$. Hence, when $\mathcal{D}$ is projectively normal, we have that:

$$\left(\frac{y_i}{y_j}\right)^{\cdot} = \sum_{k,l} a_{k,l}\left(\frac{y_k}{y_j}\right)\left(\frac{y_l}{y_j}\right),$$

i.e. the ratios y_i/y_j form a closed system of coordinates under differentiation. At the bad points, the concept of projective normality plays an important role: this enables one to show that y_i/y_j is a bona fide Taylor series starting from every point in a neighborhood of the point in question.

Naively, to compactify the affine manifold M, we might be tempted to take the compactified $\overline{M}$ of M (by projectivizing the equations defining the variety M) into the complex projective space $\mathbb{P}^m(\mathbb{C})$. This can never work for a general reason: an Abelian variety $\mathcal{A}$ of dimension greater than or equal to two is never a complete intersection, i.e. it can never be described in a projective space $\mathbb{P}^m(\mathbb{C})$ by $m-n$ global homogeneous polynomial equations. In other words, for M to be the affine subset of an Abelian variety, $\overline{M}$ must be singular at infinity. By blowing up the singularity along the locus reached by the flow and blowing down the part of the locus that is not reached by the flow, we show that $\overline{M}$ transforms into an Abelian variety $\mathcal{A}$ and the locus at infinity transforms into a subvariety of codimension 1.

To prove the algebraic complete integrability of a given Hamiltonian system, the main steps of the method are as follows. The first step is to show the existence of the Laurent solutions, which requires an argument precisely every time k is an integer eigenvalue of L, and therefore, $L - kI$ is not invertible. We show the existence of the remaining constants of the motion in involution so as to reach the number $n + k$. For given $c_1, ..., c_m$, the set:

$$\mathcal{D} \equiv \left\{ \begin{array}{l} x_i(t) = t^{-\nu_i}\left(x_i^{(0)} + x_i^{(1)}t + x_i^{(2)}t^2 + \cdots\right), 1 \leq i \leq m, \\ \text{Laurent solutions such that } : H_j\left(x_i(t)\right) = c_j + \text{ Taylor part }, \end{array} \right\}$$

defines one or several $n - 1$ dimensional algebraic varieties (divisor) having the property that $\bigcap_{i=1}^{n+k} \{H_i = c_i, z \in \mathbb{C}^m\} \cup \mathcal{D}$ is a smooth, compact, connected variety with n commuting vector fields independent at every point. In other words, it is a complex algebraic torus (i.e. Abelian variety) $\mathcal{A} = \mathbb{C}^n/Lattice$. The flows $J\frac{\partial H_{k+i}}{\partial z}, ..., J\frac{\partial H_{k+n}}{\partial z}$ are straight-line motions on $\mathcal{A}$. From the divisor $\mathcal{D}$, a lot of information can be obtained with regard to the periods and the action-angle variables. As we have pointed out, the majorant method (exercise 8.1) and the concept of projective normality or normally generated line bundle will play a crucial role in the study of the convergence of all the formal Laurent series solutions.

Then, we proceed to the compactification of M into an Abelian variety $\mathcal{A}$. If the system of these equations is algebraically completely integrable, then it has a family of Laurent series solutions depending on a sufficient number of free parameters. These solutions evolve according to rectilinear motions in complex time on Abelian varieties. We have seen previously that these varieties have a whole series of remarkable properties. For example, the Kodaira embedding theorem combined with the Lefschetz theorem asserts that meromorphic functions on the torus having a pole of order one, two or at most three along a positive divisor allow the torus to be smoothly embedded in a projective space. A crucial step is to find this positive divisor; it turns out that it is defined by the relations between the free parameters obtained by confining the Laurent series solutions to invariant manifolds. From a divisor on an Abelian variety, we can indeed have several pieces of information about the torus, which ultimately allows us to write Abelian integrals and thus linearize differential equations. We will explain such situations explicitly with the help of important problems.

8.2. The Hénon–Heiles system

Consider the system:

$$\begin{aligned} \dot{y}_1 &= x_1, \qquad \dot{x}_1 = -ay_1 - 2y_1y_2, \\ \dot{y}_2 &= x_2, \qquad \dot{x}_2 = -by_2 - y_1^2 - 6y_2^2, \end{aligned} \qquad [8.5]$$

corresponding to a generalized Hénon–Heiles Hamiltonian:

$$H = \frac{1}{2}(x_1^2 + x_2^2) + \frac{1}{2}(ay_1^2 + by_2^2) + y_1^2 y_2 + 2y_2^3, \qquad [8.6]$$

where a, b, are the constant parameters, and y_1, y_2, x_1, x_2 are the canonical coordinates and momenta, respectively. The second integral of motion is:

$$H_2 = y_1^4 + 4y_1^2 y_2^2 - 4x_1^2 y_2 + 4x_1 x_2 y_1 + 4ay_1^2 y_2 + (4a - b)x_1^2 + a(4a - b)y_1^2. \qquad [8.7]$$

The system [8.5] can be written in the form:

$$\dot{z} = f(z) = J\frac{\partial H}{\partial z}, \quad z = (y_1, y_2, x_1, x_2)^\intercal, \qquad [8.8]$$

where $H = H_1$ [8.6],

$$\frac{\partial H}{\partial z} = \left(\frac{\partial H}{\partial y_1}, \frac{\partial H}{\partial y_2}, \frac{\partial H}{\partial x_1}, \frac{\partial H}{\partial x_2}\right)^\intercal, \qquad J = \begin{pmatrix} O & I \\ -I & O \end{pmatrix}$$

The second flow commuting with the first is regulated by the equations

$$\dot{z} = J\frac{\partial H_2}{\partial z}, \qquad z = (y_1, y_2, x_1, x_2)^\intercal,$$

with H_2 defined by equation [8.7], and is written explicitly as:

$$\begin{aligned} \dot{y}_1 &= 2\,(4a - b)\,x_1 - 8x_1y_2 + 4x_2y_1, \\ \dot{y}_2 &= 4x_1y_1, \\ \dot{x}_1 &= -2a\,(4a - b)\,y_1 - 4x_1x_2 - 8ay_1y_2 - 8y_1y_2^2 - 4y_1^3, \\ \dot{x}_2 &= 4x_1^2 - 4ay_1^2 - 8y_1^2 y_2. \end{aligned}$$

The invariant (or level) manifold:

$$M = \bigcap_{i=1}^{2} \{z : H_i(z) = c_i\} \subset \mathbb{C}^4, \qquad [8.9]$$

is a smooth affine surface for generic $c = (c_1, c_2) \in \mathbb{C}^2$. The question I address is how does one find the compactification of $\mathcal{A}$ into an Abelian surface? Now, the system [8.8] is integrable in the sense of Liouville but how does one effectively integrate the problem?

We check the existence near each movable singularity of a Laurent series, which represents the general solution. In fact, the Laurent decomposition of such asymptotic solutions has the following form $z(t) \equiv (y_1(t), y_2(t), x_1(t), x_2(t))$, where

$$y_1 = \frac{y_1^{(0)}}{t} + y_1^{(1)} + y_1^{(2)}t + y_1^{(3)}t^2 + \cdots, \qquad x_1 = \dot{y}_1,$$

$$y_2 = \frac{y_2^{(0)}}{t^2} + \frac{y_2^{(1)}}{t} + y_2^{(2)} + y_2^{(3)}t + y_2^{(4)}t^2 + \cdots, \qquad x_2 = \dot{y}_2,$$

which depend on $\dim(\textit{phase space}) - 1 = 3$ free parameters. Putting these expressions into [8.8], solving inductively for the $z^{(k)}$, we find at the 0th step a nonlinear equation,

$$z^{(0)} + f(z^{(0)}) = 0, \qquad [8.10]$$

and at the kth step, a linear system of equations,

$$(L - kI)\, z^{(k)} = \begin{cases} 0 \text{ for } k = 1, \\ \text{quadratic polynomial in } z^{(1)}, ..., z^{(k-1)} \text{ for } k > 1, \end{cases} \qquad [8.11]$$

where L denotes the Jacobian map of [8.10]. One parameter appear at the 0th step, i.e. in the resolution of [8.10], and the two remaining ones at the kth step, $k = 3$, $k = 6$. Equation [8.11] has the following explicit form:

$$\begin{cases} y_1^{(0)} + y_1^{(0)}y_2^{(0)} = 0 \\ y_2^{(0)} + \left(y_2^{(0)}\right)^2 = 0 \end{cases}$$

and obviously:

$$y_1^{(0)} = \alpha = \text{free parameter}, \qquad y_2^{(0)} = -1.$$

The system [8.11] for $k = 1, 2, ..., 6$ is explicitly given by the following equations with the corresponding solutions:

(a) For $k = 1$, we have:

$$\begin{cases} y_1^{(0)}y_2^{(1)} + y_1^{(1)}y_2^{(0)} = 0 \\ y_2^{(1)} + 6y_2^{(0)}y_2^{(1)} = 0 \end{cases}$$

from where, $y_1^{(1)} = y_2^{(1)} = 0$.

(b) For $k = 2$, we have:

$$\begin{cases} 2y_1^{(2)}y_2^{(0)} + 2y_1^{(0)}y_2^{(2)} + 2y_1^{(1)}y_2^{(1)} + ay_1^{(0)} = 0 \\ \left(y_1^{(0)}\right)^2 + 6\left(y_2^{(1)}\right)^2 + 12y_2^{(0)}y_2^{(2)} + by_2^{(0)} = 0 \end{cases}$$

whose solution is:

$$y_1^{(2)} = \frac{\alpha^3}{12} + \frac{\alpha a}{2} - \frac{\alpha b}{12}, \quad y_2^{(2)} = \frac{\alpha^2}{12} - \frac{b}{12}.$$

(c) For $k = 3$, we have

$$\begin{cases} 2y_1^{(1)}y_2^{(2)} + 2y_1^{(2)}y_2^{(1)} + 2y_1^{(3)}y_2^{(0)} + 2y_1^{(0)}y_2^{(3)} + 2y_1^{(3)} + ay_1^{(1)} = 0 \\ 2y_1^{(0)}y_1^{(1)} + 12y_2^{(0)}y_2^{(3)} + 12y_2^{(1)}y_2^{(2)} + by_2^{(1)} = 0 \end{cases}$$

from where,

$$y_1^{(3)} = \beta = \text{free parameter}, \qquad y_2^{(3)} = 0.$$

(d) For $k = 4$, we have:

$$\begin{cases} 6y_1^{(4)} + 2y_1^{(3)}y_2^{(1)} + 2y_1^{(4)}y_2^{(0)} + 2y_1^{(1)}y_2^{(3)} + 2y_1^{(2)}y_2^{(2)} \\ \qquad +2y_1^{(0)}y_2^{(4)} + ay_1^{(2)} = 0 \\ 2y_2^{(4)} + 2y_1^{(0)}y_1^{(2)} + \left(y_1^{(1)}\right)^2 + 6\left(y_2^{(2)}\right)^2 + 12y_2^{(0)}y_2^{(4)} \\ \qquad +12y_2^{(1)}y_2^{(3)} + by_2^{(2)} = 0 \end{cases}$$

whose solution is:

$$y_1^{(4)} = \frac{\alpha a b}{24} - \frac{\alpha^5}{72} + \frac{11\alpha^3 b}{720} - \frac{11\alpha^3 a}{120} - \frac{\alpha b^2}{720} - \frac{\alpha a^2}{8},$$

$$y_2^{(4)} = \frac{\alpha^4}{48} + \frac{\alpha^2 a}{10} - \frac{\alpha^2 b}{60} - \frac{b^2}{240}.$$

(e) For $k = 5$, we have:

$$\begin{cases} 12y_1^{(5)} + 2y_1^{(3)}y_2^{(2)} + 2y_1^{(4)}y_2^{(1)} + 2y_1^{(1)}y_2^{(4)} + 2y_1^{(2)}y_2^{(3)} \\ \qquad +2y_1^{(0)}y_2^{(5)} + 2y_1^{(5)}y_2^{(0)} + ay_1^{(3)} = 0 \\ 6y_2^{(5)} + 2y_1^{(0)}y_1^{(3)} + 2y_1^{(1)}y_1^{(2)} + 12y_2^{(1)}y_2^{(4)} + 12y_2^{(2)}y_2^{(3)} \\ \qquad +12y_2^{(0)}y_2^{(5)} + by_2^{(3)} = 0 \end{cases}$$

whose solution is:

$$y_1^{(5)} = -\frac{\beta\alpha^2}{12} + \frac{\beta b}{60} - \frac{a\beta}{10}, \qquad y_2^{(5)} = \frac{\alpha\beta}{3}.$$

(f) For $k = 6$, we have:

$$\begin{cases} 20y_1^{(6)} + 2y_1^{(0)}y_2^{(6)} + 2y_1^{(2)}y_2^{(4)} + 2y_1^{(3)}y_2^{(3)} + 2y_1^{(6)}y_2^{(0)} + 2y_1^{(5)}y_2^{(1)} \\ \qquad +2y_1^{(1)}y_2^{(5)} + 2y_1^{(4)}y_2^{(2)} + ay_1^{(4)} = 0 \\ 12y_2^{(6)} + 2y_1^{(0)}y_1^{(4)} + \left(y_1^{(2)}\right)^2 + 6\left(y_2^{(3)}\right)^2 + 2y_1^{(1)}y_1^{(3)} + 12y_2^{(0)}y_2^{(6)} \\ \qquad +12y_2^{(1)}y_2^{(5)} + 12y_2^{(2)}y_2^{(4)} + by_2^{(4)} = 0 \end{cases}$$

from where,

$$y_1^{(6)} = -\frac{\alpha\gamma}{9} - \frac{\alpha^7}{15552} - \frac{\alpha^5 a}{2160} + \frac{\alpha^5 b}{12960} + \frac{\alpha^3 b^2}{25920} + \frac{\alpha^3 a^2}{1440} - \frac{\alpha^3 ab}{4320} + \frac{\alpha ab^2}{1440}$$
$$-\frac{\alpha b^3}{19440} - \frac{\alpha a^2 b}{288} + \frac{\alpha a^3}{144},$$

and

$$y_2^{(6)} = \gamma = \text{free parameter.}$$

Using the majorant method (exercise 8.1), we can show that the formal Laurent series solution is convergent. Consequently, we have the following proposition.

PROPOSITION 8.1.– *The non-identically zero solutions of [8.10] define a line. The system [8.11] has* 1 *degree of freedom for* $k = 3$ *and* $k = 6$. *The flow* X_{H_1} *[8.8] possesses Laurent series solutions which depend on three free parameters* α, β *and* γ. *The first free parameter* α *appears in the resolution of [8.10] and the two remaining ones* β, γ *at the* k^{th} *step,* $k = 3$, $k = 6$ *of [8.11]. These Laurent series solutions are explicitly given by:*

$$\begin{aligned} y_1 &= \frac{y_1^{(0)}}{t} + y_1^{(2)}t + y_1^{(3)}t^2 + y_1^{(4)}t^3 + y_1^{(5)}t^4 + y_1^{(6)}t^5 + \cdots \\ y_2 &= \frac{y_2^{(0)}}{t^2} + y_2^{(2)} + y_2^{(4)}t^2 + y_2^{(5)}t^3 + y_2^{(6)}t^4 + \cdots \\ x_1 &= -\frac{y_1^{(0)}}{t^2} + y_1^{(2)} + 2y_1^{(3)}t + 3y_1^{(4)}t^2 + 4y_1^{(5)}t^3 + 5y_1^{(6)}t^4 + \cdots \\ x_2 &= -2\frac{y_2^{(0)}}{t^3} + 2y_2^{(4)}t + 3y_2^{(5)}t^2 + 4y_2^{(6)}t^3 + \cdots \end{aligned} \qquad [8.12]$$

with leading terms given above.

Consider points at infinity, which are limit points of trajectories of the flow. To be precise, we search for the set of Laurent solutions which remain confined to the fixed affine invariant surface M [8.9] related to specific values of generic constants c_1 and c_2.

PROPOSITION 8.2.– *The pole solutions [8.12] restricted to the surface M [8.9] are a smooth genus 3 hyperelliptic curve $\mathcal{D}$ [8.13], which is a double ramified cover of an elliptic curve $\mathcal{E}$ [8.14].*

PROOF.– By substituting the solutions [8.12] in the constants of the motion $H_1 = c_1$ and $H_2 = c_2$, we eliminate the parameter γ linearly, leading to an algebraic relation between the two remaining parameters, which is nothing but the equation of the divisor $\mathcal{D}$ along which the $z(t) = (y_1(t), y_2(t), x_1(t), x_2(t))$ blow up. To be more precise, $\mathcal{D}$ is the closure of the continuous components of {Laurent series solutions $z(t)$ such that $H_i(z(t)) = c_i,\ 1 \le i \le 2$}, i.e.

$$\mathcal{D} = t^0 - \text{coefficient of} \bigcap_{k=1}^{2} \left\{z \in \mathbb{C}^4 : H_k(z(t)) = c_k\right\}.$$

Thus, we find an algebraic curve defined by:

$$\mathcal{D} = \overline{\{(\beta,\alpha) : \beta^2 = P_8(\alpha)\}}, \qquad [8.13]$$

where

$$\begin{aligned} P_8(\alpha) &= -\frac{7}{15552}\alpha^8 - \frac{1}{432}\left(5a - \frac{13}{18}b\right)\alpha^6 \\ &\quad -\frac{1}{36}\left(\frac{671}{15120}b^2 + \frac{17}{7}a^2 - \frac{943}{1260}ab\right)\alpha^4 \\ &\quad -\frac{1}{36}\left(4a^3 - \frac{1}{2520}b^3 - \frac{13}{6}a^2b + \frac{2}{9}ab^2 - \frac{10}{7}c_1\right)\alpha^2 + \frac{1}{36}c_2. \end{aligned}$$

The curve $\mathcal{D}$ determined by an eight-order equation is smooth and hyperelliptic, and its genus is three. Moreover, the map:

$$\sigma : \mathcal{D} \longrightarrow \mathcal{D}, \quad (\beta,\alpha) \longmapsto (\beta,-\alpha), \qquad [8.14]$$

is an involution on $\mathcal{D}$ and the quotient $\mathcal{E} = \mathcal{D}/\sigma$ is an elliptic curve defined by:

$$\mathcal{E} = \overline{\{(\beta,\zeta) : \beta^2 = P_4(\zeta)\}}, \qquad [8.15]$$

where $P_4(\zeta)$ is the degree-four polynomial in $\zeta = \alpha^2$ obtained from the polynomial $P_8(\alpha)$ above. The hyperelliptic curve $\mathcal{D}$ is thus a two-sheeted ramified covering of the elliptic curve $\mathcal{E}$ [8.15],

$$\rho : \mathcal{D} \longrightarrow \mathcal{E}, \quad (\beta,\alpha) \longmapsto (\beta,\zeta), \qquad [8.16]$$

ramified at the four points covering $\zeta = 0$ and ∞. This ends the proof of the proposition. ■

Let $\mathcal{A}$ be a smooth surface compactifying M [8.9]. Consider a basis $1, f_1, ..., f_N$ of the vector space:

$$\mathcal{L}(\mathcal{D}) \equiv \{f : f \text{ meromorphic on } \mathcal{A}, (f) \geq -\mathcal{D}\},$$

of meromorphic functions on $\mathcal{A}$ with at worst a simple pole along $\mathcal{D}$ and the map:

$$\mathcal{A} \to \mathbb{P}^N(\mathbb{C}),\ p \curvearrowright [1, f_1(p), ..., f_N(p)],$$

considered projectively, because if at p some $f_i(p) = \infty$, we divide by f_i having the highest order pole near p, which makes every element finite. The Kodaira embedding theorem tells us that if the line bundle associated with the divisor is positive, then for $k \in \mathbb{N}$, the functions of $\mathcal{L}(k\mathcal{D})$ embed smoothly $\mathcal{A}$ into $\mathbb{P}^N(\mathbb{C})$ and then by the Chow theorem, $\mathcal{A}$ can be realized as an algebraic variety, i.e.

$$\mathcal{A} = \bigcap_i \{Z \in \mathbb{P}^N : P_i(Z) = 0\},$$

where $P_i(Z)$ are the homogeneous polynomials. In fact, in our case, $k = 2$ and $N = 7$ suffice, i.e. the divisor $2\mathcal{D}$ provides a smooth embedding into $\mathbb{P}^7(\mathbb{C})$, via the meromorphic section of $\mathcal{L}(2\mathcal{D})$.

It is easy to find a set of polynomial functions $\{1, f_1, ..., f_N\}$ in $\mathcal{L}(2\mathcal{D})$ such that the embedding of $2D$ with those functions into $\mathbb{P}^N(\mathbb{C})$ yields a curve of genus $N+2$. Straightforward calculation, using asymptotic expansions, shows that the space $\mathcal{L}(2\mathcal{D})$ is spanned by the following functions:

$$\begin{aligned}
\mathcal{L}(2\mathcal{D}) &= \{f_0, f_1, ..., f_7\}, \\
&= \{1, y_1, y_1^2, y_2, x_1, x_1^2 + y_1^2 y_2, x_2 y_1 - 2x_1 y_2, x_1 x_2 + 2a y_1 y_2 + 2y_1 y_2^2\}, \\
&= \left\{1, \frac{\alpha}{t}, \frac{\alpha^2}{t^2}, -\frac{1}{t^2}, -\frac{\alpha}{t^2}, -\frac{\alpha^2(\alpha^2 - 8a - b)}{t^2}, \frac{\alpha(\alpha^2 + 4a - b)}{2t^2}, \frac{6\beta}{t^2}\right\} \\
&\qquad + \text{higher order terms in } t.
\end{aligned}$$

The map

$$2\mathcal{D} \longrightarrow \mathbb{P}^7(\mathbb{C}),\ p \longmapsto \lim_{t\to 0} t^2 [f_0(p), f_1(p), ..., f_7(p)] = \left[0, 0, f_2^{(0)}(p), ..., f_7^{(0)}(p)\right],$$

maps the curve $2\mathcal{D}$ into $\widetilde{2\mathcal{D}} \subseteq \mathbb{P}^7(\mathbb{C})$, and the genus of $2\mathcal{D}$ is 9.

PROPOSITION 8.3.– *(a) The orbits of the vector field [8.8] running through $2\mathcal{D}$ form a smooth surface Π near $2\mathcal{D}$ such that $\Pi\backslash 2\mathcal{D} \subseteq \mathcal{A}$. Moreover, the variety $\mathcal{A} = M \cup \Pi$, is smooth, compact and connected.*

(b) The variety $\mathcal{A}$ comes equipped with two everywhere independent commuting vector fields, which extend holomorphically on $\mathcal{A}$.

(c) The system [8.8] is a.c.i. and the corresponding flow evolves on an Abelian surface $\mathcal{A} \simeq \mathbb{C}^2/Lattice$.

PROOF.– *(a)* Let:

$$\Phi(t,p) = \{z(t) = (y_1(t), y_2(t), x_1(t), x_2(t)) : t \in \mathbb{C}, 0 < |t| < \varepsilon\},$$

be the orbit of the vector field [8.8] going through the point $p \in 2\mathcal{D}$. Let $\Pi_p \subset \mathbb{P}^7(\mathbb{C})$ be the surface element formed by the divisor $2\mathcal{D}$ and the orbits going through p, and set $\Pi \equiv \bigcup_{p\in 2\mathcal{D}} \Pi_p$. Consider the curve $\mathcal{S} = H \cap \Pi$ where $H \subset \mathbb{P}^7(\mathbb{C})$ is a hyperplane transversal to the direction of the flow. If $\mathcal{S}$ is smooth, then using the implicit function theorem the surface Π is smooth. But if $\mathcal{S}$ is singular at 0, then Π would be singular along the trajectory ($t-$axis) which goes immediately into the affine part M. So, M would be singular, which is absurd because M is the fiber of a morphism from $\mathbb{C}^4$ to $\mathbb{C}^2$ and so smooth for almost all the two constants of the motion c_i. Let $\overline{M}$ be the projective closure of M into $\mathbb{P}^4(\mathbb{C})$, let $Z = [Z_0, Z_1 = z_1 Z_0, ..., Z_4 = z_4 Z_0] \in \mathbb{P}^4(\mathbb{C})$, and let $M_\infty = \overline{M} \cap \{Z_0 = 0\}$, be the locus at infinity. Consider the map,

$$\overline{M} \subseteq \mathbb{P}^4(\mathbb{C}) \longrightarrow \mathbb{P}^7(\mathbb{C}),\ Z \longmapsto f(Z),$$

where $f = (f_0, f_1, ..., f_7) \in \mathcal{L}(2D)$ and let $\mathcal{A} = f(\overline{M})$. In a neighborhood $V(p) \subseteq \mathbb{P}^7(\mathbb{C})$ of p, we have $\Pi_p = \mathcal{A}$ and $\Pi_p\backslash 2D \subseteq M$. Otherwise there would exist an element of surface $\Pi_p' \subseteq T$ such that:

$$\Pi_p \cap \Pi_p' = t - \text{axis}, \qquad \Phi(t,p) = t - \text{axis}\backslash p \subseteq M,$$

and hence M would be singular along the t-axis, which is impossible. Since the variety $\overline{M} \cap \{Z_0 \neq 0\}$ is irreducible and since the generic hyperplane section $H_{gen.}$ of $\overline{M}$ is also irreducible, all hyperplane sections are connected, and hence, M_∞ is also connected. Now, consider the graph $\Gamma_f \subseteq \mathbb{P}^4(\mathbb{C}) \times \mathbb{P}^7(\mathbb{C})$ of the map f, which is irreducible together with $\overline{M}$. It follows from the irreducibility of M_∞ that a generic hyperplane section $\Gamma_f \cap \{H_{gen.} \times \mathbb{P}^7(\mathbb{C})\}$ is irreducible, hence the special hyperplane section $\Gamma_f \cap \{\{Z_0 = 0\} \times \mathbb{P}^7(\mathbb{C})\}$ is connected, and therefore, the projection map:

$$proj_{\mathbb{P}^7(\mathbb{C})}[\Gamma_f \cap \{\{Z_0 = 0\} \times \mathbb{P}^7(\mathbb{C})] = f(M_\infty) \equiv 2\mathcal{D},$$

is connected. Hence, the variety $M \cup \Pi = \mathcal{A}$ is compact, connected and embeds smoothly into $\mathbb{P}^7(\mathbb{C})$ via f.

(b) Let g^{t_1} and g^{t_2} be the flows generated respectively by vector fields X_{H_1} and X_{H_2}. For $p \in 2\mathcal{D}$ and for small $\varepsilon > 0$, $g^{t_1}(p), \forall t_1, 0 < |t_1| < \varepsilon$, is well defined and $g^{t_1}(p) \in M$. Then, we may define g^{t_2} on M by:

$$g^{t_2}(q) = g^{-t_1}g^{t_2}g^{t_1}(q),\ q \in U(p) = g^{-t_1}(U(g^{t_1}(p))),$$

where $U(p)$ is a neighborhood of p. By commutativity, we can see that g^{t_2} is independent of t_1:

$$g^{-t_1-\varepsilon_1}g^{t_2}g^{t_1+\varepsilon_1}(q) = g^{-t_1}g^{-\varepsilon_1}g^{t_2}g^{t_1}g^{\varepsilon_1} = g^{-t_1}g^{t_2}g^{t_1}(q).$$

We affirm that $g^{t_2}(q)$ is holomorphic away from $2\mathcal{D}$. This is because $g^{t_2}g^{t_1}(q)$ is holomorphic away from $2\mathcal{D}$ and g^{t_1} is holomorphic in $U(p)$ and maps bi-holomorphically $U(p)$ onto $U(g^{t_1}(p))$.

(c) Since the flows g^{t_1} and g^{t_2} are holomorphic and independent on $2\mathcal{D}$, we can show along the same lines as in the Arnold–Liouville theorem (Arnold 1989) that $\mathcal{A}$ is a torus. And that will done, by considering the holomorphic map:

$$\mathbb{C}^2 \longrightarrow T,\ (t_1, t_2) \longmapsto g^{t_1}g^{t_2}(p),$$

for a fixed origin $p \in \mathcal{A}$. Then, the additive subgroup:

$$Lattice = \{(t_1, t_2) \in \mathbb{C}^2 : g^{t_1}g^{t_2}(p) = p\},$$

is a lattice of $\mathbb{C}^2$, and hence, $\mathbb{C}^2/lattice \longrightarrow \mathcal{A}$ is a biholomorphic diffeomorphism. Therefore, $\mathcal{A} \subseteq \mathbb{P}^7(\mathbb{C})$ is conformal to a complex torus $\mathbb{C}^2/Lattice$ and an Abelian surface as a consequence of Chow's theorem. This completes the proof of the proposition. ■

REMARK 8.1.– *Recall that a Kähler manifold (Definition 5.9) is a manifold with a Kähler metric, i.e. a Hermitian metric whose associated differential two-form of type* $(1,1)$ *is closed. The complex torus* $\mathbb{C}^2/lattice$ *with the Euclidean metric* $\sum dz_i \otimes d\overline{z}_i$ *is a Kähler manifold, and any compact complex manifold that can be embedded in projective space is also a Kähler manifold. Now, a compact complex Kähler manifold having as many independent meromorphic functions as its dimension is a projective variety. We have shown that* $\mathcal{A}$ *is a complex torus* $\mathbb{C}^2/lattice$ *and so in particular* $\mathcal{A}$ *is a Kähler variety with Kähler metric given by* $dt_1 \otimes d\overline{t}_1 + dt_2 \otimes d\overline{t}_2$*. As mentioned above, a compact complex Kähler variety having the required number (as its dimension) of independent meromorphic functions is a projective variety. Thus,* $\mathcal{A}$ *is both a projective variety and a complex torus* $\mathbb{C}^2/Lattice$*, and hence, an Abelian surfaces as a consequence of Chow theorem.*

Now, we need to describe what Abelian surfaces arise this way and how to construct these embedding in theta functions.

PROPOSITION 8.4.– *The flow [8.8] evolves on Abelian surface $\mathcal{A} \subseteq \mathbb{P}^7(\mathbb{C})$ of period matrix:*

$$\begin{pmatrix} 2 & 0 & a & c \\ 0 & 4 & c & b \end{pmatrix}, \quad Im\begin{pmatrix} a & c \\ c & b \end{pmatrix} > 0, \quad (a, b, c \in \mathbb{C}),$$

and it will be expressed in terms of Abelian integrals, involving the differentials (see [8.18]).

PROOF.– Note that the affine invariant surface M [8.9], has the following involution $\sigma : (y_1, y_2, x_1, x_2) \longmapsto (y_1, y_2, -x_1, -x_2)$, which maps X_{H_j} into $-X_{H_j}$, $j = 1, 2$, where X_{H_1} is the flow [8.8] and X_{H_2} the other flow commuting with the first, thus σ amounts to a reflection about some appropriately chosen origin on $\mathcal{A}$. This map acts on the parameters of the Laurent solution [8.12] as follows $(t, \beta, \alpha) \longmapsto (-t, \beta, -\alpha)$. Since $\mathcal{L}$ is symmetric ($\sigma^*\mathcal{L} \simeq \mathcal{L}$), σ can be lifted to $\mathcal{L}$ as an involution $\widetilde{\sigma}$ in two ways differing in sign and for each section (theta function) $s \in H^0(\mathcal{L})$, we therefore have $\widetilde{\sigma}s = \pm s$. Recall that a section $s \in H^0(\mathcal{L})$ is called even (respectively odd) if $\widetilde{\sigma}s = +s$ (respectively $\widetilde{\sigma}s = -s$). Under $\widetilde{\sigma}$, the vector space $H^0(\mathcal{L})$ splits into an even and odd subspace:

$$H^0(L) = H^0(\mathcal{L})^{even} \oplus H^0(\mathcal{L})^{odd},$$

with $H^0(\mathcal{L})^{even}$ containing all the even sections and $H^0(\mathcal{L})^{odd}$ all odd ones. By the classification theory of ample line bundles on Abelian varieties and proposition 8.3 *(c)*, $\mathcal{A} \simeq \mathbb{C}^2/Lattice$ with the period lattice given by the columns of the matrix:

$$\begin{pmatrix} \delta_1 & 0 & a & c \\ 0 & \delta_2 & c & b \end{pmatrix}, \quad \text{Im}\begin{pmatrix} a & c \\ c & b \end{pmatrix} > 0, \quad (a, b, c \in \mathbb{C}),$$

with

$$\delta_1\delta_2 = \dim H^0\left(\mathcal{L}^{\otimes 2}\right) = \dim \mathcal{L}(2\mathcal{D}) = g(2\mathcal{D}) - 1 = 8.$$

Hence, we have two possibilities: *(i)* $\delta_1 = 1$, $\delta_2 = 8$ and *(ii)* $\delta_1 = 2, \delta_4$. From the formulas (see section 6.7), the corresponding line bundle $L^{\otimes 2}$ has in case (i) five even sections and three odd ones, and in case (ii), six even sections and two odd ones. The space $L^{\otimes 2}$ splits into two subspaces $\left(\mathcal{L}^{\otimes 2}\right)^{even}$ and $\left(\mathcal{L}^{\otimes 2}\right)^{odd}$ of even and odd functions for the σ-involution:

$$\begin{aligned} \mathcal{L}^{\otimes 2} &= \left(\mathcal{L}^{\otimes 2}\right)^{even} \oplus \left(\mathcal{L}^{\otimes 2}\right)^{odd} = \{f_0, f_1, f_2, f_3, f_5, f_7\} \oplus \{f_4, f_6\}, \\ &= \{1, y_1, y_1^2, y_2, x_1^2 + y_1^2 y_2, x_1 x_2 + 2a y_1 y_2 + 2 y_1 y_2^2\} \oplus \{x_1,\ x_2 y_1 - 2 x_1 y_2\}. \end{aligned}$$

Among the functions of $\mathcal{L}^{\otimes 2}$, there are six even and two odd functions for the involution σ, showing that case (ii) is the only alternative and the period matrix has the form:

$$\begin{pmatrix} 2\ 0\ a\ c \\ 0\ 4\ c\ b \end{pmatrix}, \ \mathrm{Im}\begin{pmatrix} a & c \\ c & b \end{pmatrix} > 0, \quad (a, b, c \in \mathbb{C}).$$

Consider on $\mathcal{A}$ the holomorphic 1-forms dt_1 and dt_2 defined by:

$$dt_i \left(X_{H_j}\right) = \delta_{ij}.$$

We can check that the 1-forms:

$$\omega_1 = \frac{\alpha^2 d\alpha}{\beta}, \qquad \omega_2 = \frac{d\alpha}{\beta}, \qquad \omega_3 = \frac{\alpha d\alpha}{\beta}, \tag{8.17}$$

yield a basis of $H^0\left(\mathcal{D}, \Omega^1_{\mathcal{D}}\right)$. Next, we show, using the Laurent solutions, that the differentials dt_1 and dt_2 corresponding to the flows generated respectively by $H = H_1$ and H_2, restricted to the curve $\mathcal{D}$ [8.13], descend to two differentials on $\mathcal{D}$:

$$dt_1 \mid_{\mathcal{D}} = \frac{\alpha^2 d\alpha}{\beta} = \omega_1, \qquad dt_2 \mid_{\mathcal{D}} = \frac{d\alpha}{\beta} = \omega_2. \tag{8.18}$$

This concludes the proof of the proposition. ■

REMARK 8.2.– *The divisor $\mathcal{D}$ defines on the Abelian surface $\mathcal{A}$ a polarization $(1, 2)$. We have seen that the eight-dimensional space of sections of the line bundle $L^{\otimes 2}$ splits into two subspaces $\left(L^{\otimes 2}\right)^{even}$ and $\left(L^{\otimes 2}\right)^{odd}$ of even and odd sections (theta functions):*

$$L^{\otimes 2} = \left(L^{\otimes 2}\right)^{even} \oplus \left(L^{\otimes 2}\right)^{odd} = \{f_0, f_1, f_2, f_3, f_5, f_7\} \oplus \{f_4, f_6\}.$$

The remarkable property is that $W\left(\left(L^{\otimes 2}\right)^{even}, \left(L^{\otimes 2}\right)^{odd}\right) \subset \left(\left(L^{\otimes 2}\right)^{even}\right)^{\otimes 2}$ where $W(,)$ denotes the Wronskian $W(f_i, f_j) \equiv f_i X(f_j) - f_j X(f_i)$ between two theta functions, with respect to an arbitrary holomorphic vector field X on the Abelian surface. Then, the Abelian surface, as embedded in $\mathbb{P}^7$ can be described by six quadratic relations between the theta functions, three of which involve even sections only and another three involving even and odd sections. From the divisor $\mathcal{D}$ and the line bundle $\mathcal{L}$, a lot of information can be obtained. For example, Adler and van Moerbeke (1988) have obtained a one-dimensional family of birational maps between the Hénon–Heiles system, Kowalewski's top and the geodesic flow on $SO(4)$ for the Manakov metric. Such birational maps are given by identifying the three eight-dimensional spaces $\mathcal{L}$ of Hénon–Heiles, Kowalewski and Manakov.

We will now show that the Abelian surface $\mathcal{A}$ can be identified as the dual of the Prym variety $Prym(\mathcal{D}/\mathcal{E})$ on which the problem linearizes.

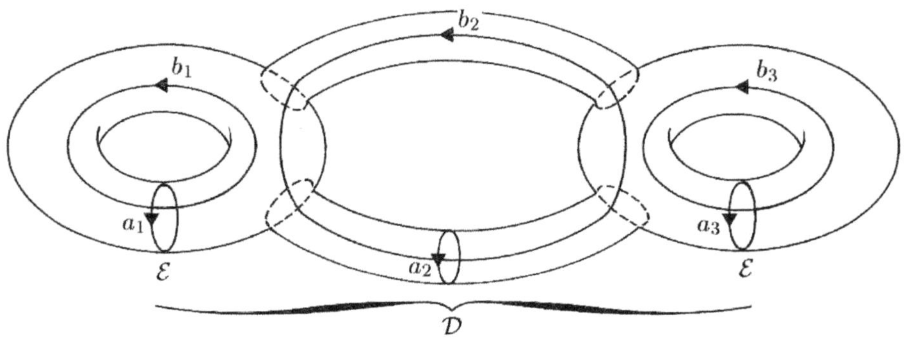

Figure 8.1. *Genus-three curve with a canonical basis of cycles*

PROPOSITION 8.5.– *The Abelian surface $\mathcal{A}$ which completes the affine surface M is the dual Prym variety $Prym^*(\mathcal{D}/\mathcal{E})$ of the genus-three hyperelliptic curve $\mathcal{D}$ [8.13] for the involution σ [8.14] interchanging the sheets of the double covering ρ [8.16], and the problem linearizes on this variety.*

PROOF.– Let $(a_1, a_2, a_3, b_1, b_2, b_3)$ be a canonical homology basis of $\mathcal{D}$ such that $\sigma(a_1) = a_3$, $\sigma(b_1) = b_3$, $\sigma(a_2) = -a_2$, $\sigma(b_2) = -b_2$, for the involution σ [7.34].

As a basis of holomorphic differentials on the curve $\mathcal{D}$, we take the differentials $\omega_1, \omega_2, \omega_3$ [8.17] and obviously $\sigma^*(\omega_1) = -\omega_1, \sigma^*(\omega_2) = -\omega_2, \sigma^*(\omega_3) = \omega_3$. Recall (see theorem 6.16) that the Prym variety $Prym(\mathcal{D}/\mathcal{E})$ is a sub-Abelian variety of the Jacobi variety $Jac(\mathcal{D}) = Pic^0(\mathcal{D}) = H^1(\mathcal{O}_{\mathcal{D}})/H^1(\mathcal{D}, \mathbb{Z})$, constructed from the double cover ρ [8.16], the involution σ on $\mathcal{D}$ interchanging sheets extends by linearity to a map $\sigma : Jac(\mathcal{D}) \to Jac(\mathcal{D})$ and up to some points of order two, with $Jac(\mathcal{D})$ splitting into an even part and an odd part: the even part is an elliptic curve (the quotient of $\mathcal{D}$ by the involution σ, i.e. $\mathcal{E}$ [8.15]) and the odd part is a two-dimensional Abelian surface $Prym(\mathcal{D}/\mathcal{E})$. We consider the period matrix Ω of $Jac(D)$:

$$\Omega = \begin{pmatrix} \int_{a_1}\omega_1 & \int_{a_2}\omega_1 & \int_{a_3}\omega_1 & \int_{b_1}\omega_1 & \int_{b_2}\omega_1 & \int_{b_3}\omega_1 \\ \int_{a_1}\omega_2 & \int_{a_2}\omega_2 & \int_{a_3}\omega_2 & \int_{b_1}\omega_2 & \int_{b_2}\omega_2 & \int_{b_3}\omega_2 \\ \int_{a_1}\omega_3 & \int_{a_2}\omega_3 & \int_{a_3}\omega_3 & \int_{b_1}\omega_3 & \int_{b_2}\omega_3 & \int_{b_3}\omega_3 \end{pmatrix}.$$

Then,

$$\Omega = \begin{pmatrix} \int_{a_1}\omega_1 & \int_{a_2}\omega_1 & -\int_{a_1}\omega_1 & \int_{b_1}\omega_1 & \int_{b_2}\omega_1 & -\int_{b_1}\omega_1 \\ \int_{a_1}\omega_2 & \int_{a_2}\omega_2 & -\int_{a_1}\omega_2 & \int_{b_1}\omega_2 & \int_{b_2}\omega_2 & -\int_{b_1}\omega_2 \\ \int_{a_1}\omega_3 & 0 & \int_{a_1}\omega_3 & \int_{b_1}\omega_3 & 0 & \int_{b_1}\omega_3 \end{pmatrix},$$

and therefore, the period matrices of $Jac(\mathcal{E})$ (i.e. $\mathcal{E}$), $Prym(\mathcal{D}/\mathcal{E})$ and $Prym^*(\mathcal{D}/\mathcal{E})$ are respectively:

$$\Delta = \begin{pmatrix} \int_{a_1} \omega_3 & \int_{b_1} \omega_3 \end{pmatrix}, \qquad \Gamma = \begin{pmatrix} 2\int_{a_1}\omega_1 & \int_{a_2}\omega_1 & 2\int_{b_1}\omega_1 & \int_{b_2}\omega_1 \\ 2\int_{a_1}\omega_2 & \int_{a_2}\omega_2 & 2\int_{b_1}\omega_2 & \int_{b_2}\omega_2 \end{pmatrix},$$

and

$$\Gamma^* = \begin{pmatrix} \int_{a_1}\omega_1 & \int_{a_2}\omega_1 & \int_{b_1}\omega_1 & \int_{b_2}\omega_1 \\ \int_{a_1}\omega_2 & \int_{a_2}\omega_2 & \int_{b_1}\omega_2 & \int_{b_2}\omega_2 \end{pmatrix}.$$

Let

$$L_\Omega = \left\{ \sum_{i=1}^{3} m_i \int_{a_i} \begin{pmatrix} \omega_1 \\ \omega_2 \\ \omega_3 \end{pmatrix} + n_i \int_{b_i} \begin{pmatrix} \omega_1 \\ \omega_2 \\ \omega_3 \end{pmatrix} : m_i, n_i \in \mathbb{Z} \right\},$$

be the period lattice associated with Ω. Let us denote also by L_Δ, the period lattice associated with Δ. We have the following diagram:

$$\begin{array}{ccccccc}
 & & 0 & & & & \\
 & & \downarrow & & & & \\
 & & \mathcal{E} & & \mathcal{D} & & \\
 & & \downarrow \rho^* & \swarrow & \downarrow \rho & & \\
0 \longrightarrow \ker N_\rho & \longrightarrow & Prym(\mathcal{D}/\mathcal{E} \oplus \mathcal{E} = Jac(\mathcal{D}) & \xrightarrow{N_\rho} & \mathcal{E} & \longrightarrow 0 \\
 & \searrow & \downarrow & & & & \\
 & & \mathcal{A} = M \cup 2\mathcal{D} \simeq \mathbb{C}^2/Lattice & & & & \\
 & & \downarrow & & & & \\
 & & 0 & & & &
\end{array}$$

where N_ρ is the norm mapping (surjective). The polarization map:

$$\tau : Prym(\mathcal{D}/\mathcal{E}) \longrightarrow \mathcal{A} = Prym^*(\mathcal{D}/\mathcal{E}),$$

has kernel $(\rho^*\mathcal{E}) \simeq \mathbb{Z}_2 \times \mathbb{Z}_2$ and the induced polarization on $Prym(\mathcal{D}/\mathcal{E})$ is of type (1,2). Let:

$$\mathcal{A} \longrightarrow \mathbb{C}^2/L_\Lambda : p \longmapsto \int_{p_0}^{p} \begin{pmatrix} dt_1 \\ dt_2 \end{pmatrix},$$

be the uniformizing map where dt_1, dt_2 [8.18] are two differentials on $\mathcal{A}$ corresponding to the flows generated respectively by H_1, H_2 such that: $dt_1|_\mathcal{D} = \omega_1$ and $dt_2|_\mathcal{D} = \omega_2$,

$$L_\Lambda = \left\{ \sum_{k=1}^{4} n_k \begin{pmatrix} \int_{\nu_k} dt_1 \\ \int_{\nu_k} dt_2 \end{pmatrix} : n_k \in \mathbb{Z} \right\},$$

is the lattice associated with the period matrix:

$$\Lambda = \begin{pmatrix} \int_{\nu_1} dt_1 \int_{\nu_2} dt_1 \int_{\nu_3} dt_1 \int_{\nu_4} dt_1 \\ \int_{\nu_1} dt_2 \int_{\nu_2} dt_2 \int_{\nu_3} dt_2 \int_{\nu_4} dt_2 \end{pmatrix},$$

and $(\nu_1, \nu_2, \nu_3, \nu_4)$ is a basis of $H_1(\mathcal{A}, \mathbb{Z})$. By the Lefschetz theorem on a hyperplane section, the map $H_1(\mathcal{D}, \mathbb{Z}) \longrightarrow H_1(\mathcal{A}, \mathbb{Z})$ induced by the inclusion $\mathcal{D} \hookrightarrow \mathcal{A}$ is surjective, and consequently, we can find four cycles $\nu_1, \nu_2, \nu_3, \nu_4$ on the curve $\mathcal{D}$ such that:

$$\Lambda = \begin{pmatrix} \int_{\nu_1} \omega_1 \int_{\nu_2} \omega_1 \int_{\nu_3} \omega_1 \int_{\nu_4} \omega_1 \\ \int_{\nu_1} \omega_2 \int_{\nu_2} \omega_2 \int_{\nu_3} \omega_2 \int_{\nu_4} \omega_2 \end{pmatrix},$$

and

$$L_\Lambda = \left\{ \sum_{k=1}^{4} n_k \begin{pmatrix} \int_{\nu_k} \omega_1 \\ \int_{\nu_k} \omega_2 \end{pmatrix} : n_k \in \mathbb{Z} \right\}.$$

The cycles $\nu_1, \nu_2, \nu_3, \nu_4$ in $\mathcal{D}$ which we look for are a_1, a_2, b_1, b_2, and they generate $H_1(T, \mathbb{Z})$ such that:

$$\Lambda = \begin{pmatrix} \int_{a_1} \omega_1 \int_{a_2} \omega_1 \int_{b_1} \omega_1 \int_{b_2} \omega_1 \\ \int_{a_1} \omega_2 \int_{a_2} \omega_2 \int_{b_1} \omega_2 \int_{b_2} \omega_2 \end{pmatrix},$$

is a Riemann matrix. So, $\Lambda = \Gamma^*$, i.e. the period matrix of $Prym^*(\mathcal{D}/\mathcal{E})$ dual of $Prym(\mathcal{D}/\mathcal{E})$. Consequently, $\mathcal{A}$ and $Prym^*(\mathcal{D}/\mathcal{E})$ are two Abelian varieties analytically isomorphic to the same complex torus $\mathbb{C}^2/L_\Lambda$. By Chow's theorem, $\mathcal{A}$ and $Prym^*(\mathcal{D}/\mathcal{E})$ are then algebraically isomorphic. This ends the proof of the proposition. ■

We will now show that $\mathcal{A}$ can also be seen as a double unramified cover of the Jacobian variety $Jac(\mathcal{C})$ of the two-genus hyperelliptic curve $\mathcal{C}$ [8.19]. The remarkable property of this curve is that the flow of solutions of the equations of motion is linearized on its Jacobian $Jac(\mathcal{C})$ and so, the solutions can be expressed in terms of theta functions of two variables. We also discuss the Kummer surface associated with $\mathcal{A}$; it will be constructed explicitly.

PROPOSITION 8.6.– *The torus can be regarded as a double unramified cover of the Jacobian variety $Jac(\mathcal{C})$ of the two-genus hyperelliptic curve $\mathcal{C}$ [8.19] and the system [8.8] can be integrated in terms of genus-two hyperelliptic functions of time. The involution σ [8.14] on the Jacobian $Jac(\mathcal{C})$ leads to a singular surface $K_\mathcal{C}$ which after desingularization defines a K-3 surface $\widetilde{K_\mathcal{C}}$.*

PROOF.– Remember that the Laurent solutions [8.12] restricted to the surface M [8.9] are parameterized by a hyperelliptic curve $\mathcal{D}$ [8.13] of genus three. The latter is a double ramified cover of an elliptic curve $\mathcal{E}$ [8.15] and can also be seen as a two-sheeted unramified cover $\pi : \mathcal{D} \longrightarrow \mathcal{C}$, $(\beta, \alpha) \longmapsto (\eta, \zeta)$ of the following hyperelliptic curve $\mathcal{C}$ of genus two:

$$\mathcal{C} : \eta^2 = \zeta P_4(\zeta). \qquad [8.19]$$

Here, we will show, using the results obtained above and applying the method explained in Vanhaecke (2001), that the linearized flow can be realized on the Jacobian variety $Jac(\mathcal{C})$ of the two-genus hyperelliptic curve $\mathcal{C}$. The torus $\mathcal{A}$ can be regarded as a double unramified cover of the Jacobian variety $Jac(\mathcal{C})$ and the system [8.8] can be integrated in terms of genus-two hyperelliptic functions of time. The differentials dt_1 and dt_2 [8.18], corresponding to the flows X_{H_1} and X_{H_2}, restricted to the curve $\mathcal{D}$, go down to $\mathcal{C}$. Indeed, we have:

$$dt_1\,|_{\mathcal{D}} = \omega_1 = \frac{\alpha^2 d\alpha}{\beta} = \frac{\zeta d\zeta}{\eta}, \qquad dt_2\,|_{\mathcal{D}} = \omega_2 = \frac{d\alpha}{\beta} = \frac{d\zeta}{\eta},$$

yielding the two hyperelliptic differentials on $\mathcal{C}$. The map ρ [8.16] extends to a map $\widetilde{\rho} : \mathcal{A} \rightarrow Jac(\mathcal{C})$. We denote by $|\mathcal{C}|$ the linear system, i.e. the set of all effective divisors linearly equivalent to $\mathcal{C}$. We have $|\mathcal{C}| = \mathbb{P}(L(\mathcal{D}))$; associating with each non-zero function $f \in \mathcal{L}(\mathcal{C})$. An Abelian variety $A \simeq \mathbb{C}^n/Lattice$ has a natural involution σ, induced by the sign flip $(z_1, \ldots, z_n) \longmapsto (-z_1, \ldots, -z_n)$ in $\mathbb{C}^n$. Its fixed points are exactly the 2^{2n} half-periods of $\mathcal{A}$. The quotient $\mathcal{A}/\sigma$ is called the Kummer surface. In our case, $n = 2$, and the involution σ [8.14] on the Jacobian $Jac(\mathcal{C})$ leads to a singular surface $K_{\mathcal{C}}$, which, after desingularization, at the 16 fixed points of the involution σ, defines a $K3$ surface $\widetilde{K_{\mathcal{C}}}$. It has no holomorphic 1-forms, and it has a trivial canonical divisor. The Kummer surface $K_{\mathcal{C}}$ of $Jac(\mathcal{C})$ can be considered as the image of $\psi_{\widetilde{\rho}^*|2\mathcal{C}|} : \mathcal{A} \rightarrow \mathbb{P}^3$ with $\widetilde{\rho}^*|2\mathcal{C}| \subset |2\mathcal{D}|$. Since $\mathcal{C}$ is hyperelliptic of genus two, it has six distinct Weierstrass points (Indeed, let Γ a smooth hyperelliptic curve of genus $g \geq 2$ with $\phi : \Gamma \rightarrow \mathbb{P}^1$ the two-sheeted map, then all the branch points of ϕ are the only Weierstrass points of Γ. By the Riemann–Hurwitz formula, the number of these points is equal to $2g + 2$.). Choose a Weierstrass point P on the curve $\mathcal{C}$ and coordinates $[Z_0, Z_1, Z_2, Z_3]$ for $\mathbb{P}^3$ such that $\psi_{\widetilde{\varphi}^*|2\mathcal{C}|}(P) = [0, 0, 0, 1]$. This point will be a singular point for the Kummer surface K of equation:

$$a\,(Z_0, Z_1, Z_2)\, Z_3^2 + b\,(Z_0, Z_1, Z_2)\, Z_3 + c\,(Z_0, Z_1, Z_2) = 0,$$

where a, b and c are the polynomials of degree $2, 3$ and 4, respectively. After a projective transformation which fixes $[0, 0, 0, 1]$, we may assume that.

$$a\,(Z_0, Z_1, Z_2) = Z_1^2 - 4Z_0Z_2.$$

We can construct an algebraic map from M to the Jacobi variety $Jac\,(C)$:

$$M \longrightarrow Jac(\mathcal{C}), \quad p \in M \longmapsto (\zeta_1 + \zeta_2) \in Jac(\mathcal{C}),$$

and the flows generated by the constants of the motion are straight lines on $Jac(\Gamma)$, i.e. the linearizing equations are given by:

$$\int_{\zeta_1(0)}^{\zeta_1(t)} \omega_1 + \int_{\zeta_2(0)}^{\zeta_2(t)} \omega_1 = c_1 t, \qquad \int_{\zeta_1(0)}^{\zeta_1(t)} \omega_2 + \int_{\zeta_2(0)}^{\zeta_2(t)} \omega_2 = c_2 t,$$

where ω_1, ω_2 [8.17] span the two-dimensional space of holomorphic differentials on the curve C and ζ_1, ζ_2 are two appropriate variables given by:

$$\zeta_1 = \frac{-Z_1 + \sqrt{a(Z_0, Z_1, Z_2)}}{2Z_0}, \qquad \zeta_2 = \frac{-Z_1 - \sqrt{a(Z_0, Z_1, Z_2)}}{2Z_0},$$

algebraically related to the originally given ones, for which the Hamilton–Jacobi equation could be solved by separation of variables. This finishes the proof of the proposition. ■

8.3. Kowalewski's spinning top

Kowalewski's top is a solid body rotating about a fixed point, whose principal moments of inertia $\lambda = \text{diag}(A, B, C)$ (with regard to the fixed point) satisfy the relation $A = B = 2C$ and its center of mass belongs to the equatorial plane (AB plane) through the fixed point. The motion of this body is governed by the equations:

$$\dot{m} = m \wedge \lambda m + \mu g \gamma \wedge l, \qquad \dot{\gamma} = \gamma \wedge \lambda m,$$

where m, l and γ denote respectively the angular momentum, the center of mass and the unit vector in the direction of gravity, μ is the mass of the solid, and g is the acceleration of gravity. By properly picking the axes of inertia in the equatorial plane and after some rescaling and normalization, they may be taken as $l = (1, 0, 0)$, $\lambda m = \left(\frac{m_1}{2}, \frac{m_2}{2}, m_3\right)$. The equations of motion take on the form:

$$\begin{aligned} \dot{m}_1 &= m_2 m_3, & \dot{\gamma}_1 &= 2m_3\gamma_2 - m_2\gamma_3, \\ \dot{m}_2 &= -m_1 m_3 + 2\gamma_3, & \dot{\gamma}_2 &= m_1\gamma_3 - 2m_3\gamma_1, \\ \dot{m}_3 &= -2\gamma_2, & \dot{\gamma}_3 &= m_2\gamma_1 - m_1\gamma_2, \end{aligned}$$

where, without restricting generality, we chose $\mu g = 1$, and we used the substitution $t \longrightarrow 2t$. This system has four first integrals (constants of motion):

$$H_1 = \frac{1}{2}\left(m_1^2 + m_2^2\right) + m_3^2 + 2\gamma_1,$$

$$H_2 = m_1\gamma_1 + m_2\gamma_2 + m_3\gamma_3, \qquad H_3 = \gamma_1^2 + \gamma_2^2 + \gamma_3^2,$$

$$H_4 = \left(\left(\frac{m_1 + im_2}{2}\right)^2 - (\gamma_1 + i\gamma_2)\right)\left(\left(\frac{m_1 - im_2}{2}\right)^2 - (\gamma_1 - i\gamma_2)\right),$$

and is written in the form of the Hamiltonian vector field:

$$\dot{x} = J\frac{\partial H}{\partial x}, \quad H \equiv H_1, \quad x = (m_1, m_2, m_3, \gamma_1, \gamma_2, \gamma_3)^\intercal, \tag{8.20}$$

where

$$J = \begin{pmatrix} 0 & -m_3 & m_2 & 0 & -\gamma_3 & \gamma_2 \\ m_3 & 0 & -m_1 & \gamma_3 & 0 & -\gamma_1 \\ -m_2 & m_1 & 0 & -\gamma_2 & \gamma_1 & 0 \\ 0 & -\gamma_3 & \gamma_2 & 0 & 0 & 0 \\ \gamma_3 & 0 & -\gamma_1 & 0 & 0 & 0 \\ -\gamma_2 & \gamma_1 & 0 & 0 & 0 & 0 \end{pmatrix}.$$

A second flow commuting with the first is regulated by the equations:

$$\dot{x} = J\frac{\partial H_4}{\partial x}, \quad x = (m_1, m_2, m_3, \gamma_1, \gamma_2, \gamma_3)^\intercal, \tag{8.21}$$

The first integrals H_1 and H_4 are in involution, while H_2 and H_3 are trivial. Let:

$$M = \bigcap_{k=1}^{4} \{x \in \mathbb{C}^6 : H_k(x) = c_k\}, \quad c_3 = 1, \tag{8.22}$$

be the affine variety defined by the four generic constants of motion. The invariant variety M is a smooth affine surface for generic values of c_1, c_2 and c_3. So, the question I address is how does one find the compactification of M into an Abelian surface? As in the previous problem, the idea of the direct proof is closely related to the geometric spirit of the (real) Arnold–Liouville theorem (Arnold 1989). Namely, a compact complex n-dimensional variety on which there exist n holomorphic commuting vector fields which are independent at every point is analytically isomorphic to a n-dimensional complex torus $\mathbf{C}^n/Lattice$, and the complex flows generated by the vector fields are straight lines on this complex torus. Now, the main problem will be to complete M into a non-singular compact complex algebraic variety $\mathcal{A} = M \cup \mathcal{D}$, in such a way that the vector fields X_{H_1} and X_{H_4} generated respectively by H_1 and H_4, extend holomorphically along a divisor $\mathcal{D}$ and remain independent there. If this is possible, $\mathcal{A}$ is an algebraic complex torus (an Abelian variety) and the coordinates restricted to M are Abelian functions. A naive guess would be to take the natural compactification $\overline{M}$ of M by projectivizing the equations in $\mathbb{P}^6(\mathbb{C})$. Indeed, this can never work for a general reason: an Abelian variety $\mathcal{A}$ of dimensions greater than or equal than two is never a complete smooth intersection, i.e. it can never be described in some projective space $\mathbb{P}^n(\mathbb{C})$ by n-dim $\mathcal{A}$ global polynomial homogeneous equations. In other words, if M is to be the affine part of an Abelian surface, $\overline{M}$ must have a singularity somewhere along the locus at infinity $\overline{M} \cap \{Z = 0\}$. In fact, what happens in this specific case? Let Z, $M_j = \frac{m_j}{Z}$, $\Gamma_j = \frac{\gamma_j}{Z}$, be the projective coordinates and consider the transformation:

$$\overline{M} \longrightarrow \overline{M}, \quad (Z, M_1, M_2, M_3, \Gamma_1, \Gamma_2, \Gamma_3) \longmapsto (Z, X_1, X_2, M_3, U_1, U_2, \Gamma_3),$$

where $2X_1 = M_1 + iM_2$, $2X_2 = M_1 - iM_2$, $U_1 = \Gamma_1 + i\Gamma_2$, $U_2 = \Gamma_1 - i\Gamma_2$. In these new variables, the equations defining $\overline{M}$ are written:

$$\begin{aligned}
F_1 &\equiv 2X_1X_2 + M_3^2 + (U_1 + U_2)Z - c_1Z^2 = 0,\\
F_2 &\equiv X_1U_2 + X_2U_1 + M_3\Gamma_3 - c_2Z^2 = 0,\\
F_3 &\equiv U_1U_2 + \Gamma_3^2 - Z^2 = 0,\\
F_4 &\equiv (X_1^2 - U_1Z)(X_2^2 - U_2Z) - c_4Z^4.
\end{aligned}$$

PROPOSITION 8.7.– *The projective variety $\overline{M}$ on $\mathbb{P}^6(\mathbb{C})$ defined by the above equations is not an Abelian surface. Furthermore, $\overline{M}$ intersects the hyperplane at infinity $Z = 0$ according to the two straight lines $d_1 = (0, X_1, 0, 0, U_1, 0, 0)$, $d_2 = (0, 0, X_2, 0, 0, U_2, 0)$ and the circle $S^1 = (0, 0, 0, 0, U_1, U_2, \Gamma_3)$, where $U_1U_2 + \Gamma_3^2 = 0$. Moreover, $\overline{M}_c$ is singular along the lines d_1, d_2 (singularity of type $Y^4 + U^3Z^3 = 0$, Y, Z small) and along the circle S^1 (singularity of type $\Gamma_3^4X^2 = (\Gamma_3^4 + \Gamma_3^2 - 1)Z^2$, X, Z small).*

PROOF.– Suppose that $\overline{M}$ is an Abelian surface. Thus, this surface is isomorphic to a two-dimensional complex algebraic torus $\mathbb{C}^2/lattice$, where the lattice is generated by four independent vectors in $\mathbb{C}^2$. Let (φ_1, φ_2) be the natural coordinates of this torus. The canonical divisor of this Abelian surface is empty as, up to scalars, the only 2-form is given by $d\varphi_1 \wedge d\varphi_2$, which clearly does not vanish anywhere on $\overline{M}$. The canonical bundle $K_{\overline{M}}$ of $\overline{M}$ is given in terms of the canonical bundle $K_{\mathbb{P}^6(\mathbb{C})}$ of $\mathbb{P}^6(\mathbb{C})$ by the adjunction formula $K_{\overline{M}} = \left(K_{\mathbb{P}^6(\mathbb{C})} \otimes \left[\overline{M}\right]\right)_{\overline{M}}$, where $\left[\overline{M}\right]$ denote the line bundle associated with the divisor $\overline{M}$ in $\mathbb{P}^6(\mathbb{C})$. Let:

$$x_1 = \frac{M_1}{Z},\quad x_2 = \frac{M_2}{Z},\quad x_3 = \frac{M_3}{Z},\quad y_1 = \frac{\Gamma_1}{Z},\quad y_2 = \frac{\Gamma_2}{Z},\quad y_3 = \frac{\Gamma_3}{Z},$$

be affine coordinates on $\mathbb{P}^6(\mathbb{C})\backslash\{Z = 0\}$ and consider the form:

$$\omega = dx_1 \wedge dx_2 \wedge dx_3 \wedge dy_1 \wedge dy_2 \wedge dy_3.$$

This form has no zeroes or poles in $\{Z \neq 0\}$. Let:

$$u = \frac{Z}{\Gamma_3},\quad u_1 = \frac{M_1}{\Gamma_3},\quad u_2 = \frac{M_2}{\Gamma_3},\quad u_3 = \frac{M_3}{\Gamma_3},\quad v_1 = \frac{\Gamma_1}{\Gamma_3},\quad v_2 = \frac{\Gamma_2}{\Gamma_3},$$

be affine coordinates on $\mathbb{P}^6(\mathbb{C})\backslash\{Z = 0\}$. Then:

$$u = \frac{1}{y_3},\quad u_1 = \frac{x_1}{y_3},\quad u_2 = \frac{x_2}{y_3},\quad u_3 = \frac{x_3}{y_3},\quad v_1 = \frac{y_1}{y_3},\quad v_2 = \frac{y_2}{y_3}$$

and so:

$$\omega = \frac{1}{u^7} du \wedge du_1 \wedge du_2 \wedge du_3 \wedge dv_1 \wedge dv_2.$$

Therefore, the divisor of ω is $(\omega) = -7\mathbf{H}$ where $\mathbf{H}$ is a hyperplan in $\mathbb{P}^6(\mathbb{C})$ and, consequently, $K_{\mathbb{P}^6(\mathbb{C})} = [-7\mathbf{H}]$. Since every line bundle on projective space is a multiple of $[\mathbf{H}]$, we have $K_{\overline{M}} = [\mathbf{H}]_{\overline{M}}$ and this implies that every holomorphic 2-form on $\overline{M}$ would have a zero on M, which is absurd. The second part of the proposition is straightforward. We will describe the singularities of $\overline{M}$ at infinity. First, let us analyze the locus at infinity $C = \overline{M} \cap \{Z = 0\}$. It is shown that $C = d_1 \cup d_2 \cup S^1$, where $d_1 = (0, X_1, 0, 0, U_1, 0, 0)$ and $d_2 = (0, 0, X_2, 0, 0, U_2, 0)$ are two straight lines tangent to the circle $S^1 = (0, 0, 0, 0, U_1, U_2, \Gamma_3)$, $U_1 U_2 + \Gamma_3^2 = 0$. In the neighborhoods of these lines and of this circle, we have rank $\left.\frac{\partial(F_1, F_2, F_3, F_4)}{\partial(Z, X_1, X_2, M_3, U_1, U_2, \Gamma_3)}\right|_{Z=0} = 3$, and the variety $\overline{M}$ is singular along the straight lines d_1, d_2 (singularity of type $Y^4 + U^3 Z^3 = 0$, Y, Z small) and along the circle S^1 (singularity of type $\Gamma_3^4 X^2 = (\Gamma_3^4 + \Gamma_3^2 - 1)Z^2$, X, Z small). ■

The solutions to the system of differential equations [8.20] intersect d_1 and d_2; however, the solutions to equation [8.20] ignore S^1 and the same facts hold for the other vector field [8.21] commuting with the first. To regularize (i.e. make the trajectories parallel) the flow at infinity, it is necessary to solve the singularities of the projective variety $\overline{M}$. We have just seen that these singularities are complicated, and we must (probably) blow up $\overline{M}$ many times along the lines and blow down $\overline{M}$ along the circle. Generally, the method of blowing up and blowing down sub-varieties is considered very delicate, and its use in the present case is difficult. In fact, we will show that the existence of meromorphic solutions to the differential equations [8.20] depending on four free parameters can be used to manufacture the tori, without ever going through the delicate procedure of blowing up and down. Information about the tori can then be gathered from the divisor. The study of convergence of these Laurent series solutions will be carried out via the majorant method and the concept of projective normality.

PROPOSITION 8.8.– *The differential equations [8.20] admit two distinct families of Laurent solutions depending on five free parameters. The first meromorphic solution, with the Laurent series, is:*

$$m_1(t) = \frac{\alpha}{t} + i(\alpha^2 - 2)\beta + o(t), \qquad \gamma_1(t) = \frac{1}{2t^2} + o(t),$$

$$m_2(t) = \frac{i\alpha}{t} - \alpha^2\beta + o(t), \qquad \gamma_2(t) = \frac{i}{2t^2} + o(t),$$

$$m_3(t) = \frac{i}{t} + \alpha\beta + o(t), \qquad \gamma_3(t) = \frac{\beta}{t} + o(t).$$

The second meromorphic solution, with the Laurent series, is:

$$m_1(t) = \frac{\alpha}{t} - i\left(\alpha^2 - 2\right)\beta + o(t), \qquad \gamma_1(t) = \frac{1}{2t^2} + o(t),$$

$$m_2(t) = -\frac{i\alpha}{t} - \alpha^2\beta + o(t), \qquad \gamma_2(t) = -\frac{i}{2t^2} + o(t),$$

$$m_3(t) = -\frac{i}{t} + \alpha\beta + o(t), \qquad \gamma_3(t) = \frac{\beta}{t} + o(t).$$

PROOF.– The proof of this proposition is a straightforward computation. ■

The problem of convergence of these series is solved using the majorant method (around all points where $\alpha, \beta \neq 0$)) and the techniques of projective normality (around bad points). Consider the set of Laurent solution which remain confined to a fixed affine invariant surface, i.e.

$$\begin{aligned} \mathcal{D}_\epsilon &= \left\{ \begin{array}{c} \text{the Laurent solutions } m_j(t), \gamma_j(t), 1 \leq j \leq 3, \text{ such that:} \\ H_k(m_j(t), \gamma_j(t)) = c_k + \text{Taylor part}, 1 \leq k \leq 4 \end{array} \right\}, \\ &= \left\{ \begin{array}{c} (\alpha, \beta, \epsilon) \text{ such that: } (\alpha^2 - 1)((\alpha^2 - 1)\beta^4 - P(\beta) + c_4 = 0, \\ P(\beta) \equiv c_1\beta^2 - 2\epsilon c_2 \beta - 1, \quad \epsilon \equiv \pm i \end{array} \right\}, \end{aligned}$$

where $c_3 = 1$ and c_1, c_2, c_4 are the generic constants. Note that the map:

$$\sigma_\epsilon : \mathcal{D}_\epsilon \longrightarrow \mathcal{D}_\epsilon, \quad (\alpha, \beta, \epsilon) \longmapsto (-\alpha, \beta, \epsilon), \qquad [8.23]$$

is an involution on $\mathcal{D}_\epsilon$, and the quotient $\mathcal{D}_\epsilon^0 = \mathcal{D}_\epsilon / \sigma_\epsilon$ by this involution is an elliptic curve defined by:

$$\mathcal{D}_\epsilon^0 : u^2 = P^2(\beta) - 4c_4\beta^4.$$

The curve $\mathcal{D}_\epsilon$ is a two-sheeted ramified covering of $\mathcal{D}_\epsilon^0$,

$$\varphi_\epsilon : \mathcal{D}_\epsilon \longrightarrow \mathcal{D}_\epsilon^0, (\alpha, u, \beta, \epsilon) \longmapsto (u, \beta, \epsilon), \qquad [8.24]$$

$$\mathcal{D}_\epsilon : \left\{ \begin{array}{l} \alpha^2 = \dfrac{2\beta^4 + P(\beta) + u}{2\beta^4} \\ u^2 = P^2(\beta) - 4c_4\beta^4. \end{array} \right. \qquad [8.25]$$

The curve $\mathcal{D}_\epsilon$ has four points at infinity and four branch points on the elliptic curve $\mathcal{D}_\epsilon^0$ and according to the Riemann–Hurwitz formula (see Appendix, Chapter 9), the genus of $\mathcal{D}_\epsilon$ is $g(\mathcal{D}_\epsilon) = 3$.

PROPOSITION 8.9.– *The divisor on which the solutions of [8.20] blow up consists of two irreducible components $\mathcal{D}_{\epsilon=i}$ and $\mathcal{D}_{\epsilon=-i}$, which are both non-singular curves of genus 3. Each of these curves $\mathcal{D}_\epsilon$ is a double cover of an elliptic curve $\mathcal{D}_\epsilon^0$ ramified at four points.*

Let $\mathcal{D} \equiv \mathcal{D}_{\epsilon=i} + \mathcal{D}_{\epsilon=-i}$. For positive integers k, define the space $\mathcal{L}(k\mathcal{D})$ of meromorphic functions having a k-fold pole at worst along $\mathcal{D}$.

PROPOSITION 8.10.– *The space $\mathcal{L}(\mathcal{D})$ is spanned by the following functions:*

$$f_0 = 1, f_1 = m_1, f_2 = m_2, f_3 = m_3, f_4 = \gamma_3, f_5 = f_1^2 + f_2^2,$$

$$f_6 = 4f_1f_4 - f_3f_5,\ f_7 = (f_2\gamma_1 - f_1\gamma_2)\, f_3 + 2f_4\gamma_2.$$

In addition, $\mathcal{D}$ is embedded into $\mathbb{P}^7(\mathbb{C})$ according to:

$$p = (\alpha, u, \beta) \longmapsto \lim_{t\to 0} t(1, f_1(p), ..., f_7(p)) = \left(0, f_1^{(0)}(p), ..., f_7^{(0)}(p)\right),$$

In a way that $\mathcal{D}_{\epsilon=i}$ and $\mathcal{D}_{\epsilon=-i}$ intersect each other transversally in four points at infinity $\left(\alpha = \pm 1,\ u = \pm\beta^2\sqrt{c_1^2 - 4c_4},\ \beta = \infty\right)$ and that the geometric genus of $\mathcal{D}$ is 9.

PROOF.– Let $L^{(r)}$ be the set of polynomial functions $f = f(x(t))$ of degree $\leq r$ modulo the constants $H_k = c_k, 1 \leq k \leq 4$ and such that:

$$f(x(t)) = t^{-1}(x^{(0)} + x^{(1)}t + \cdots),$$

$x^{(0)} \neq 0$ on $\mathcal{D}$ with $x(t) = (m_1, m_2, m_3, \gamma_1, \gamma_2, \gamma_3)$ given explicitly above. We look for r such that the geometric genus of $\mathcal{D}^{(r)} = N_r + 2$, $\mathcal{D}^{(r)} \subset \mathbb{P}^{N_r}(\mathbb{C})$ and we will show that it is unnecessary to go beyond $r = 4$. Indeed, using the Laurent series obtained previously, we obtain that the spaces $L^{(r)}$, nested according to weighted degree, are generated as follows:

$$L^{(1)} = \{f_0 = 1, f_1 = m_1, f_2 = m_2, f_3 = m_3\},$$

$$L^{(2)} = L^{(1)} \oplus \left\{f_4 = \gamma_3, f_5 = -\frac{1}{4}(f_1^2 + f_2^2)\right\},$$

$$L^{(3)} = L^{(2)} \oplus \{f_6 = f_3f_5 + f_1f_4\},$$

$$L^{(4)} = L^{(3)} \oplus \{f_7 = (f_2\gamma_1 - f_1\gamma_2)f_3 + 2f_4\gamma_2\}.$$

The spaces $L^{(1)}$, $L^{(2)}$ and $L^{(3)}$ do not provide the embedding because $g(\mathcal{D}^{(r)}) \neq \dim L^{(r)} + 1$, $r = 1, 2, 3$. More precisely, since the embedding in $\mathbb{P}^3(\mathbb{C})$ via $L^{(1)}$ does not separate the sheets, we then pass to a $L^{(2)}$ and we also find that the embedding in $\mathbb{P}^6(\mathbb{C})$ is invalid because $g\left(\mathcal{D}^{(2)} \text{ embedded in } \mathbb{P}^5(\mathbb{C})\right) - 2 > 5$, which contradicts the fact that $N_r + 1 = g(\mathcal{D}^{(2)}) - 1$. In the same way, using space $L^{(3)}$, we obtain $g\left(\mathcal{D}^{(3)} \text{ embedded in } \mathbb{P}^6(\mathbb{C})\right) - 2 > 6$, and the contradiction persists. We must therefore consider $L^{(4)}$ and study the embedding of $\mathcal{D}$ into $\mathbb{P}^7(\mathbb{C})$ using the functions of $L^{(4)}$. Let:

$$\left(\frac{f_0}{f_4}, \ldots, \frac{f_7}{f_4}\right) \equiv F_k = F_k^{(0)} + F_k^{(1)}t + \ldots, \quad 0 \leq k \leq 7,$$

and $p_j \equiv \left(\alpha = \pm 1, u = \pm\beta^2\sqrt{c_1^2 - 4c_4}, \beta = \infty\right)$, $1 \leq j \leq 4$, be the four points at ∞ of $\mathcal{D}_\epsilon$. Therefore:

$$\begin{aligned} F_k^{(0)}(p_j) &= \left(0, \frac{\alpha}{\beta}, \frac{\epsilon\alpha}{\beta}, \frac{\epsilon}{\beta}, 1, \epsilon\alpha, \epsilon(\alpha^2 - 1)\beta, -\frac{c_2}{\beta} + \frac{\epsilon(1-u)}{\beta^2}\right)(p_j), \\ &= \left(0, 0, 0, 0, 1, \pm\epsilon, 0, \mp\epsilon\sqrt{c_1^2 - 4c_4}\right) = 4 \text{ distinct points.} \end{aligned}$$

The four points at ∞ are separated on each curve $\mathcal{D}_\varepsilon$ but using the transformation $\mathcal{D}_{\epsilon=i} \longrightarrow \mathcal{D}_{\epsilon=-i}$, $(\alpha, u, \beta) \longmapsto (-\alpha, -u, \beta)$, we show:

$$F_k^{(0)}(p_j)\Big|_{\mathcal{D}_{\epsilon=i}} = F_k^{(0)}(p_j)\Big|_{\mathcal{D}_{\epsilon=-i}},$$

i.e. the four points at ∞ are identified pairwise with the four points at ∞ on the other curve. Let $s = \frac{1}{\beta}$ be a local parameter for p_j. Since $\frac{\partial F_k^{(0)}}{\partial s}(p_j) = (0, \pm 1, \pm\epsilon, \epsilon, 0, -c_2)$, then:

$$\frac{\partial F_k^{(0)}}{\partial s}(p_j)\Bigg|_{\mathcal{D}_{\epsilon=i}} \neq \frac{\partial F_k^{(0)}}{\partial s}(p_j)\Bigg|_{\mathcal{D}_{\epsilon=-i}},$$

and the curve $\mathcal{D}_{\epsilon=i}$ intersects transversely the curve $\mathcal{D}_{\epsilon=-i}$ in four points at infinity $b_1, \ldots, b_4 = (\alpha = \infty, \beta = 0)$, $p_1, \ldots, p_4 = (\alpha = \pm 1, u = \pm\beta^2\sqrt{c_1^2 - 4c_4}, \beta = \infty)$. In a neighborhood of the points $b_j = (\alpha = \infty, \beta = 0)$, $1 \leq j \leq 4$, we divide by f_1. By setting,

$$\left(\frac{f_0}{f_1}, \ldots, \frac{f_7}{f_1}\right) \equiv G_k = G_k^{(0)} + G_k^{(1)}t + \ldots, \quad 0 \leq k \leq 7,$$

we obtain $G_k^{(0)}(p_j) = (0, 1, \epsilon, 0, 0, 0, \mp 1, 0)$ which are the coordinates of four different points in $\mathbb{P}^7(\mathbb{C})$. Therefore, $g\left(\mathcal{D}^{(4)}\right)$ embedded in $\mathbb{P}^7(\mathbb{C})) - 2 = 7$, i.e. $g\left(\mathcal{D}^{(4)}\right) = 9$. ■

To avoid overburdening the text, we will keep the notation $\mathcal{D}$ instead of $\mathcal{D}^{(4)}$ as being embedded in $\mathbb{P}^7(\mathbb{C})$. At all points where a, $\alpha, \beta \neq 0$, the Laurent solutions are nicely convergent (by the majorant method). Therefore, at most points of $\mathcal{D}$, there is a transversal fiber to the curve. Hence, this defines a smooth surface strip around $\mathcal{D}$

except at the bad points. Now, we need to construct a surface strip around $\mathcal{D}$ at the bad points as well. For doing that, we must use the concept of normally generated line bundle or projective normality. Ultimately, we wish to prove that in the various charts:

$$\left(\frac{f_j}{f_k}\right)^{\cdot} = \text{polynomial}\left(\frac{f_j}{f_k}\right), \quad 1 \le j \le 7, k \text{ fixed} \qquad [8.26]$$

This enables us to show that (f_j/f_k), is a bona fide Taylor series starting from every point in a neighborhood of the point in question $\subseteq \mathbb{P}^7(\mathbb{C})$. As mentioned above, let $\mathcal{L} \equiv L^{(4)}$ and $\mathcal{D} \equiv \mathcal{D}^{(4)}$. In fact, we will prove a somewhat stronger statement than equation [8.26], namely that equation [8.26] is satisfied with quadratic polynomials. By inspection, we see that the $f_0, \ldots, f_7$ do not satisfy that property. For example,

$$\left(\frac{f_0}{f_4}\right)^{\cdot} = \frac{f_1\gamma_2 - f_2\gamma_1}{f_4^2} \ne \text{polynomial}\left(\frac{f_j}{f_4}\right), \quad 1 \le j \le 7$$

The problem is that $\mathcal{D}$ although very ample, will not be projectively normal (more precisely, the line bundle $[\mathcal{D}]$ is not normally generated) which forces us to look at the divisor $2\mathcal{D}$ or maybe simply $2\mathcal{D}_{\epsilon=i} + \mathcal{D}_{\epsilon=-i}$. Hence, we must take functions with higher-order poles. Let us consider, for instance, the space $\mathcal{L}(2\mathcal{D}_{\epsilon=i} + \mathcal{D}_{\epsilon=-i})$ of functions which have double poles at worst along $\mathcal{D}_{\epsilon=i}$ and simple ones at worst along $\mathcal{D}_{\epsilon=-i}$. By the Riemann–Roch theorem, if $2\mathcal{D}_{\epsilon=i} + \mathcal{D}_{\epsilon=-i}$ is a divisor on an Abelian variety, this space must be 18-dimensional. Indeed, let:

$$\mathcal{L}(2\mathcal{D}_{\epsilon=i} + \mathcal{D}_{\epsilon=-i}) = L^{(4)} \oplus \{g_8, \ldots, g_{17}\} \equiv \{g_0, \ldots, g_{17}\},$$

where $g_0 = 1$, $g_1 = \frac{1}{2}(m_1 + im_2)$, $g_2 = \frac{1}{2}(m_1 - im_2)$, $g_3 = f_3$, $g_4 = f_4$, $g_5 = f_5$, $g_6 = f_6$, $g_7 = f_7$, $g_8 = g_2^2$, $g_9 = g_2g_3$, $g_{10} = g_2g_4$, $g_{11} = g_2^2 - \gamma_1 - i\gamma_2$, $g_{12} = g_1g_{11}$, $g_{13} = g_1^2g_{11}$, $g_{14} = g_1^3g_{11}$, $g_{15} = g_2g_5$, $g_{16} = g_2g_6$, $g_{17} = g_2g_7$, where we read off the behavior of the g_k along $\mathcal{D}_\epsilon$ from the Laurent series solutions; here, (g_k) denotes the divisor restricted to $\mathcal{D}_\epsilon$: $(g_1) = -\mathcal{D}_{\epsilon=-i}$, $(g_2) = -\mathcal{D}_{\epsilon=i}$, $(g_k) = -\mathcal{D}_{\epsilon=i} - \mathcal{D}_{\epsilon=-i}$, $3 \le k \le 7$, $(g_8) = -2\mathcal{D}_{\epsilon=i}$, $(g_k) = -2\mathcal{D}_{\epsilon=i} - \mathcal{D}_{\epsilon=-i}$, $k = 9, 10, 15, 16, 17$, $(g_{11}) = -2\mathcal{D}_{\epsilon=i} + 2\mathcal{D}_{\epsilon=-i}$, $(g_{12}) = -2\mathcal{D}_{\epsilon=i} + \mathcal{D}_{\epsilon=-i}$, $(g_k) = -2\mathcal{D}_{\epsilon=-i}$, $k = 13, 14$. Note that all g_k have the property $(g_k) \ge -2\mathcal{D}_{\epsilon=i} - \mathcal{D}_{\epsilon=-i}$. Consider now the embedding of $2\mathcal{D}_{\epsilon=i} + \mathcal{D}_{\epsilon=-i}$ via these functions:

$$2\mathcal{D}_{\epsilon=i} + \mathcal{D}_{\epsilon=-i} \to \mathbb{P}^{17}(\mathbb{C}), p \mapsto \lim_{t \to 0} t\,(g_0(p), \ldots, g_{17}(p)) = \left(0, g_1^{(0)}(p), \ldots, g_{17}^{(0)}(p)\right).$$

For the points $b_j = (\alpha = \infty, \beta = 0)$, it is appropriate to divide by g_8, which implies that $\left(\frac{g_k}{g_8}\right)(b_j)$, $0 \le j \le 17$, is finite. In a neighborhood of the points at infinity $p_j = \left(\alpha = \pm 1, u = \pm\beta^2\sqrt{c_1^2 - 4c_4}, \beta = \infty\right)$, $1 \le j \le 4$, we divide $g_0, \ldots, g_{17}$ by g_{10}, which makes $\left(\frac{g_k}{g_{10}}\right)(p_j)$, $0 \le j \le 17$, finite. Next, using differential equations [8.20], we show that in a neighborhood of the points p_j and modulo linear combination of the constants of motion:

$$\left(\frac{g_k}{g_{10}}\right)^{\cdot} = \frac{\dot{g}_k g_{10} - g_k \dot{g}_{10}}{g_{10}^2} = \text{quadratic polynomial of} \left(\frac{g_0}{g_{10}}, ..., \frac{g_{17}}{g_{10}}\right).$$

Also, in a neighborhood of the points b_j, we have:

$$\left(\frac{g_k}{g_8}\right)^{\cdot} = \text{quadratic polynomial of} \left(\frac{g_0}{g_8}, ..., \frac{g_{17}}{g_8}\right).$$

It follows that at the bad points p_j, b_j, the series $\left(\frac{g_k}{g_{10}}\right)(p_j)$, $\left(\frac{g_k}{g_8}\right)(b_j)$, $0 \leq k \leq 17$, converges as a consequence of Picard's theorem applied to the system of ordinary differential equations $\left(\frac{g_k}{g_l}\right)^{\cdot}$, $l = 8, 10$. Therefore, we have the following proposition:

PROPOSITION 8.11.– *The divisor $2\mathcal{D}_{\epsilon=i} + \mathcal{D}_{\epsilon=-i}$ is projectively normal and has a smooth embedding in $\mathbb{P}^{17}(\mathbb{C})$, which shows that in particular the Laurent series solutions converge everywhere.*

In summary, the use of the Laurent series (and the study of convergence of various series intervening in the problem via the majorant method and the concept of a normally generated line bundle) shows that the orbits of the vector field [8.20] going through the curve $\mathcal{D}$ form a smooth surface Π near $\mathcal{D}$ such that $\Pi \backslash M \subseteq \mathcal{A}$ and the variety $\mathcal{A} = M \cup \Pi$ is smooth, compact and connected. More precisely, for generic c, the affine surface M completes into an Abelian surface $\mathcal{A}$ of polarization $(1, 2)$, by adding a singular divisor $\mathcal{D}$ consisting of two isomorphic genus-three curves $\mathcal{D}_{\epsilon=i}$ and $\mathcal{D}_{\epsilon=-i}$ intersecting in four points. Each $\mathcal{D}_{\epsilon=\pm i}$ is a double cover of an elliptic curve $\mathcal{D}_\epsilon^0$ ramified at four points. It defines a line bundle and a polarization $(1, 2)$ on $\mathcal{A}$. The divisors $2\mathcal{D}_{\epsilon=i}$, $2\mathcal{D}_{\epsilon=-i}$ or $\mathcal{D}_{\epsilon=i}+\mathcal{D}_{\epsilon=-i}$ (of geometric genus 9) are all very ample and define a polarization $(2, 4)$; and moreover, the eight-dimensional space of sections of the corresponding line bundle embeds the Abelian surface into $\mathbb{P}^7(\mathbb{C})$. In other words, the line bundle $[\mathcal{D}]$ defines a polarization of type $(2, 4)$ on $\mathcal{A}$ and leads to an embedding of $\mathcal{A}$ in $\mathbb{P}^7(\mathbb{C})$; it is not normally generated, but the line bundle $[2\mathcal{D}_{\epsilon=i}+\mathcal{D}_{\epsilon=-i}]$ is. The Abelian surface $\mathcal{A}$ is equipped with two everywhere independent commuting vector fields. Furthermore, it is shown that the Abelian surface $\mathcal{A}$ can also be identified as the dual of the Pym variety $(\mathcal{D}_\epsilon/\mathcal{D}_\epsilon^0)$, and the problem linearizes on this variety. More precisely, we have:

PROPOSITION 8.12.– *The Abelian surface $\mathcal{A}$ which completes the affine surface M [8.22] is the dual Prym variety Prym* $(\mathcal{D}_\epsilon/\mathcal{D}_\epsilon^0)$ of the genus-three curve $\mathcal{D}_\epsilon$ [8.25] for the involution σ_ϵ [8.23] interchanging the sheets of the double covering φ_ϵ [8.24].*

PROOF.– Simply follow the same reasoning as in proposition 8.5, while taking account of the notations used here. ■

8.4. Kirchhoff's equations of motion of a solid in an ideal fluid

We consider the motion of a solid in a perfect fluid described using the Kirchhoff equations:

$$\dot{p} = p \wedge \frac{\partial H}{\partial l}, \qquad \dot{l} = p \wedge \frac{\partial H}{\partial p} + l \wedge \frac{\partial H}{\partial l}, \qquad [8.27]$$

where (p_1, p_2, p_3) is the velocity of a point fixed relatively to the solid, (l_1, l_2, l_3) is the angular velocity of the body expressed with regard to a frame of reference also fixed relatively to the solid, and H is the Hamiltonian (a differentiable function to be specified below). Explicitly, these questions are written in the form:

$$\begin{aligned}
\dot{p}_1 &= p_2\frac{\partial H}{\partial l_3} - p_3\frac{\partial H}{\partial l_2}, & \dot{l}_1 &= p_2\frac{\partial H}{\partial p_3} - p_3\frac{\partial H}{\partial p_2} + l_2\frac{\partial H}{\partial l_3} - l_3\frac{\partial H}{\partial l_2},\\
\dot{p}_2 &= p_3\frac{\partial H}{\partial l_1} - p_1\frac{\partial H}{\partial l_3}, & \dot{l}_2 &= p_3\frac{\partial H}{\partial p_1} - p_1\frac{\partial H}{\partial p_3} + l_3\frac{\partial H}{\partial l_1} - l_1\frac{\partial H}{\partial l_3},\\
\dot{p}_3 &= p_1\frac{\partial H}{\partial l_2} - p_2\frac{\partial H}{\partial l_1}, & \dot{l}_3 &= p_1\frac{\partial H}{\partial p_2} - p_2\frac{\partial H}{\partial p_1} + l_1\frac{\partial H}{\partial l_2} - l_2\frac{\partial H}{\partial l_1},
\end{aligned}$$

where (p_1, p_2, p_3) is the velocity of a point fixed relatively to the solid, (l_1, l_2, l_3) is the angular velocity of the body expressed with regard to a frame of reference also fixed relatively to the solid, and H is the Hamiltonian. We distinguish two integrable cases: the case of Clebsch and the case of Lyapunov–Steklov.

In Clebsch's case, equations [8.27] have the four invariants:

$$\begin{aligned}
H_1 &= H = \frac{1}{2}\left(a_1p_1^2 + a_2p_2^2 + a_3p_3^2 + b_1l_1^2 + b_2l_2^2 + b_3l_3^2\right),\\
H_2 &= p_1^2 + p_2^2 + p_3^2, \qquad\qquad H_3 = p_1l_1 + p_2l_2 + p_3l_3,\\
H_4 &= \frac{1}{2}\left(b_1p_1^2 + b_2p_2^2 + b_3p_3^2 + \varrho\left(l_1^2 + l_2^2 + l_3^2\right)\right),
\end{aligned}$$

with $\frac{a_2-a_3}{b_1} + \frac{a_3-a_1}{b_2} + \frac{a_1-a_2}{b_3} = 0$, and the constant ϱ satisfies the conditions $\varrho = \frac{b_1(b_2-b_3)}{a_2-a_3} = \frac{b_2(b_3-b_1)}{a_3-a_1} = \frac{b_3(b_1-b_2)}{a_1-a_2}$. We give a sketch of the study of this problem, and we refer the reader to Adler et al. (2004) and Lesfari (2022) for details. Consider points at infinity which are limit points of trajectories of the flow. In fact, there is a Laurent decomposition of such asymptotic solutions,

$$x(t) = t^{-1}\left(x^{(0)} + x^{(1)}t + x^{(2)}t^2 + \ldots\right), \qquad x \equiv (p_1, p_2, p_3, l_1, l_2, l_3),$$

which depend on dim($phase\ space$) $-1=5$ free parameters. Taking into account only solution trajectories lying on the invariant surface:

$$M=\bigcap_{i=1}^{4}\{x:H_i(x)=c_i\}\subset\mathbb{C}^6,$$

we obtain one-parameter families which are parameterized by a curve:

$$\mathcal{D}:\theta^2+c_1\beta^2\gamma^2+c_2\alpha^2\gamma^2+c_3\alpha^2\beta^2+c_4\alpha\beta\gamma=0,$$

where θ is an arbitrary parameter and where $\alpha=x_4^{(0)},\beta=x_5^{(0)},\gamma=x_6^{(0)}$ parameterizes the elliptic curve:

$$\mathcal{E}:\beta^2=d_1^2\alpha^2-1,\ \gamma^2=d_2^2\alpha^2+1,$$

with d_1,d_2 such that: $d_1^2+d_2^2+1=0$. The curve $\mathcal{D}$ is a two-sheeted ramified covering of the elliptic curve $\mathcal{E}$. The branch points are defined by the 16 zeroes of $c_1\beta^2\gamma^2+c_2\alpha^2\gamma^2+c_3\alpha^2\beta^2+c_4\alpha\beta\gamma$ on $\mathcal{E}$. The curve $\mathcal{D}$ is unramified at infinity, and by Hurwitz's formula, the genus of $\mathcal{D}$ is 9. Upon putting $\zeta\equiv\alpha^2$, the curve $\mathcal{D}$ can also be seen as a four-sheeted unramified covering of the following curve of genus three:

$$\mathcal{C}:\left(\theta^2+c_1\beta^2\gamma^2+\left(c_2\gamma^2+c_3\beta^2\right)\zeta\right)^2-c_4^2\zeta\beta^2\gamma^2=0.$$

Moreover, the map $\tau:\mathcal{C}\longrightarrow\mathcal{C},\ (\theta,\zeta)\longmapsto(-\theta,\zeta)$, is an involution on $\mathcal{C}$ and the quotient $\mathcal{C}_0=\mathcal{C}/\tau$ is an elliptic curve defined by:

$$\mathcal{C}_0:\eta^2=c_4^2\zeta\left(d_1^2d_2^2\zeta^2+\left(d_1^2-d_2^2\right)\zeta-1\right).$$

The curve $\mathcal{C}$ is a double ramified covering of $\mathcal{C}_0$, $\mathcal{C}\longrightarrow\mathcal{C}_0,\ (\theta,\eta,\zeta)\longmapsto(\eta,\zeta)$,

$$\mathcal{C}:\begin{cases}\theta^2=-c_1\beta^2\gamma^2-\left(c_2\gamma^2+c_3\beta^2\right)\zeta+\eta\\ \eta^2=c_4^2\zeta\left(d_1^2d_2^2\zeta^2+\left(d_1^2-d_2^2\right)\zeta-1\right).\end{cases}$$

The affine surface M defined by making the invariants of the system equal to generic constants can be completed into a non-singular compact complex algebraic variety (Abelian surface) $\mathcal{A}=M\cup\mathcal{D}$ by adjoining at infinity a smooth curve $\mathcal{D}$ of genus nine. The latter is a double cover of an elliptic curve $\mathcal{E}$ ramified over 16 points. The flow evolves on an Abelian surface $\mathcal{A}\subseteq\mathbb{P}^7(\mathbb{C})$, whose period matrix reads:

$$\begin{pmatrix}2&0&a&c\\0&4&c&b\end{pmatrix},\qquad \operatorname{Im}\begin{pmatrix}a&c\\c&b\end{pmatrix}>0.$$

The functions of $\mathcal{L}(\mathcal{D})$ (the space of meromorphic functions with at worst a simple pole along $\mathcal{D}$) embed $\widetilde{M_c}$ smoothly into $\mathbb{P}^7(\mathbb{C})$ and the divisor $\mathcal{D}$ defines on $\widetilde{M_c}$ a polarization $(2,4)$. The divisor $\mathcal{D}$, although very ample, will not be projectively normal (i.e. its line bundle $[\mathcal{D}]$ is not normally generated) because the eight functions spanning $\mathcal{L}(\mathcal{D})$ do not form a closed system of quadratic differential equations, which forces us to look at the divisor $2\mathcal{D}$. Indeed, a direct calculation using the Laurent series shows that the 32 functions spanning $\mathcal{L}(2\mathcal{D})$ form a closed system of quadratic differential equations and the divisor $2\mathcal{D}$ is projectively normal (i.e. the line bundle $[2\mathcal{D}]$ is normally generated) and has a smooth embedding in $\mathbb{P}^{31}(\mathbb{C})$, and in particular, the Laurent series solutions converge everywhere. The system of differential equations [8.27] has one single five-parameter family of Laurent. Also, $\mathcal{A}$ can be identified as $Prym(\mathcal{C}/\mathcal{C}_0)$. The map:

$$\mathcal{A} \longrightarrow \mathbb{P}^7(\mathbb{C}), \quad (t_1, t_2) \longmapsto [1, X_1(t_1, t_2), ..., X_7(t_1, t_2)],$$

is an embedding of $\mathcal{A}$ into $\mathbb{P}^7(\mathbb{C})$ where $(1, X_1, ..., X_7)$ forms a base of the space $\mathcal{L}(\mathcal{D})$ of meromorphic functions with at worst a simple pole along $\mathcal{D}$ (the functions $X_1, ..., X_7$ are expressed in a simple way as a function of $(p_1, p_2, p_3, l_1, l_2, l_3)$. The solutions of the differential system in question in terms of theta functions are given by:

$$X_k(t) = \frac{\theta_k[(t_1^0, t_2^o) + t(n_1, n_2)]}{\theta_0[(t_1^0, t_2^o) + t(n_1, n_2)]}, \quad k = 1, ..., 7$$

where $(\theta_0, ..., \theta_7)$ forms a base of the space of theta functions associated with $\mathcal{D}$. The two theta functions θ_0, θ_7 are odd, while the six theta functions θ_1, ..., θ_6 are even. Moreover, the integration of this problem can be done using hyperelliptic functions of genus two. In Lyapunov–Steklov's case, we have

$$H_1 = H = \frac{1}{2}\sum_{k=1}^{3}(a_k p_k^2 + b_k l_k^2) + \sum_{k=1}^{3} c_k p_k l_k,$$

$a_1 = A^2 b_1 (b_2 - b_3)^2 + B$, $a_2 = A^2 b_2 (b_3 - b_1)^2 + B$, $a_3 = A^2 b_3 (b_1 - b_2)^2 + B$, $c_1 = Ab_2 b_3 + C$, $c_2 = Ab_1 b_3 + C$, $c_3 = Ab_1 b_2 + C$, where A, B, C are the constants. A fourth first integral is given by

$$H_4 = \frac{1}{2}\sum_{k=1}^{3}(d_k p_k^2 + l_k^2) - A\sum_{k=1}^{3} b_k p_k l_k,$$

where $d_1 = A^2 (b_2 - b_3)^2$, $d_2 = A^2 (b_3 - b_1)^2$ and $d_3 = A^2 (b_1 - b_2)^2$. A long and delicate calculation shows that Prym varieties also come into play when studying this problem (see Adler et al. 2004), and in addition, the integration is done using hyperelliptic functions of genus two and the solutions can be expressed in terms of theta functions.

8.5. Exercises

EXERCISE 8.1.– *Consider the system of nonlinear differential equations:*

$$\dot{x}_1 = f_1(x_1, ..., x_m), ..., \dot{x}_m = f_m(x_1, ..., x_m), \qquad [8.28]$$

where $f_1, ..., f_m$ are the rational functions of m complex variables $x_1, ..., x_m$ and which map a domain from $\mathbb{C}^m$ to $\mathbb{C}$. We assume that this system is quasi-homogeneous, i.e. that there exist positive integers $s_1, ..., s_m$ such that:

$$f_j(\alpha^{s_1}x_1, ..., \alpha^{s_m}x_m) = \alpha^{s_j+1} f_j(x_1, ..., x_m), \quad 1 \leq j \leq m,$$

for each non-zero constant α. Suppose that:

$$x_j = t^{-s_j} \sum_{k=0}^{\infty} c_j^{(k)} t^k, \quad 1 \leq j \leq m, \qquad [8.29]$$

where $c^{(0)} \neq 0$ is the formal solution in the form of the Laurent series, obtained by the method of undetermined coefficients, of the quasi-homogeneous system [8.28]. Show that the coefficients $c_j^{(0)}$ satisfy the nonlinear equations:

$$s_j c_j^{(0)} + f_j\left(c_1^{(0)}, ..., c_m^{(0)}\right) = 0, \qquad [8.30]$$

where $1 \leq j \leq m$, while $c_j^{(1)}, c_j^{(2)}, ...$ each satisfy a system of linear equations of the form:

$$(\mathcal{M} - k\mathcal{I})\, c^{(k)} = \textit{polynomial in} \quad c_j^{(0)}, ..., c_j^{(k-1)}, 1 \leq j \leq m, k \geq 1, \qquad [8.31]$$

where

$$c^{(k)} = \left(c_1^{(k)}, ..., c_m^{(k)}\right)^{\mathsf{T}}, \qquad \mathcal{M} \equiv \left(\frac{\partial f_j}{\partial x_l}\left(c_1^{(0)}, ..., c_m^{(0)}\right) + \delta_{jl} s_j\right)_{1 \leq j,l \leq m},$$

is the Jacobian matrix of equation [8.30]. Study the convergence of the series [8.29].

Solution: Substituting equation [8.29] into equation [8.28], while taking into account the quasi-homogeneity of the system, we obtain:

$$\sum_{k=0}^{\infty} (k - s_j) c_j^{(k)} t^{k-s_j-1} = f_j\left(\sum_{k=0}^{\infty} c_1^{(k)} t^{k-s_1}, ..., \sum_{k=0}^{\infty} c_m^{(k)} t^{k-s_m}\right),$$

$$= f_j\left(t^{-s_1}\left(c_1^{(0)} + \sum_{k=1}^{\infty} c_1^{(k)} t^k\right), ..., t^{-s_m}\left(c_m^{(0)} + \sum_{k=1}^{\infty} c_m^{(k)} t^k\right)\right),$$

$$= t^{-s_j-1} f_j \left(c_1^{(0)} + \sum_{k=1}^{\infty} c_1^{(k)} t^k, ..., c_m^{(0)} + \sum_{k=1}^{\infty} c_m^{(k)} t^k \right).$$

Then, we develop the second member as follows:

$$\sum_{k=0}^{\infty} (k - s_j) c_j^{(k)} t^k = f_j \left(c_1^{(0)}, ..., c_n^{(0)} \right) + \sum_{l=1}^{m} \frac{\partial f_j}{\partial x_l} \left(c_1^{(0)}, ..., c_m^{(0)} \right) \sum_{k=1}^{\infty} c_l^{(k)} t^k$$

$$+ \sum_{k=2}^{\infty} t^k \sum_{(\alpha,\tau)\in\Delta_k} \frac{1}{\alpha!} \frac{\partial^\alpha f_j}{\partial x^\alpha} \left(c_1^{(0)}, ..., c_m^{(0)} \right) \prod_{l=1}^{m} \left(c_l^{(\tau_l)} \right)^{\alpha_l},$$

where $\alpha = (\alpha_1, ..., \alpha_m)$, $\tau = (\tau_1, ..., \tau_m)$, $|\alpha| = \sum_{l=1}^{m} \alpha_l$, $\alpha! = \prod_{l=1}^{m} \alpha_l!$,

$$\Delta_k = \left\{ (\alpha, \tau) : \tau_l > 0, \forall l, |\alpha| > 2, \sum_{l=1}^{m} \alpha_l \tau_l = k \right\}.$$

By identifying the terms having the same power on the first and second members, we successively obtain for $k = 0$ the expression [8.30], for $k = 1$, $(\mathcal{M} - \mathcal{I})\, c^{(1)} = 0$ and for $k \geq 2$,

$$\left((\mathcal{M} - k\mathcal{I})\, c^{(k)} \right)_j = - \sum_{(\alpha,\tau)\in D_k} \frac{1}{\alpha!} \frac{\partial^\alpha f_j}{\partial x^\alpha} \left(c_1^{(0)}, ..., c_m^{(0)} \right) \prod_{l=1}^{m} \left(c_l^{(\tau_l)} \right)^{\alpha_l}, \quad [8.32]$$

where $\tau_l > 0$, $\sum_{l=1}^{m} \alpha_l \tau_l = k$, which leads to the explicit expressions of equation [8.31]. The solution obtained by the method of undetermined coefficients is formal because we obtain it by performing on the series, which we assume a priori to be convergent, various operations whose validity remains to be justified. The theorem will therefore be established as soon as we have verified that these series are convergent. For this purpose, we use the method of majorization functions. First, note that free parameters appear either in the system [8.28] of m equations with m unknowns, when it admits a continuous set of solutions, or by the fact that $\lambda_j \equiv k \in \mathbb{N}^*, 1 \leq j \leq m$, is an Eigenvalue of the matrix $\mathcal{M}$. Therefore, the coefficients can be seen as being rational functions on an affine variety $\mathbb{V}$, with fiber the locus $\bigcap_{j=1}^{m} \left\{ s_j c_j^{(0)} + f_j \left(c_1^{(0)}, ..., c_m^{(0)} \right) = 0 \right\}$. Let $m_0 \in \mathbb{V}$ and let K be a compact subset of $\mathbb{V}$, containing an open neighborhood of m_0. Note that K can be equipped with the topology of the complex plane. Let $A = 1 + \max \left\{ \left| c_j^{(\tau_j)}(m_0) \right| \right\}$, $1 \leq \tau_j \leq \lambda_m$, $1 \leq j \leq m$, where λ_m denotes the largest eigenvalue of the matrix $\mathcal{M}$. Let B and C be the two constants with $C > A$ such that in the compact K, we have:

$$\left| \frac{\partial^\alpha f_j}{\partial x^\alpha}(m_0) \right| \leq \alpha! B^{|\alpha|}, \qquad \left| (\mathcal{M}(m_0) - k\mathcal{I}_m)^{-1} \right| \leq C, \quad k \geq \lambda_m + 1.$$

From [8.32], we deduce that:

$$\left|c_j^{(k)}(m_0)\right| \leq C \sum_{(\alpha,\tau)\in D} B^{|\alpha|} \prod_{l=1}^{m} \left|c_l^{(\tau_l)}\right|^{\alpha_l}, \qquad k \geq \lambda_m + 1.$$

Now consider the series $\Phi(t) = At + \sum_{k=2}^{\infty} \beta_k t^k$, where β_k are the real numbers defined inductively by:

$$\beta_1 \equiv A, \qquad \beta_k \equiv C \sum_{(\alpha,\tau)\in D} B^{|\alpha|} \prod_{l=1}^{m} \beta_{\tau_l}^{\alpha_l}, \quad k \geq 2.$$

We check by recurrence that $\Phi(t)$ is a majorant for $\sum_{k=1}^{\infty} c_j^{(k)} t^k$, where $1 \leq j \leq m$. Indeed, we have $\left|c_j^{(1)}\right| \leq A$. Suppose $\left|c_j^{(l)}\right| \leq \beta_l, l < k, \forall j$. Then, for $k \geq \lambda_m + 1$,

$$\left|c_j^{(k)}(m_0)\right| \leq C \sum_{(\alpha,\tau)\in D} B^{|\alpha|} \prod_{l=1}^{m} \left|c_l^{(\tau_l)}\right|^{\alpha_l} \leq C \sum_{(\alpha,\tau)\in D} B^{|\alpha|} \prod_{l=1}^{m} |\beta_{\tau_l}^{\alpha_l}| = \beta_k.$$

On the contrary, it follows from the definition of the numbers β_k that:

$$\Phi(t) = At + CB^2 \frac{(m\Phi(t))^2}{1 - Bm\Phi(z)}.$$

The root

$$\Phi(t) = \frac{1 + mABt - \sqrt{(1 - 2mAB(1 + 2mBC)t + m^2A^2B^2t^2)}}{2mB(1 + mBC)},$$

provides the desired majorant. Hence, the possibility of a power series expansion near $t = 0$.

EXERCISE 8.2.– *We return to the problem of Kowalewski's spinning top (section 8.3).*

(a) Show that there exist on the Abelian surface $\mathcal{A}$ two holomorphic differentials dt_1 and dt_2 such that:

$$dt_{1|\mathcal{D}_{\pm i}} = \omega_1 = \frac{k_1\left(\alpha^2 - 1\right)\beta^2 d\beta}{\alpha u}, \qquad dt_{2|\mathcal{D}_{\pm i}} = \omega_2 = \frac{k_2 d\beta}{\alpha u},$$

where $k_1, k_2 \in \mathbb{C}$ and ω_1, ω_2 are the holomorphic differentials on $\mathcal{D}_{\pm i}$.

(b) Show that the vector field X_H (respectively X_{H_4}) is regular along the divisor $\mathcal{D}$, transversal at all points $\beta \neq 0$ (respectively $\beta \neq \infty$) and doubly tangent at $\beta = 0$ (respectively $\beta = \infty$).

(c) Determine the space of holomorphic differentials on the divisor $\mathcal{D}$. What can we say about the embedding of $\mathcal{D}$ into $\mathbb{P}^7(\mathbb{C})$?

Solution: *(a)* Let $p \in \mathcal{D}_{\pm i} \cap \{\alpha u \neq 0\}$ where u is given in equation [8.25]. Around the point p, we consider two coordinates τ and x on $\mathcal{A}$ by setting:

$$\tau = \frac{1}{m_3} = \mp it + o(t), \quad x = (m_1 \pm im_2)/2 = \mp i\beta + o(t) \quad \text{on} \quad \mathcal{D}_{\pm i}.$$

We denote by $\frac{\partial}{\partial t_1}$ (respectively $\frac{\partial}{\partial t_2}$) the derivation operator according to the vector field X_H (respectively X_{H_4}). We have:

$$dt_1 = \frac{1}{\triangle(\tau, x)}\left(\frac{\partial x}{\partial t_2}d\tau - \frac{\partial \tau}{\partial t_2}dx\right), \quad dt_2 = \frac{1}{\triangle(\tau, x)}\left(-\frac{\partial x}{\partial t_1}dt + \frac{\partial \tau}{\partial t_1}dx\right),$$

where $\triangle(\tau, x) \equiv \frac{\partial \tau}{\partial t_1}\frac{\partial x}{\partial t_2} - \frac{\partial \tau}{\partial t_2}\frac{\partial x}{\partial t_1}$. Using Laurent's series (proposition 8.8), we obtain:

$$\frac{\partial \tau}{\partial t_1} = \mp i + o(t), \quad \frac{\partial x}{\partial t_1} = -2\alpha\beta^2 + o(t), \quad \frac{\partial \tau}{\partial t_2} = \mp 4i(\alpha^2 - 1)\beta^2 + o(t),$$

$$\frac{\partial x}{\partial t_2} = 8\alpha(\alpha^2 - 1)\beta^4 - c_1\beta^2 \pm 2ic_2\beta + 1 + o(t).$$

Hence,

$$\omega_1 \equiv dt_1|_{\mathcal{D}_\epsilon} = k_1\frac{(\alpha^2 - 1)\beta^2}{\alpha u}d\beta, \quad \omega_2 \equiv dt_2|_{\mathcal{D}_\epsilon} = k_2\frac{d\beta}{\alpha u}, \quad k_1, k_2 \in \mathbb{C}.$$

(b) Note that:

$$\exists F_k^{(0)}, F_l^{(1)} : \det\begin{pmatrix} \frac{\partial}{\partial s}F_k^{(0)}(p_j) & \frac{\partial}{\partial s}F_l^{(0)}(p_j) \\ F_k^1(p_j) & F_l^1(p_j) \end{pmatrix} \neq 0,$$

and therefore the vector field X_H is transversal to $\mathcal{D}$, at the points $p_j \in \mathcal{D}_i \cap \mathcal{D}_{-i}$. The function:

$$\frac{\omega_1}{\omega_2} = \frac{k_1}{k_2}(\alpha^2 - 1)\beta^2 \sim \frac{1}{\beta^2},$$

is meromorphic in a neighborhood of $b_j = (\alpha = \infty, \beta = 0)$, $1 \leq j \leq 4$ and provides the tangent to the curve $\mathcal{D}$ in the coordinates t_1 and t_2. The function $\frac{\omega_1}{\omega_2}$ vanishes when the vector field X_{H_4} is tangent to $\mathcal{D}$ and has a pole when X_H is tangent to $\mathcal{D}$. Therefore, the zeros b_j of ω_2 provide the four points at which the vector field X_H is tangent to $\mathcal{D}$. We find that:

$$\forall \mathcal{F}_k^{(0)}, \mathcal{F}_l^{(1)} : \det\begin{pmatrix} \frac{\partial}{\partial \beta}\mathcal{F}_k^{(0)}(b_j) & \frac{\partial}{\partial \beta}\mathcal{F}_l^{(0)}(b_j) \\ \mathcal{F}_k^1(b_j) & \mathcal{F}_l^1(b_j) \end{pmatrix} = 0,$$

so the vector field X_H is doubly tangent to $\mathcal{D}$ at points b_j.

(*c*) Considering the addition formula concerning the Poincaré residue (see Appendix, Chapter 9), applied to the 2-form $\omega = f_1 dt_1 \wedge dt_2$ where $f_j \in \mathcal{L}(\mathcal{D})$, we have:

$$\omega = \frac{dt_1 \wedge dt_2}{\frac{1}{f_j}} \longrightarrow \operatorname{Res} \omega|_{\mathcal{D}} = -\left.\frac{dt_1}{\frac{\partial}{\partial t_2}\left(\frac{1}{f_j}\right)}\right|_{\mathcal{D}} = \left.\frac{dt_2}{\frac{\partial}{\partial t_1}\left(\frac{1}{f_j}\right)}\right|_{\mathcal{D}},$$

$$= \left.\frac{dt_2}{\frac{\partial}{\partial t_1}\left(\frac{t_1}{f_j^{(0)}} + \circ(t_1^2)\right)}\right|_{\mathcal{D}} = f_j^{(0)} dt_2\Big|_{\mathcal{D}} = f_j^{(0)}\omega_2.$$

Therefore, the space of holomorphic differentials on $\mathcal{D}$ is:

$$\left\{f_1^{(0)}\omega_2, f_2^{(0)}\omega_2, ..., f_7^{(0)}\omega_2\right\} \oplus \{\omega_1, \omega_2\},$$

where $f_1^{(0)}, f_2^{(0)}, ..., f_7^{(0)}$ are the first coefficients of the functions $f_1, f_2, ..., f_7$ (proposition 8.10). The embedding of $\mathcal{D}$ into $\mathbb{P}^7(\mathbb{C})$ is up to two holomorphic differentials the canonical embedding:

$$p = (\alpha, u, \beta) \in \mathcal{D} \longmapsto \left\{\omega_2, f_1^{(0)}\omega_2, f_2^{(0)}\omega_2, ..., f_7^{(0)}\omega_2\right\} \in \mathbb{P}^7(\mathbb{C}).$$

EXERCISE 8.3.– *We consider the group $SO(4)$, its Lie algebra $so(4)$ and the Killing form $\langle X, Y\rangle = -\frac{1}{2}$ tr $(X.Y)$ in $so(4)$ where:*

$$X = (X_{ij})_{1\leq i,j\leq 4} = \sum_{i=1}^{6} x_i e_i = \begin{pmatrix} 0 & -x_3 & x_2 & -x_4 \\ x_3 & 0 & -x_1 & -x_5 \\ -x_2 & x_1 & 0 & -x_6 \\ x_4 & x_5 & x_6 & 0 \end{pmatrix} \in so(4).$$

A left-invariant metric on $SO(4)$ is defined by a non-singular symmetric linear map $\Lambda : so(4) \longrightarrow so(4)$, $X \longmapsto \Lambda.X$, with $\langle gX, gY\rangle = \langle X, \Lambda^{-1}.Y\rangle$ where $g \in SO(4)$. Therefore, the geodesic flow for this metric is written in the form (Euler–Arnold equations):

$$\dot{X} = [X, \Lambda.X], \qquad [8.33]$$

where

$$\Lambda.X = (\lambda_{ij}X_{ij})_{1\leq i,j\leq 4} = \sum_{i=1}^{6} \lambda_i x_i e_i = \begin{pmatrix} 0 & -\lambda_3 x_3 & \lambda_2 x_2 & -\lambda_4 x_4 \\ \lambda_3 x_3 & 0 & -\lambda_1 x_1 & -\lambda_5 x_5 \\ -\lambda_2 x_2 & \lambda_1 x_1 & 0 & -\lambda_6 x_6 \\ \lambda_4 x_4 & \lambda_5 x_5 & \lambda_6 x_6 & 0 \end{pmatrix}.$$

(a) Briefly explain the relationship between Kirchhoff equations and geodesic flow on SO(4).

(b) Verify that the Euler–Arnold equations [8.33] can be written explicitly in the form:

$$\begin{aligned}
\dot{x}_1 &= (\lambda_3 - \lambda_2)\, x_2 x_3 + (\lambda_6 - \lambda_5)\, x_5 x_6,\\
\dot{x}_2 &= (\lambda_1 - \lambda_3)\, x_1 x_3 + (\lambda_4 - \lambda_6)\, x_4 x_6,\\
\dot{x}_3 &= (\lambda_2 - \lambda_1)\, x_1 x_2 + (\lambda_5 - \lambda_4)\, x_4 x_5,\\
\dot{x}_4 &= (\lambda_3 - \lambda_5)\, x_3 x_5 + (\lambda_6 - \lambda_2)\, x_2 x_6,\\
\dot{x}_5 &= (\lambda_4 - \lambda_3)\, x_3 x_4 + (\lambda_1 - \lambda_6)\, x_1 x_6,\\
\dot{x}_6 &= (\lambda_2 - \lambda_4)\, x_2 x_4 + (\lambda_5 - \lambda_1)\, x_1 x_5,
\end{aligned} \qquad [8.34]$$

and form a Hamiltonian vector field:

$$\dot{x}(t) = J\frac{\partial H}{\partial x}, \quad x \in \mathbb{R}^6, \qquad [8.35]$$

with

$$H = \frac{1}{2}\langle X, \Lambda X\rangle = \frac{1}{2}\left(\lambda_1 x_1^2 + \lambda_2 x_2^2 + \cdots + \lambda_6 x_6^2\right),$$

and

$$J = \begin{pmatrix}
0 & -x_3 & x_2 & 0 & -x_6 & x_5\\
x_3 & 0 & -x_1 & x_6 & 0 & -x_4\\
-x_2 & x_1 & 0 & -x_5 & x_4 & 0\\
0 & -x_6 & x_5 & 0 & -x_3 & x_2\\
x_6 & 0 & -x_4 & x_3 & 0 & -x_1\\
-x_5 & x_4 & 0 & -x_2 & x_1 & 0
\end{pmatrix} \in so(6).$$

(c) Show that the system [8.34] admits the following trivial first integrals:

$$H_2 = \frac{1}{2}\left(x_1^2 + x_2^2 + \cdots + x_6^2\right), \qquad H_3 = x_1 x_4 + x_2 x_5 + x_3 x_6.$$

We assume that:

$$\begin{aligned}
&\lambda_1\lambda_6\lambda_4 + \lambda_1\lambda_2\lambda_5 - \lambda_1\lambda_2\lambda_4 + \lambda_3\lambda_6\lambda_5 - \lambda_3\lambda_6\lambda_4 - \lambda_3\lambda_2\lambda_5,\\
&+\lambda_4\lambda_2\lambda_5 + \lambda_4\lambda_1\lambda_3 - \lambda_4\lambda_1\lambda_5 + \lambda_6\lambda_2\lambda_3 - \lambda_6\lambda_2\lambda_5 - \lambda_1\lambda_6\lambda_3 = 0.
\end{aligned}$$

This relation is invariant by cyclic translation: $1 \to 4$, $2 \to 5$, $3 \to 6$, *and it is verified if (Manakov condition):*

$$\lambda_1 = \frac{\beta_2 - \beta_3}{\alpha_2 - \alpha_3}, \qquad \lambda_2 = \frac{\beta_1 - \beta_3}{\alpha_1 - \alpha_3}, \qquad \lambda_3 = \frac{\beta_1 - \beta_2}{\alpha_1 - \alpha_2},$$

$$\lambda_4 = \frac{\beta_1 - \beta_4}{\alpha_1 - \alpha_4}, \qquad \lambda_5 = \frac{\beta_2 - \beta_4}{\alpha_2 - \alpha_4}, \qquad \lambda_6 = \frac{\beta_3 - \beta_4}{\alpha_3 - \alpha_4},$$

with $\alpha_j, \beta_j \in \mathbb{C}$, $\prod_{j<k} (\alpha_j - \alpha_k) \neq 0$. *Show that there exists a fourth first integral of the form:*

$$H_4 = \frac{1}{2}\left(\mu_1 x_1^2 + \mu_2 x_2^2 + \cdots + \mu_6 x_6^2\right).$$

where the parameters $\mu_1, ..., \mu_6$ *can be parametrized (like* $\lambda_1, ..., \lambda_6$*) by*

$$\lambda_1 = \frac{\mu_2 - \mu_3}{\alpha_2 - \alpha_3}, \qquad \lambda_2 = \frac{\mu_1 - \mu_3}{\alpha_1 - \alpha_3}, \qquad \lambda_3 = \frac{\mu_1 - \mu_2}{\alpha_1 - \alpha_2},$$

$$\lambda_4 = \frac{\mu_1 - \mu_4}{\alpha_1 - \alpha_4}, \qquad \lambda_5 = \frac{\mu_2 - \mu_4}{\alpha_2 - \alpha_4}, \qquad \lambda_6 = \frac{\mu_3 - \mu_4}{\alpha_3 - \alpha_4}.$$

Verify that these integrals are functionally independent, in involution, and deduce that the system in question is completely integrable.

(d) Show that the system of differential equations [8.34] has solutions in the form of Laurent series depending on five free parameters. Show that the invariant manifold:

$$\mathbb{A} = \bigcap_{j=1}^{4} \left\{x \equiv (x_1, ..., x_6) \in \mathbb{C}^6 : H_j(x) = c_j\right\},$$

where $c_j, j = 1, ..., 4$ *are the generic constants is the affine part of an Abelian surface* $\mathcal{A}$. *The system in question is algebraically completely integrable, and the vector fields generated by* H *and* H_4 *linearize on the Abelian surface* $\mathcal{A}$. *The latter is isomorphic to the complex torus* $\mathbb{C}^2/L_\Omega$ *where the lattice* L_Ω *is generated by the period matrix:*

$$\Omega = \begin{pmatrix} 2 & 0 & a & c \\ 0 & 4 & c & b \end{pmatrix}, \quad Im\begin{pmatrix} a & c \\ c & b \end{pmatrix} > 0, \quad (a, b, c \in \mathbb{C}),$$

with $\mathcal{A}\backslash\mathbb{A}$ = *algebraic curve* $\mathcal{D}$ *of genus* 9, *which is a ramified covering along an elliptic curve* $\mathcal{E}$. *(*$\mathcal{D}$ *and* $\mathcal{E}$ *are to be determined explicitly).*

(e) Show that the eight functions $1, x_1, ..., x_6, x_7 = (\alpha_3 - \alpha_1)(\alpha_4 - \alpha_2)x_1x_4 + (\alpha_3 - \alpha_2)(\alpha_4 - \alpha_1)x_2x_5$, *form a basis of the space* $\mathcal{L}(\mathcal{D})$) *of meromorphic functions on* $\mathcal{A}$ *having at most one simple pole along* $\mathcal{D}$, *and that furthermore the map:*

$$\mathcal{A} \approx \mathbb{C}^2/L_\Omega \longrightarrow \mathbb{P}^7(\mathbb{C}), \quad (t_1, t_2) \longmapsto [1, x_1(t_1, t_2) : ... : x_6(t_1, t_2)],$$

is an embedding of $\mathcal{A}$ *into projective space* $\mathbb{P}^7(\mathbb{C})$.

(f) Consider the quadratic polynomial:

$$Q_s(x) = \sum_{j=1}^{4} s_j H_j(x), \quad s = [s_1, s_2, s_3, s_4] \in \mathbb{P}^3(\mathbb{C}),$$

where we denote by the same letters $H_1, ..., H_4$ the matrices of the quadratic forms $H_1, ..., H_4$ with respect to the canonical basis. Let:

$$\Delta = \left\{ s = [s_1 : s_2 : s_3 : s_4] \in \mathbb{P}^3(\mathbb{C}) : \det Q_s = \det \sum_{j=1}^{4} s_j H_j = 0 \right\},$$

be the discriminant variety. Show that $\Delta = K_1 \cup K_2 \cup K_3$ is a surface of degree 6 in $\mathbb{P}^3(\mathbb{C})$ where K_j, $1 \leq j \leq 3$ are quadratic cones in $\mathbb{P}^3(\mathbb{C})$, and that furthermore Δ contains an elliptic curve $\mathcal{C}_0$ of quadrics of rank 3. For an appropriate choice of $\lambda_1, ..., \lambda_6, \mu_1, ..., \mu_6$, the curve $\mathcal{C}_0$ can be parameterized as follows:

$$\mathcal{C}_0 = \left\{ s = [2y : z^2 : -z : 1] \in \mathbb{P}^3(\mathbb{C}) | y^2 = \prod_{j=1}^{4} (z - a_j) \right\},$$

so that for each $s \in \mathcal{C}_0$, the quadric $Q_s = z^2 H_2 - zH + H_4 + 2yH_3$ is of rank three.

(g) Show that the curve $\mathcal{D}$ obtained in question (c) *is an unramified covering of degree four along the curve:*

$$\mathcal{C} : w^2 + c_2 z^2 - c_1 z + c_4 + 2c_3 y - 0, \quad y^2 = \prod_{j=1}^{4} (z - a_j),$$

of genus three, the latter being a double covering $\mathcal{C} \to \mathcal{C}_0, (w, y, z) \mapsto (y, z)$ ramified along the elliptic curve $\mathcal{C}_0$ with branch points on $\mathcal{C}_0 \cap \left\{ \sum_{j=1}^{4} s_j H_j(x) \right\} =$ {four points}. Show that, in addition, the Abelian surface $\mathcal{A}$ is characterized as being the variety $Prym(\mathcal{C}/\mathcal{C}_0)$ for the involution $(w, y, z) \to (-w, y, z)$.

Hint: The Kirchhoff equations studied previously (section 8.4) are a special case of the Euler–Arnold equations associated with geodesic flows on Lie groups for a left-invariant metric. This is the geodesic flow on the group $E(3) = SO(3) \times \mathbb{R}^3$ of displacements in three-dimensional Euclidean space. In fact, the problem of the motion of a solid in a perfect fluid is a limiting case of the geodesic flow on SO(4), which suggests that the integrable Clebsch and Lyapunov–Steklov cases are limits of integrable situations on SO(4). Indeed, let the differential system be:

$$\dot{x} = x \wedge \frac{\partial H}{\partial x} + y \wedge \frac{\partial H}{\partial y},$$

$$\dot{y} = y \wedge \frac{\partial H}{\partial x} + x \wedge \frac{\partial H}{\partial y},$$

where $x = (x_1, x_2, x_3) \in \mathbb{R}^3$, $y = (y_1, y_2, y_3) \in \mathbb{R}^3$ and $\wedge$ is the vector product in $\mathbb{R}^3$. To obtain the Kirchhoff equations, simply set $(x, y) = \left(l, \frac{p}{\varepsilon}\right)$ and consider the limit $\varepsilon \to 0$. The above equations are those of the geodesic flow on the group $SO(4)$ for a left-invariant metric defined by the quadratic form H and can be written in the form:

$$\dot{X} = \left[X, \frac{\partial H}{\partial X}\right],$$

with $X \in so(4)$. For further information on the study of this problem and the solution to all the questions raised, we can profitably consult Adler et al. (2004) and Lesfari (2022).

EXERCISE 8.4.– *We consider the Hamiltonian of the Toda lattice in the form:*

$$H = \frac{1}{2}\sum_{j=1}^{3} y_j^2 + \sum_{j=1}^{3} e^{x_j - x_{j+1}}, \quad x_{3+j} = x_j.$$

Using the Flaschka transformations: $z_j = e^{x_j - x_{j+1}}$, $z_{j+j} = -y_j$, *the Toda system can be written explicitly:*

$$\begin{aligned} \dot{z}_1 &= z_1(z_5 - z_4), & \dot{z}_4 &= z_3 - z_1, \\ \dot{z}_2 &= z_2(z_6 - z_5), & \dot{z}_5 &= z_1 - z_2, \\ \dot{z}_3 &= z_3(z_4 - z_6), & \dot{z}_6 &= z_2 - z_3, \end{aligned} \qquad [8.36]$$

and admits the following four first integrals:

$$\begin{aligned} G_1 &= z_1 z_2 z_3 = a_1, \\ G_2 &= z_4 + z_5 + z_6 = a_2, \\ G_3 &= -z_1 - z_2 - z_3 + \frac{1}{2}(z_4^2 + z_5^2 + z_6^2) = a_3, \\ G_4 &= z_1 z_6 + z_2 z_4 + z_3 z_5 + z_4 z_5 z_6 = a_4. \end{aligned} \qquad [8.37]$$

Briefly study the complete algebraic integrability of this system.

Solution: Note that without restricting the generality, we can choose $a_2 = 0$. The system [8.36] is completely integrable and the Hamiltonian structure is defined by the Poisson bracket $\{F, H\} = \left\langle \frac{\partial F}{\partial z}, J\frac{\partial H}{\partial z}\right\rangle = \sum_{k,l=1}^{6} J_{kl}\frac{\partial F}{\partial z_k}\frac{\partial H}{\partial z_l}$, where:

$$J = \begin{pmatrix} 0 & 0 & 0 & -z_1 & z_1 & 0 \\ 0 & 0 & 0 & 0 & -z_2 & z_2 \\ 0 & 0 & 0 & z_3 & 0 & -z_3 \\ z_1 & 0 & -z_3 & 0 & 0 & 0 \\ -z_1 & z_2 & 0 & 0 & 0 & 0 \\ 0 & -z_2 & z_3 & 0 & 0 & 0 \end{pmatrix},$$

an antisymmetric matrix with polynomial elements satisfying the Jacobi identity. The system [8.36] is algebraically integrable and admits Laurent solutions depending on five free parameters:

– First family of solutions in meromorphic Laurent series:

$$z_1 = \frac{1}{t^2} + w - \frac{u+v}{4}t + \cdots, \qquad z_2 = ut - u(\beta-\gamma)t^2 + \cdots,$$

$$z_3 = vt + v(\beta-\gamma)t^2 + \cdots, \qquad z_4 = \frac{1}{t} + \beta - wt + \frac{u+5v}{8}t^2 + \cdots,$$

$$z_5 = -\frac{1}{t} + \beta + wt - \frac{5u+v}{8}t^2 + \cdots,$$

$$z_6 = \gamma + \frac{u-v}{2}t^2 + \frac{(u+v)(\gamma-\beta)}{3}t^3 + \cdots$$

– Second family of solutions in meromorphic Laurent series:

$$z_1 = vt + v(\beta-\gamma)t^2 + \cdots, \qquad z_2 = \frac{1}{t^2} + w - \frac{u+v}{2}t + \cdots,$$

$$z_3 = ut - u(\beta-\gamma)t^2 + \cdots, \qquad z_4 = \gamma + \frac{u-v}{2}t^2 + \cdots,$$

$$z_5 = \frac{1}{t} + \beta - wt + \cdots, \qquad z_6 = -\frac{1}{t} + \beta + wt + \cdots$$

– Third family of solutions in meromorphic Laurent series:

$$z_1 = ut - u(\beta-\gamma)t^2 + \cdots, \qquad z_2 = vt + v(\beta-\gamma)t^2 + \cdots,$$

$$z_3 = \frac{1}{t^2} + w + \cdots, \qquad z_4 = -\frac{1}{t} + \beta + wt + \cdots,$$

$$z_5 = \gamma + \frac{u-v}{2}t^2 + \cdots, \qquad z_6 = \frac{1}{t} + \beta - wt + \cdots$$

Substituting these developments into equations [8.37], we obtain three copies $\mathcal{D}_j$, $j = 1, 2, 3$ (each corresponding to a family of solutions above) of a hyperelliptic curve of genus two, with:

$$u = \frac{c_1}{u} - P(\gamma), \qquad P(\gamma) = \gamma^3 - 2c_2\gamma^2 + (2c_2^2 - c_3)\gamma - c_4, \qquad v = \frac{c_1}{u},$$

$$\beta = c_2 - \frac{1}{2}\gamma, \qquad w = \frac{1}{4}\gamma^2 - \frac{c_2}{3}\gamma + \frac{c_2^2 - c_3}{3}, \qquad u + v = \pm\sqrt{P^2(\gamma) + 4c_1}.$$

The Laurent series above are therefore parameterized by three copies $\mathcal{D}_1$, $\mathcal{D}_2$ and $\mathcal{D}_3$ of a hyperelliptic curve of genus two. It is not difficult to show that equations [8.36] linearize on the Jacobian variety of a hyperelliptic curve of genus two and that the affine manifold defined by the constants [8.37] is completed in this Jacobian variety by the addition of the three copies $\mathcal{D}_j$ of this hyperelliptic curve of genus two. These curves each have a double point in common (tacnode) with each other, and the flow [8.36] is tangent at this point to each curve $\mathcal{D}_j$. In addition, the involution $(z_1, z_2, z_3, z_4, z_5, z_6) \longmapsto (z_2, z_1, z_3, z_6, z_5, z_4)$ on the invariant manifold acts on the free parameters as follows: $(u, v, w, t, \beta, \gamma) \longmapsto (-v, -u, w, -t, \beta, \gamma)$, $(\mathcal{D}_1, \mathcal{D}_2, \mathcal{D}_3) \longmapsto (\mathcal{D}_2, \mathcal{D}_1, \mathcal{D}_3)$ and comes down to a reflection on the origin of the Jacobian variety. The embedding of this Jacobian variety is done in the projective space $\mathbb{P}^8(\mathbb{C})$ using the functions of $\mathcal{L}(\mathcal{D}_1 + \mathcal{D}_2 + \mathcal{D}_3)$ for which a base is provided using the Laurent solutions by:

$$(1, z_5, z_6, z_1 + z_4 z_5, z_2 + z_5 z_6, z_2 z_4 - z_5(z_1 + z_4 z_5), z_1 z_2, z_2 z_3, z_1 z_3).$$

9

Appendix: Riemann Surfaces and Algebraic Curves

In this appendix, we have collected some notions on compact Riemann surfaces or complex algebraic curves. For details and further information see for example Lesfari (2023). Complex algebraic curves $\mathcal{C}$ or compact Riemann surfaces are complex manifolds of complex dimension *1* equipped with atlases whose coordinate changes are holomorphic maps. We define $\mathcal{C}$ by the equation:

$$P(w,z) \equiv p_0(z)w^n + p_1(z)w^{n-1} + \cdots + p_n(z) = 0,$$

where $P(w,z)$ is a polynomial with two complex variables w and z, of degree n in w and irreducible, and $p_0(z) \neq 0$, $p_1(z)$,...,$p_n(z)$ are polynomials in z. The curve $\mathcal{C}$ is homeomorphic to a torus with g holes. The number g is called the genus of the curve $\mathcal{C}$. An important particular case is represented by the elliptic curves ($g = 1$) and hyperelliptic curves of genus g whose affine equation is $P(w,z) = w^2 - p_n(z) = 0$, where $p_n(z)$ is a polynomial without multiple roots, of degree $n = 2g + 1$ or $2g + 2$. Throughout this section, $\mathcal{C}$ denotes a non-singular g complex algebraic curve, that is, such that $gradP \equiv \left(\frac{\partial P}{\partial w}, \frac{\partial P}{\partial z}\right) \neq 0$.

A differential form on $\mathcal{C}$ is written in the form $\omega = f(\tau)d\tau$, where τ is a local parameter and f a complex function of τ. We say that ω : (i) is an Abelian differential if $f(\tau)$ is a meromorphic function on $\mathcal{C}$; (ii) is an holomorphic differential if $f(\tau)$ is a holomorphic function on $\mathcal{C}$; (iii) has a pole of order k or a zero of order k at a point p if $f(\tau)$ has a pole of order k or a zero of order k at this point. An Abelian differential is said to be of the first kind if it is holomorphic on $\mathcal{C}$, of the second kind if its residue is zero at each point of $\mathcal{C}$ and finally of the third kind if its residue is non-zero in at least one point of $\mathcal{C}$.

Let $(a_1, \ldots, a_g, b_1, \ldots, b_g)$ be a basis of cycles in the homology group $H_1(C, \mathbb{Z})$ in such a way that the intersection indices of cycles two by two are written: $(a_j, a_j) = (b_j, b_j) = 0$, $(a_j, b_k) = \delta_{jk}$, $1 \leq j, k \leq g$. We'll say that $(a_1, \ldots, a_g, b_1, \ldots, b_g)$ is a symplectic basis of the homology group $H_1(X, \mathbb{Z})$. We call periods of ω following the cycles a_i, b_j, the numbers $\int_{a_i} \omega$ and $\int_{b_i} \omega$.

THEOREM 9.1.– *The dimension of the set of holomorphic differentials on the curve C is equal to its genus g.*

The formula for the Poincaré residue asserts that all holomorphic differentials on C are of type $\frac{z^i w^j}{\partial P/\partial w} dz$, for some i and j. For example,

$$\omega_k = \frac{z^{k-1} dz}{\sqrt{\prod_{j=1}^{2g+1} (z - z_j)}}, \quad k = 1, 2, \ldots, g,$$

form a basis of holomorphic differentials on the hyperelliptic curve of genus g associated with the equation:

$$P(w, z) = w^2 - \prod_{j=1}^{2g+1} (z - z_j) = 0.$$

THEOREM 9.2.– *Let ω and ω' be two holomorphic differentials on C. Then,*

$$\sum_{k=1}^{g} \left(\int_{a_k} \omega \int_{b_k} \omega' - \int_{b_k} \omega \int_{a_k} \omega' \right) = 0.$$

If moreover ω is non-zero, then

$$-\frac{1}{2\sqrt{-1}} \sum_{k=1}^{g} \left(\int_{a_k} \omega \int_{b_k} \overline{\omega} - \int_{b_k} \omega \int_{a_k} \overline{\omega} \right) > 0.$$

(These two expressions are called Riemann bilinear relations.)

Let $(\omega_1, \ldots, \omega_g)$ be a holomorphic differential basis on the curve C The period matrix of C is defined by $\Omega = (E, F)$, where

$$E = \begin{pmatrix} \int_{a_1} \omega_1 & \cdots & \int_{a_g} \omega_1 \\ \vdots & \ddots & \vdots \\ \int_{a_1} \omega_g & \cdots & \int_{a_g} \omega_g \end{pmatrix}, \qquad F = \begin{pmatrix} \int_{b_1} \omega_1 & \cdots & \int_{b_g} \omega_1 \\ \vdots & \ddots & \vdots \\ \int_{b_1} \omega_g & \cdots & \int_{b_g} \omega_g \end{pmatrix}.$$

Riemann bilinear relations are equivalent respectively to the following relations:

$$\Omega J \Omega^{\top} = 0, \qquad \frac{1}{2\sqrt{-1}} \Omega J \overline{\Omega}^{\top} > 0,$$

where $J = \begin{pmatrix} 0 & I \\ -I & 0 \end{pmatrix}$ is a matrix of order $2g$, I is the unit matrix of order g.

Let us also note that we can find a new basis $(\eta_1, \ldots, \eta_g)$ where

$$\eta_k = \sum_{j=1}^{g} c_{kj} \omega_j, \quad 1 \leq k \leq g,$$

such that: $\int_{a_l} \eta_l = \delta_{lk}$ $1 \leq k, l \leq g$, i.e. such that the period matrix associated with this basis is of the form (I, Z) where I is the unit matrix and $Z = E^{-1}F$. The matrix Z is symmetric and has a positive definite imaginary part.

A divisor $\mathcal{D}$ on $\mathcal{C}$ is an element of the form

$$\mathcal{D} = \sum_{j=1}^{k} n_j p_j, \quad n_j \in \mathbb{Z}, \quad p_j \in \mathcal{C}.$$

The set of divisors is an Abelian group noted $Div(\mathcal{D})$. The addition of the divisors is defined by the addition of the coefficients. To a meromorphic function f on $\mathcal{C}$ corresponds a divisor denoted by (f), the divisor of the zeros and the poles of f. More precisely, we have

$$(f) = \sum_{j=1}^{k} m_j a_j - \sum_{j=1}^{l} n_j b_j,$$

where $a_1, \ldots, a_k$ are the zeros of f of multiplicity $m_1, \ldots, m_k$ respectively and $b_1, \ldots, b_l$ are poles of f of multiplicity $n_1, \ldots, n_l$ respectively. The degree of the divisor $\mathcal{D}$ is an integer denoted $\deg \mathcal{D}$ and is defined by $\deg \mathcal{D} = \sum_{j=1}^{k} n_j$. The map $\deg : Div(\mathcal{C}) \longrightarrow \mathbb{Z}$ is a group homomorphism. Every divisor of a meromorphic function is called a principal divisor. The set of principal divisors forms a subgroup of $\mathrm{Div}^{\circ}(X)$. On any compact Riemann surface, a meromorphic function $f \not\to 0$ has the same number of zeros as poles, so $\deg (f) = 0$. That is, any principal divisor has degree 0 but in general the divisors with degree 0 are not all principal. We say that the divisor $\mathcal{D}$ is positive (or effective) and we write $\mathcal{D} \geq 0$, if the integers n_j are positive. Two divisors $\mathcal{D}_1$ and $\mathcal{D}_2$ are said to be linearly equivalent (that we denote $\mathcal{D}_1 \sim \mathcal{D}_2$) if $\mathcal{D}_1 - \mathcal{D}_2$ is the principal, that is, if $\mathcal{D}_1 - \mathcal{D}_2 = (f)$, where f is a meromorphic function.

THEOREM 9.3 (ABEL).– *Let* $p_1, ..., p_k, q_1, ...q_k$ *be points of* C. *There is a meromorphic function such that:*

$$(f) = \sum_{j=1}^{k} q_j - \sum_{j=1}^{k} p_j,$$

if and only if, there exists a closed path γ *such that:*

$$\forall \omega \in \Omega(C), \quad \sum_{j=1}^{k} \int_{p_j}^{q_j} \omega = \int_\gamma \omega.$$

Let $\mathcal{D}$ be a divisor on C,

$$\mathcal{L}(\mathcal{D}) = \{f \text{ meromorphic function on } C : (f) + \mathcal{D} \geq 0\},$$

and

$$\Omega(\mathcal{D}) = \{\omega \text{ meromorphic differential form on } C : (\omega) + \mathcal{D} \geq 0\}.$$

THEOREM 9.4 (RIEMANN-ROCH).– *We have*

$$\dim \mathcal{L}(\mathcal{D}) - \dim \Omega(-\mathcal{D}) = \deg \mathcal{D} - g + 1.$$

COROLLARY 9.1 (RIEMANN–HURWITZ FORMULA).– *Let* C_1 *and* C_2 *be two algebraic curves of genus* $g(C_1)$ *and* $g(C_2)$ *respectively and* h *a non-constant holomorphic map. Then,*

$$2g(C_1) - 2 = \deg(h)(2g(C_2) - 2) + V(h),$$

where $V(h)$ *is the branching index of* h.

THEOREM 9.5 (ADDITION THEOREM).– *Let* $\mathcal{D}_1$ *and* $\mathcal{D}_2$ *be two positive divisors of degree* n, $\mathcal{E}_1$ *a positive divisor of degree* g *and* $\omega \in \Omega(C)$. *then there exists a positive divisor* $\mathcal{E}_2$ *of degree* g *such that:* $\int_{\mathcal{D}_1}^{\mathcal{D}_2} \omega = \int_{\mathcal{E}_1}^{\mathcal{E}_2} \omega.$

Let $S^g(C)$ be the set of positive divisors on the curve C and define an equivalence relation on this set: $\mathcal{D}_1 \equiv \mathcal{D}_2$ if and only if there exists a meromorphic function f such that: $(f) = \mathcal{D}_1 - \mathcal{D}_2$, or what amounts to the same, if and only if $\int_{\mathcal{D}_1}^{\mathcal{D}_2} \omega = \int_\gamma \omega$, $\forall \omega \in \Omega(C)$, with γ a closed path on C. The Jacobian variety of the curve C, is defined by $Jac(C) = S^g/\equiv$.

THEOREM 9.6.– $S^g/\equiv$ *is homeomorphic to the torus* $\mathbb{C}^g/\Lambda$ *where* Λ *is the lattice defined by the period matrix.*

THEOREM 9.7 (JACOBI INVERSION PROBLEM).– *Let $\mathcal{D}_1$ be a positive divisor of degree g and $(\omega_1, \dots, \omega_g)$ a normalized basis of $\Omega(\mathcal{C})$. Then, for all $(s_1, \dots, s_g) \in \mathbb{C}^g$, there exists a positive divisor $\mathcal{D}_2$ of degree g such that:* $\displaystyle\int_{\mathcal{D}_1}^{\mathcal{D}_2} \omega = s_k$.

10

Appendix: Elliptic Functions and Elliptic Integrals

This appendix presents the basic ideas and properties of elliptic functions and elliptic integrals as an expository essay. The Weierstrass $\wp$ function is studied; it is an elliptic function of order two which has a double pole at the origin in every point of the fundamental parallelogram. Then we introduce the other two Weierstrass functions: the function ζ and the function σ. These functions are connected by interesting relations to the function $\wp$, but unlike the function $\wp$, the function ζ is a meromorphic function with a single pole in the fundamental parallelogram while the function σ is a holomorphic function everywhere. The Weierstrass functions, and especially the elliptic function $\wp$, often intervene when solving many problems. An overview of Jacobi's functions is given. These are elliptic functions of the second order which have two simple poles in the parallelogram of the periods. These functions are often used when solving practical problems. For details and further information see, for example, Lesfari (2023).

10.1. Elliptic functions

Let ω_1 and ω_2 be two complex numbers, $\mathbb{R}$-linearly independent (that is, such that ω_2 is not zero and the quotient $\dfrac{\omega_1}{\omega_2}$ is not real). We consider the lattice $\Lambda = \mathbb{Z}\omega_1 \oplus \mathbb{Z}\omega_2 = \{\omega \equiv m\omega_1 + n\omega_2 : m, n \in \mathbb{Z}\}$, it is a discrete subgroup of $\mathbb{C}$ and it forms a set of parallelograms.

DEFINITION 10.1.– *We call the fundamental parallelogram generated by ω_1 and ω_2 any parallelogram $\Pi = \{z_0 + \alpha\omega_1 + \beta\omega_2 \ : z_0 \in \mathbb{C}, \alpha, \beta \in [0,1]\}$, having affixes z_0, $z_0 + \alpha\omega_1$, $z_0 + \alpha\omega_2$, $z_0 + \alpha\omega_1 + \beta\omega_2$ with $z_0 \in \mathbb{C}$, $0 \leq \alpha, \beta \leq 1$.*

The quotient $\mathbb{C}/\Lambda$ of $\mathbb{C}$ by the equivalence relation determined by Λ as follows, $z_1, z_2 \in \mathbb{C}$, $z_1 \sim z_2 \bmod.\Lambda \iff z_1 - z_2 \in \Lambda$, is a torus. This is homeomorphic to $S^1 \times S^1$, visualizable by gluing both pairs of sides in a square or in a parallelogram.

DEFINITION 10.2.– *A function f is called doubly periodic (of periods ω_1 and ω_2) if $f(z+\omega_1) = f(z)$, $f(z+\omega_2) = f(z)$. In other words, if $f(z+\omega) = f(z)$, $\forall \omega \in \Lambda$. It is also said that f is Λ-periodic.*

REMARK 10.1.– *A non-constant meromorphic function may have at most two linearly independent periods. Moreover, the elements ω_1 and ω_2 are not unique. If ω_1 and ω_2 are two periods of f, then $-\omega_1$ and $-\omega_2$ are also two periods of f and any period of f is written: $\omega = m\omega_1 + n\omega_2$ where $m, n \in \mathbb{Z}$.*

DEFINITION 10.3.– *A meromorphic doubly periodic function is called an elliptic function.*

The set of periods of an elliptic function is a closed subgroup of the additive group of complex numbers. More precisely, in the case of a non-constant elliptic function, this set is a lattice.

PROPOSITION 10.1.– *There is no non-constant elliptic function f that is holomorphic over $\mathbb{C}$. In other words, any elliptic function f having no poles is a constant.*

REMARK 10.2.– *It follows that a non-constant elliptic function has at least one pole in the fundamental parallelogram.*

PROPOSITION 10.2.– *Any non-constant elliptic function has a finite number of poles and a finite number of zeros in a fundamental parallelogram.*

PROPOSITION 10.3.– *Let f be an elliptic function and denote by $b_1, \ldots, b_m$ the poles of f (each pole being counted with multiplicity), then $\sum_{k=1}^{m} Res(f, b_k) = 0$.*

REMARK 10.3.– *The set of elliptic functions with respect to Λ is usually identified with the field of meromorphic functions (the sum, the product and the quotient of two elliptic functions of the same periods is an elliptic function). The nth derivative of an elliptic function is also an elliptic function*

PROPOSITION 10.4.– *Let f be a non-constant elliptic function. Let us denote by $a_1,\ldots,a_l$ the zeros of f of multiplicity $n_1,\ldots,n_l$ respectively and by $b_1,\ldots,b_m$ the poles of f of multiplicity $p_1,\ldots,p_m$ respectively. Then $\sum_{k=1}^{l} n_k = \sum_{k=1}^{m} p_k$. In other words, the* *number of zeros of a non-constant elliptic function is equal to the number of its poles in the fundamental parallelogram.*

PROPOSITION 10.5.– *Let f be an elliptic function. Then*

$$\sum_{k=1}^{l} n_k a_k - \sum_{k=1}^{m} p_k b_k = \textit{period},$$

where $a_1,...,a_l$ are the zeros of f of multiplicity $n_1, ..., n_l$ respectively and $b_1,...,b_m$ the poles of f of multiplicity $p_1,...,p_m$ respectively.

PROPOSITION 10.6.– *Let f and g be two elliptic functions with the same periods. Then, there exists an algebraic relation of the form $P(f(z), g(z)) = 0$, where P is a polynomial with two indeterminate and constant coefficients.*

COROLLARY 10.1.– *Every elliptic function $f(z)$ satisfies a differential equation of the form $P(f(z), f'(z)) = 0$, where P is a polynomial with two indeterminate and constant coefficients.*

10.2. Weierstrass functions

DEFINITION 10.4.– *The Weierstrass $\wp$ function is defined as the series*

$$\wp(z) = \frac{1}{z^2} + \sum_{\omega \in \Lambda \setminus \{0\}} \left(\frac{1}{(z-\omega)^2} - \frac{1}{\omega^2} \right), \qquad [10.1]$$

where $\Lambda = \mathbb{Z}\omega_1 \oplus \mathbb{Z}\omega_2$, as above.

THEOREM 10.1.– *The series [10.1] converges absolutely and uniformly on compact subsets of $\mathbb{C} \setminus \Lambda$.*

PROPOSITION 10.7.– *$\wp(z)$ is an elliptic function with ω_1 and ω_2 as periods. It is even and admits double poles at the points $\omega \in \Lambda$, whose residue is zero. In addition, $\wp'(z)$ is a doubly periodic function and it is odd.*

PROPOSITION 10.8.– *$\wp(z)$ has a Laurent series expansion near the point 0 of the form*

$$\wp(z) = \frac{1}{z^2} + \sum_{k=1}^{\infty} (2k+1) G_{2k+2} z^{2k},$$

where $G_k \equiv G_k(\Lambda) = \displaystyle\sum_{\omega \in \Lambda \setminus \{0\}} \frac{1}{\omega^k}$, $k \geq 4$ and $G_k = 0$ for odd k.

DEFINITION 10.5.– *The two complex numbers g_2 and g_3 defined by the series (called the Eisenstein series)*

$$g_2 = 60 \sum_{\omega \in \Lambda \setminus \{0\}} \frac{1}{\omega^4}, \qquad g_3 = 140 \sum_{\omega \in \Lambda \setminus \{0\}} \frac{1}{\omega^6},$$

are called Weierstrass invariants of the function $\wp(z)$.

THEOREM 10.2.– *The function $\wp(z)$ is a solution in Λ of the differential equation:*

$$(\wp'(z))^2 = 4(\wp(z))^3 - g_2\wp(z) - g_3, \qquad [10.2]$$

where g_2 and g_3 are the Weierstrass invariants of the function $\wp(z)$.

REMARK 10.4.– *Let $y^2 = 4x^3 - ax - b$, be a cubic without singularities (the discriminant of the polynomial $4x^3 - ax - b$ is non-zero). There is a lattice Λ such that: $a = g_2$ and $b = g_3$ where g_2, g_3 are the Weierstrass invariants.*

PROPOSITION 10.9.– *The function $\wp'(z)$ vanishes at the point $a \in \mathbb{C}$ with $-a \equiv a$ mod. Λ, i.e. $2a \in \Lambda$, $a \notin \Lambda$. In other words, modulo Λ, $\wp'(z)$ has three simple zeros: $\frac{\omega_1}{2}, \frac{\omega_2}{2}, \frac{\omega_1+\omega_2}{2}$. By setting $e_1 = \wp\left(\frac{\omega_1}{2}\right)$, $e_2 = \wp\left(\frac{\omega_2}{2}\right)$, $e_3 = \wp\left(\frac{\omega_1+\omega_2}{2}\right)$, we obtain*

$$e_1 \neq e_2 \neq e_3, \quad e_1 + e_2 + e_3 = 0, \quad e_1e_2 + e_2e_3 + e_3e_1 = -\frac{g_2}{4}, \quad e_1e_2e_3 = \frac{g_3}{4}.$$

REMARK 10.5.– *By setting $w = \wp(z)$, equation [10.2] is written*

$$\left(\frac{dw}{dz}\right)^2 = 4w^3 - g_2w - g_3.$$

Since $z \to 0$ when $w \to \infty$, then

$$z = \int_\infty^w \frac{dw}{\sqrt{4w^3 - g_2w - g_3}}. \qquad [10.3]$$

In other words, the function $w = \wp(z)$ is obtained by inversion of the integral [10.3] (called the Weierstrass elliptic integral). Conversely, if the polynomial $4w^3 - g_2w - g_3$ does not have multiple zeros (i.e. its discriminant is non-zero: $g_2^3 - 27g_3^2 \neq 0$), then the inversion of the integral [10.3] leads to the Weierstrass function $\wp(z)$.

PROPOSITION 10.10.– *The map*

$$\mathbb{C}/\Lambda \longrightarrow \mathbb{P}^2(\mathbb{C}), \quad z \longmapsto [1, \wp(z), \wp'(z)], z \neq 0, \quad 0 \longmapsto [0,0,1],$$

is an isomorphism between the complex torus $\mathbb{C}/\Lambda$ and the elliptic curve $\mathcal{E}$ of affine equation: $y^2 = 4x^3 - g_2x - g_3$.

REMARK 10.6.– *Let Λ a lattice in $\mathbb{C}$. It is shown that on $\mathbb{C}\backslash\Lambda$, we have $\overline{\wp(z)} = \wp(\overline{z})$ and $\wp(z_1) = \wp(z_2)$ if and only if $z_1 - z_2 \in \Lambda$ or $z_1 + z_2 \in \Lambda$. In addition, if $z \notin \Lambda$ and $2z \in \Lambda$, then $\wp'(z) = 0$.*

PROPOSITION 10.11.– *Let $u, v \notin \Lambda$ and $u \pm v \notin \Lambda$, then the function $\wp(z)$ satisfies the addition law*

$$\wp(u) + \wp(v) + \wp(u+v) = \frac{1}{4}\left(\frac{\wp'(u) - \wp'(v)}{\wp(u) - \wp(v)}\right)^2,$$

as well as the duplication formula

$$\wp(2z) = \frac{1}{4}\left(\frac{\wp''(z)}{\wp'(z)}\right)^2 - 2\wp(z).$$

Geometrically, if P_1 and P_2 are two distinct points of the elliptic curve $\mathcal{E}$ of the affine equation above, $\mathcal{E}$ has three points of intersection with the line secant $L = P_1P_2$ that goes through P_1 and P_2. In the case where $P_1 = P_2$, L is the tangent to the curve at the point $P_1 = P_2$. The coordinates of the third intersection point are expressed as rational functions of the two others.

Let $\mathcal{E}_\Lambda$ be the set of elliptic functions. This set is a vector space (and even a field). We write $\mathbb{C}(X)$ for the set of rational functions of a variable.

PROPOSITION 10.12.– *We have $\mathcal{E}_\Lambda = \mathbb{C}(\wp, \wp')$, i.e. any elliptic function for Λ is a rational function of $\wp(z)$ and $\wp'(z)$. More precisely, the application*

$$\mathbb{C}(X) \times \mathbb{C}(X) \longrightarrow \mathcal{E}_\Lambda, \quad (g, h) \longmapsto f(z) = g(\wp(z)) + \wp'(z)h(\wp(z)),$$

is an isomorphism between vector spaces. The subfield of even elliptic functions with respect to Λ is generated by $\wp(z)$, i.e. $\mathcal{E}_\Lambda^+ = \{f \in \mathcal{E}_\Lambda : f even\} = \mathbb{C}(\wp)$.

REMARK 10.7.– *According to the previous theorem, to characterize the field of elliptic functions we form the quotient ring $\mathbb{C}[X, Y]$ by the principal ideal corresponding to the equation $Y^2 = 4X^3 - g_2X - g_3$, i.e. we have*

$$\mathcal{E}_\Lambda = \mathbb{C}(\wp, \wp') \simeq \mathbb{C}[X, Y]/\left(Y^2 - 4X^3 + g_2X + g_3\right),$$

where $\wp = X$ and $\wp'$ is identified with the image of Y in the quotient.

DEFINITION 10.6.– *The Weierstrass zeta function ζ[1] is defined by*

$$\zeta(z) = \frac{1}{z} - \int_0^z \left(\wp(z) - \frac{1}{z^2}\right) dz.$$

1 Not to be confused with Riemann zeta function.

Note that the derivative of this function is $\zeta'(z) = -\wp(z)$. Replacing $\wp(z)$ with [10.1] and after integration we get

$$\zeta(z) = \frac{1}{z} + \sum_{\omega \in \Lambda \setminus \{0\}} \left(\frac{1}{z-\omega} + \frac{1}{\omega} + \frac{z}{\omega^2} \right).$$

PROPOSITION 10.13.– *The function $\zeta(z)$ is odd, it is not an elliptic function, it is not periodic and we have $\zeta(z+\omega_k) - \zeta(z) = \tau_k$, $k = 1, 2$ where τ_k are constants. The numbers ω_k and τ_k are related by Legendre's relation: $\tau_1\omega_2 - \tau_2\omega_1 = 2\pi i$.*

PROPOSITION 10.14.– *Let Λ be a lattice in $\mathbb{C}$ and $z \in \mathbb{C}\setminus\Lambda$, fixed. Then, the above series converges absolutely and the convergence is uniform on any compact of $\mathbb{C}\setminus\Lambda$. In addition, the function $\zeta(z)$ is holomorphic on $\mathbb{C}\setminus\Lambda$ and we have $\overline{\zeta(z)} = \zeta(\overline{z})$.*

PROPOSITION 10.15.– *a) Let $u, v \notin \Lambda$ and $u \pm v \notin \Lambda$. Then*

$$\wp(u+v) + \wp(u) + \wp(v) = [\zeta(u+v) - \zeta(u) - \zeta(v)]^2.$$

b) We have

$$2\frac{\begin{vmatrix} 1 & \wp(u) & \wp^2(u) \\ 1 & \wp(v) & \wp^2(v) \\ 1 & \wp(w) & \wp^2(w) \end{vmatrix}}{\begin{vmatrix} 1 & \wp(u) & \wp'(u) \\ 1 & \wp(v) & \wp'(v) \\ 1 & \wp(w) & \wp'(w) \end{vmatrix}} = \zeta(u+v+w) - \zeta(u) - \zeta(v) - \zeta(w).$$

PROPOSITION 10.16.– *We have*

$$\zeta(z) = \frac{1}{z} + \sum_{k=2}^{\infty} \frac{c_k}{2k-1} z^{2k-1},$$

where

$$c_2 = \frac{g_2}{20}, \quad c_3 = \frac{g_3}{28}, \quad c_k = \frac{3}{(2k+1)(k-3)} \sum_{m=2}^{k-2} c_m c_{k-m}, \quad k \geq 4.$$

PROPOSITION 10.17.– *If f is an elliptic function with poles $b_1, ..., b_n$ and main parts:*

$$\frac{c_{j_1}}{z-b_j} + \frac{c_{j_2}}{(z-b_j)^2} + \cdots + \frac{c_{j_k}}{(z-b_j)^k}, \quad j = 1, ..., n$$

then

$$f(z) = C + \sum_{j=1}^{n} \left(c_{j_1}\zeta(z-b_j) + \cdots + (-1)^{k-1} \frac{c_{j_k}}{(k-1)!} \zeta^{(k-1)}(z-b_j) \right),$$

where C is a constant.

DEFINITION 10.7.– *The Weierstrass sigma function σ is defined by*

$$\sigma(z) = z e^{\int_0^z \left(\zeta(z) - \frac{1}{z}\right) dz}.$$

The logarithmic derivative of the function σ is

$$(\ln \sigma(z))' = \frac{\sigma'(z)}{\sigma(z)} = \zeta(z).$$

Replacing $\zeta(z)$ with its expression

$$\zeta(z) = \frac{1}{z} + \sum_{\omega \in \Lambda \setminus \{0\}} \left(\frac{1}{z-\omega} + \frac{1}{\omega} + \frac{z}{\omega^2} \right),$$

we obtain

$$\sigma(z) = z \prod_{\omega \in \Lambda \setminus \{0\}} \left(1 - \frac{z}{\omega}\right) e^{\frac{z}{\omega} + \frac{1}{2}\left(\frac{z}{\omega}\right)^2}.$$

PROPOSITION 10.18.– *The function $\sigma(z)$ is an odd function and we have*

$$\sigma(z + \omega_k) = -e^{\tau_k\left(z + \frac{\omega_k}{2}\right)}.\sigma(z), \quad (k = 1, 2),$$

where τ_k arc constants.

PROPOSITION 10.19.– *Let f be an elliptic function of order n. Let's designate by $a_1, ..., a_n$ (resp. $b_1, ..., b_n$) the zeros (resp., poles) of f in the parallelogram of the periods. Here all zeros and poles are counted with their multiplicity orders. Then*

$$f(z) = C\sigma\left(z + \sum_{j=2}^{n} a_j - \sum_{j=1}^{n} b_j\right) \frac{\prod_{j=2}^{n} \sigma(z - a_j)}{\prod_{j=1}^{n} \sigma(z - b_j)},$$

where C is a constant.

PROPOSITION 10.20.– *Let Λ be a lattice in $\mathbb{C}$ and $z \in \mathbb{C} \setminus \Lambda$, fixed. Then, the infinite product*

$$\sigma(z) = z \prod_{\omega \in \Lambda \setminus \{0\}} \left(1 - \frac{z}{\omega}\right) e^{\frac{z}{\omega} + \frac{1}{2}\left(\frac{z}{\omega}\right)^2},$$

converges and the convergence is uniform on any compact of $\mathbb{C} \setminus \Lambda$. Moreover the function $\sigma(z)$ is holomorphic on $\mathbb{C}$.

10.3. Elliptic integrals and Jacobi elliptic functions

In general, we use the term elliptic integral for an integral of the form

$$\int R\left(s, \sqrt{P(s)}\right) ds,$$

where R is a rational function in two variables and $P(s)$ a polynomial of degree three or four with simple roots. In general, this integral is not expressed by means of elementary functions, that is to say those obtained by applying to the variable s the algebraic operations (addition, subtraction, multiplication, division) in finite number, as well as logarithmic, trigonometric and inverse functions. We will see that the inverse functions of these elliptic integrals are elliptic functions. It is shown that with the help of elementary transformations that an elliptic integral is reduced to one of the Legendre canonical forms:

$$\int \frac{ds}{\sqrt{(1-s^2)(1-k^2s^2)}}, \qquad \int \sqrt{\frac{1-k^2s^2}{1-s^2}} ds,$$

$$\int \frac{ds}{(1+ls^2)\sqrt{(1-s^2)(1-k^2s^2)}},$$

where k and l are constants. The first of these integrals is called the elliptic integral of the first kind, the second the elliptic integral of the second kind and the third the elliptic integral of the third kind. We can write these integrals in a slightly different form, by setting $s = \sin\varphi$, and the previous integrals are written

$$\int \frac{d\varphi}{\sqrt{1-k^2\sin^2\varphi}}, \qquad \int \sqrt{1-k^2\sin^2\varphi} d\varphi,$$

$$\int \frac{d\varphi}{(1+l\sin^2\varphi)\sqrt{1-k^2\sin^2\varphi}}.$$

In general, we adopt the following notations:

$$F(k,\varphi) = \int_0^{\sin\varphi} \frac{ds}{\sqrt{(1-s^2)(1-k^2s^2)}} = \int_0^{\varphi} \frac{d\varphi}{\sqrt{1-k^2\sin^2\varphi}},$$

$$E(k,\varphi) = \int_0^{\sin\varphi} \sqrt{\frac{1-k^2s^2}{1-s^2}} ds = \int_0^{\varphi} \sqrt{1-k^2\sin^2\varphi} d\varphi,$$

$$\Pi(k,l,\varphi) = \int_0^{\sin\varphi} \frac{ds}{(1+ls^2)\sqrt{(1-s^2)(1-k^2s^2)}},$$

$$= \int_0^{\varphi} \frac{d\varphi}{(1+l\sin^2\varphi)\sqrt{1-k^2\sin^2\varphi}}.$$

Integrals are often encountered where the upper bound is $\varphi = \frac{\Pi}{2}$. In this case, we write

$$F(k) = \int_0^1 \frac{ds}{\sqrt{(1-s^2)(1-k^2s^2)}} = \int_0^{\frac{\Pi}{2}} \frac{d\varphi}{\sqrt{1-k^2\sin^2\varphi}},$$

$$E(k) = \int_0^1 \sqrt{\frac{1-k^2s^2}{1-s^2}}ds = \int_0^{\frac{\Pi}{2}} \sqrt{1-k^2\sin^2\varphi d\varphi},$$

and these integrals are called complete elliptic integrals of the first and second kind, respectively. We show that

$$F(k) = \frac{\Pi}{2}\left(1 + \left(\frac{1}{2}\right)^2 + \left(\frac{1.3}{2.4}\right)^2 k^4 + \left(\frac{1.3.5}{2.4.6}\right)^2 k^6 + \cdots\right),$$

$$E(k) = \frac{\Pi}{2}\left(1 - \left(\frac{1}{2}\right)^2 - \left(\frac{1.3}{2.4}\right)^2 \frac{k^4}{3} - \left(\frac{1.3.5}{2.4.6}\right)^2 \frac{k^6}{5} - \cdots\right).$$

Let us consider elliptic integrals of the form

$$t = \int_0^s \frac{ds}{\sqrt{(1-s^2)(1-k^2s^2)}}, \qquad 0 \leq k \leq 1$$

and see with a little more detail the properties of this integral of the first kind while knowing that the properties of the other integrals are obtained in a similar way. We have seen above that the change of variable $s = \sin\varphi$ transforms this integral to the form

$$t = \int_0^\varphi \frac{d\varphi}{\sqrt{1-k^2\sin^2\varphi}}.$$

We first consider the case where $k \neq 0$ and $k \neq 1$. The function $t(\varphi)$ defined by this integral has an inverse called the amplitude of t and is written $\varphi = \mathbf{am}t = \mathbf{am}(t;k)$. Note that if $k = 0$, then $t = \arcsin s$, and $s = \sin t$. For $k \neq 0$, we note by analogy the inverse function of the integral in question by $s = \mathbf{sn}t = \mathbf{sn}(t;k)$, which we call the Jacobi elliptic function (read s, n, t by detaching the letters). The number k is called module the of the function. When there is no ambiguity on the module k, we simply write $\mathbf{sn}t$ instead of $\mathbf{sn}(t;k)$. The function $\varphi = \mathbf{am}t$ is a strictly increasing odd function of t and we have $\mathbf{am}(0) = 0$, $\frac{\partial \mathbf{am}}{\partial t}(0) = 1$. Since $s = \sin\varphi$, we can write $s = \mathbf{sn}t = \sin(\mathbf{am}t)$. The second and third Jacobi elliptic functions are defined by

$$\mathbf{cn}t = \mathbf{cn}(t;k) = \cos \mathbf{am}t, \qquad \mathbf{dn}t = \mathbf{dn}(t;k) = \sqrt{1-k^2\mathbf{sn}^2t}.$$

For $\mathbf{cn}t$, read c, n, t by detaching the letters. Similarly, for $\mathbf{dn}t$, read d, n, t by detaching the letters. Again when there is no ambiguity on the module k, we simply write $\mathbf{cn}t$ (resp. $\mathbf{dn}t$) instead of $\mathbf{cn}(t;k)$ (resp. $\mathbf{dn}(t;k)$).

PROPOSITION 10.21.– *We have* $\mathbf{sn}^2t + \mathbf{cn}^2t = 1$ *and* $\mathbf{dn}^2t + k^2\mathbf{sn}^2t = 1$.

REMARK 10.8.– *The periods of the function* $\mathbf{sn}t$ *are* $4K$ *and* $2iK'$ *with*

$$K = \int_0^1 \frac{ds}{\sqrt{(1-s^2)(1-k^2s^2)}} = F(k),$$

and

$$K' = \int_1^{\frac{1}{k}} \frac{ds}{\sqrt{(s^2-1)(1-k^2s^2)}} = \int_0^1 \frac{dr}{\sqrt{(1-r^2)(1-k'^2r^2)}} = F(k'),$$

where $k' = \sqrt{1-k^2}$ *and* $s = \dfrac{1}{\sqrt{1-k'^2r^2}}$. *Similarly, the periods of* $\mathbf{cn}t$ *are* $4K$ *and* $2K + 2iK'$ *and those of* $\mathbf{dn}t$ *are* $2K$ *and* $4iK'$.

REMARK 10.9.– *The elliptic integral of the first kind*

$$\int_0^z \frac{dz}{\sqrt{(1-z^2)(1-k^2z^2)}},$$

realizes a conformal application of the upper half-plane $\{z \in \mathbb{C} : Im\, z > 0\}$ *on the rectangle of vertices* $\pm K$, $\pm K + iK'$.

PROPOSITION 10.22.– *We have* $\mathbf{sn}(0) = 0$, $\mathbf{cn}(0) = 1$ *and* $\mathbf{dn}(0) = 1$. *The function* $\mathbf{sn}t$ *is odd while the functions* $\mathbf{cn}t$ *and* $\mathbf{dn}t$ *are pairs. In other words, we have* $\mathbf{sn}(-t) = -\mathbf{sn}t$, $\mathbf{cn}(-t) = \mathbf{cn}t$ *and* $\mathbf{dn}(-t) = \mathbf{dn}t$.

PROPOSITION 10.23.– *The derivatives of the three Jacobi elliptic functions are given by*

$$\frac{d}{dt}\mathbf{sn}t = \mathbf{cn}t.\mathbf{dn}t, \qquad \frac{d}{dt}\mathbf{cn}t = -\mathbf{sn}t.\mathbf{dn}t, \qquad \frac{d}{dt}\mathbf{dn}t = -k^2\mathbf{sn}t.\mathbf{cn}t.$$

PROPOSITION 10.24.– *Jacobi's elliptical functions satisfy the following differential equations:*

$$\left(\frac{d}{dt}\mathbf{sn}t\right)^2 = \left(1 - \mathbf{sn}^2 t\right)\left(1 - k^2\mathbf{sn}^2 t\right),$$

$$\left(\frac{d}{dt}\mathbf{cn}t\right)^2 = \left(1 - \mathbf{cn}^2 t\right)\left(k'^2 + k^2\mathbf{cn}^2 t\right),$$

$$\left(\frac{d}{dt}\mathbf{dn}t\right)^2 = \left(1 - \mathbf{dn}^2 t\right)\left(\mathbf{dn}^2 t - k'^2\right),$$

where $k' = \sqrt{1-k^2}$.

COROLLARY 10.2.– *Jacobi's elliptical functions:* $\mathbf{sn}t$, $\mathbf{cn}t$ *and* $\mathbf{dn}t$ *are obtained by inversion respectively of the integrals:*

$$t = \int_0^w \frac{dw}{\sqrt{(1-w^2)(1-k^2w^2)}},$$

$$t = \int_0^w \frac{dw}{\sqrt{(1-w^2)(k'^2+k^2w^2)}},$$

$$t = \int_0^w \frac{dw}{\sqrt{(1-w^2)(w^2-k'^2)}}.$$

where $k' = \sqrt{1-k^2}$.

Let's finally examine the case $k = 0$ and $k = 1$.

PROPOSITION 10.25.– *a) When* $k = 0$, *we have*

$$\mathbf{am}(t;0) = t, \quad \mathbf{sn}(t;0) = \sin t, \quad \mathbf{cn}(t;0) = \cos t, \quad \mathbf{dn}(t;0) = 1.$$

b) When $k = 1$, *we have*

$$\mathbf{sn}(t;1) = \tanh t, \quad \mathbf{cn}(t;1) = \frac{1}{\cosh t}, \quad \mathbf{cn}(t;1) = \frac{1}{\cosh^2 t}.$$

PROPOSITION 10.26.– *The functions* $\mathbf{sn}t$, $\mathbf{cn}t$, $\mathbf{dn}t$ *satisfy respectively the following addition formulas:*

$$\mathbf{sn}(t+\tau) = \frac{\mathbf{sn}t\mathbf{cn}\tau\mathbf{dn}\tau + \mathbf{sn}\tau\mathbf{cn}t\mathbf{dn}t}{1-k^2\mathbf{sn}^2t\mathbf{sn}^2\tau},$$

$$\mathbf{cn}(t+\tau) = \frac{\mathbf{cn}t\mathbf{cn}\tau - \mathbf{sn}t\mathbf{sn}\tau\mathbf{dn}t\mathbf{dn}\tau}{1-k^2\mathbf{sn}^2t\mathbf{sn}^2\tau},$$

$$\mathbf{dn}(t+\tau) = \frac{\mathbf{dn}t\mathbf{dn}\tau - k^2\mathbf{sn}t\mathbf{sn}\tau\mathbf{cn}t\mathbf{cn}\tau}{1-k^2\mathbf{sn}^2t\mathbf{sn}^2\tau}.$$

REMARK 10.10.– *In a sense, the Jacobi elliptic functions* $\mathbf{sn}t$ *and* $\mathbf{cn}t$ *generalize the trigonometric functions sine and cosine.*

10.4. Application: simple pendulum

A simple pendulum is an idealization of a real pendulum. It consists of a point mass suspended at the end of a cord of zero mass, in such a manner that it can swing in an arc of a circle. We denote by l the length of the cord (i.e. the radius of the circle) and by g the acceleration of gravity. The motion of the simple pendulum is described by an angle displacement x of the pendulum from its vertical equilibrium position. The equation of motion is

$$\frac{d^2x}{dt^2} + \frac{g}{l}\sin x = 0. \qquad [10.4]$$

Let $\theta = \frac{dx}{dt}$, and equation [10.4] is written $\theta d\theta + \frac{g}{l}\sin x dx = 0$. By integrating, we get $\frac{\theta^2}{2} = \frac{g}{l}\cos x + C$, where C is a constant. To determine the latter, note that when $t = 0$, $x = x_0$ (initial angle), then $\theta = 0$ (the speed is zero), hence $C = -\frac{g}{l}\cos x_0$. Therefore

$$\frac{l}{2g}\left(\frac{dx}{dt}\right)^2 = \frac{l}{2g}\theta^2 = \cos x - \cos x_0. \qquad [10.5]$$

We will study several cases:

a) Consider the case of an oscillatory movement, ie, the case where the mass passes from $x = x_0$ (the largest angle reached by the pendulum, it corresponds to a speed $\theta = 0$) to $x = 0$ (maximum speed). Since $\cos x = 1 - 2\sin^2\frac{x}{2}$, equation [10.5] becomes

$$\frac{l}{4g}\left(\frac{dx}{dt}\right)^2 = \sin^2\frac{x_0}{2} - \sin^2\frac{x}{2}. \qquad [10.6]$$

Let $\sin\frac{x}{2} = \sin\frac{x_0}{2}\sin\varphi$, hence $\frac{1}{2}\cos\frac{x}{2}dx = \sin\frac{x_0}{2}\cos\varphi d\varphi$,

$$\frac{1}{2}\sqrt{1-\sin^2\frac{x}{2}}dx = \sin\frac{x_0}{2}\sqrt{1-\sin^2\varphi}d\varphi,$$

$$\frac{1}{2}\sqrt{1-\sin^2\frac{x_0}{2}\sin^2\varphi}dx = \sin\frac{x_0}{2}\sqrt{1-\sin^2\varphi}d\varphi,$$

and so

$$dx = \frac{2\sin\frac{x_0}{2}\sqrt{1-\sin^2\varphi}}{\sqrt{1-\sin^2\frac{x_0}{2}\sin^2\varphi}}d\varphi.$$

By substitution in [10.6], we obtain $\left(\frac{d\varphi}{dt}\right)^2 = \frac{g}{l}\left(1-k^2\sin^2\varphi\right)$, where $k = \sin\frac{x_0}{2}$ is the module and $\frac{x_0}{2}$ the modular angle. Note that for $x = 0$, we have $\varphi = 0$ and therefore

$$t = \pm\sqrt{\frac{l}{g}}\int_0^\varphi \frac{d\varphi}{\sqrt{1-k^2\sin^2\varphi}}.$$

From what precedes, we have

$$\varphi = \pm\mathbf{am}\sqrt{\frac{g}{l}}t, \qquad \sin\varphi = \pm\sin\mathbf{am}\frac{g}{l}t = \pm\mathbf{sn}\sqrt{\frac{g}{l}}t,$$

and consequently, $\sin\frac{x}{2} = \pm\sin\frac{x_0}{2}\mathbf{sn}\sqrt{\frac{g}{l}}t.$

b) Consider the case of a circular motion. Equation [10.5] is written in the form

$$\frac{l}{2g}\left(\frac{dx}{dt}\right)^2 = 1-2\sin^2\frac{x}{2} - \cos x_0 = (1-\cos x_0)\left(1-k^2\sin^2\frac{x}{2}\right),$$

where $k^2 = \frac{2}{1-\cos x_0}$, with k positive and $0 < k < 1$. Taking into account the initial condition $x(0) = 0$, we obtain

$$dt = \pm\sqrt{\frac{2l}{g(1-\cos x_0)}}\int_0^\varphi \frac{d\varphi}{\sqrt{1-k^2\sin^2\varphi}}, \qquad \varphi = \frac{x}{2}.$$

Hence

$$\varphi = \pm\mathbf{am}\sqrt{\frac{g(1-\cos x_0)}{2l}}t, \qquad x = \pm 2\mathbf{am}\sqrt{\frac{g(1-\cos x_0)}{2l}}t.$$

c) Consider finally the case of an asymptotic movement. This is the case where $x_0 = \pm\pi$ and the equation [10.5] is written

$$\frac{l}{2g}\left(\frac{dx}{dt}\right)^2 = \cos x + 1 = 2\cos^2\frac{x}{2}.$$

Hence

$$t = \pm\frac{1}{2}\sqrt{\frac{l}{g}}\int_0^x \frac{dx}{\cos\frac{x}{2}} = \pm\sqrt{\frac{l}{g}}\ln\tan\left(\frac{x}{\pi} + \frac{\pi}{4}\right),$$

and $x = 4\arctan e^{\pm\sqrt{\frac{g}{l}}t} - \pi$. We check that $x \to \pm\pi$ when $t \to \infty$.

REMARK 10.11.– *For small oscillations, equation [10.4] can be reduced to a linear equation,* $\frac{d^2x}{dt^2} + \frac{g}{l}x = 0$*, whose general solution is immediate:*

$$x(t) = C_1\cos\sqrt{\frac{g}{l}}t + C_2\sqrt{\frac{l}{g}}\sin\sqrt{\frac{g}{l}}t, \qquad C_1 = x(0), \qquad C_2 = \frac{dx}{dt}(0).$$

For small oscillations, the period of the pendulum (the time need for a complete oscillation; a round trip) is $2\pi\sqrt{\frac{l}{g}}$*. On the other hand, in the case of oscillations which are not necessarily small, the period is determined according to what precedes* $4\sqrt{\frac{l}{g}}\int_0^{\frac{\pi}{2}}\frac{dx}{\sqrt{1-k^2\sin^2 x}}$ *with* $k = \sin\frac{x_0}{2}$.

References

Adler, M. and van Moerbeke, P. (1988). The Kowalewski and Hénon-Heiles motions as Manakov geodesic flows on SO(4) – a two-dimensional family of Lax pairs. *Comm. Math. Phys.*, 113(4), 659–700.

Adler, M., van Moerbeke, P., Vanhaecke, P. (2004). *Algebraic Integrability, Painlevé Geometry and Lie Algebras*. Springer, Berlin.

Arbarello, E., Cornalba, M., Griffiths, P.A., Harris, J. (1985). *Geometry of Algebraic Curves*, vol. 1. Springer, New York.

Arnold, V.I. (1978/1989). *Mathematical Methods in Classical Mechanics*, 2nd edition. Springer-Verlag, Berlin, Heidelberg and New York.

Barth, W. (1985). Abelian surfaces with (1,2)-polarization. *Adv. Studies in Pure Math.*, 10, 41–84.

Barth, W., Hulek, K., Peters, C., van de Ven, A. (2015). *Compact Complex Surfaces*. Springer, Berlin, Heidelberg.

Beauville, A. (1977). Prym varieties and the Schottky problem. *Invent. Math.*, 41, 149–196.

Belokolos, A.I., Bobenko, V.Z., Enol'skii, V.Z., Its, A.R., Matveev, V.B. (1994). *Algebro-Geometric Approach to Nonlinear Integrable Equations*. Springer-Verlag, Berlin.

Birkenhake, C.H. and Lange, H. (2004). *Complex Abelian Varieties*, 2nd edition. Springer, Berlin.

Debarre, O. (1999). *Tores et variétés abéliennes complexes*. EDP Sciences, Paris.

Dolbeault, P. (1990). *Analyse complexe*. Masson and Dunod, Paris.

Dubrovin, B.A. (1981). Theta functions and non-linear equations. *Russian Math. Surv.*, 36(2), 11–92.

Griffiths, J. and Harris, P.A. (1978). *Principles of Algebraic Geometry*. Wiley-Interscience, New York.

Hartshorne, R. (1977). *Algebraic Geometry*. Springer, New York.

Hindry, M. and Silverman, J.H. (2000). *Diophantine Geometry: An Introduction*. Springer, New York.

Hormander, L. (1990). *An Introduction to Complex Analysis in Several Variables*, 3rd edition. North-Holland, Amsterdam.

Huybrechts, D. (2005). *Complex Geometry*. Springer, Berlin.

Iitaka, S. (2011). *Algebraic Geometry: An Introduction to Birational Geometry of Algebraic Varieties*. Springer, New York.

Iyer, J.N. (1999). Projective normality of Abelian surfaces given by primitive bundles. *Manuscr. Math.*, 98, 139–153.

Koizumi, S. (1976). Theta relations and projective normality of Abelian varieties. *Am. J. Math.*, 98, 865–889.

Lang, S. (1995). *Algebraic and Abelian Functions*, 2nd corrected printing. Springer, New York.

Lazarsfeld, R. (1990). Projectivité normale des surfaces abéliennes. Edited by O. Debarre. Prepublication 14. Europroj-C.I.M.P.A., Nice.

Lesfari, A. (1988). Abelian surfaces and Kowalewski's top. *Ann. Scient. École Norm. Sup.*, 4(21), 193–223.

Lesfari, A. (2003). Le système différentiel de Hénon-Heiles et les variétés Prym. *Pacific J. Math.*, 212(1), 125–132.

Lesfari, A. (2008). Prym varieties and algebraic completely integrable systems. *J. Geom. Phys.*, 58, 1063–1079.

Lesfari, A. (2022). *Integrable Systems*. ISTE Ltd, London and John Wiley & Sons, New York.

Lesfari, A. (2023). *Courbes algébriques complexes et/ou Surfaces de Riemann compactes*. Calvage et Mounet, Paris.

Miyaoka, Y. (1974). Kähler metrics on elliptic surfaces. *Proc. Japan Acad.*, 50, 533–536.

Moishezon, B.G. (1967). On n-dimensional compact varieties with n algebraically independent meromorphic functions. *Amer. Math. Soc. Transl.*, 63, 51–177.

Mumford, D. (1966–1967). On the equations defining Abelian varieties I, II, III. *Invent. Math.*, 1, 287–354; 3, 75–135; 3, 215–244.

Mumford, D. (1969). Varieties defined by quadratic equations. In *Questioni sulle varieta algebraiche*. Corsi dal C.I.M.E. Edizioni Cremonese Roma.

Mumford, D. (1974a). Prym varieties I. In *Contributions to Analysis*, Ahlfors, L.V., Kra, I., Maskit, B., Niremberg, L. (eds). Academic Press, New York, 325–350.

Mumford, D. (1974b). *Abelian Varieties*. Oxford University Press, Oxford.

Mumford, D. (1975a). *Curves and Their Jacobians*. University of Michigan Press, Ann Arbor.

Mumford, D. (1975b). *Algebraic Geometry I: Complex Projective Varieties*. Springer-Verlag, Berlin, Heidelberg, New York.

Mumford, D. (1983). *Tata lectures on theta I, II, Progress in Math.* Birkhaüser, Boston.

Narasimhan, R. (1995). *Several Complex Variables*. University of Chicago Press, Chicago.

Ohbuchi, A. (1988). A note on the projective normality of special line bundles on Abelian varieties. *Tsukuba J. Math.*, 12(2), 341–352.

Ortega, A. (2003). Variétés de Prym associées aux revêtements n-cycliques d'une courbe hyperelliptique. *Math. Z.*, 245, 97–103.

Ramanan, S. (1980). Ample divisors on Abelian surfaces. *Proc. London Math. Soc.*, 5(2), 609–629.

Scheidemann, V. (2023). *Introduction to Complex Analysis in Several Variables*. Birkhäuser, Basel.

Shabat, B.V. (1992). *Introduction to Complex Analysis Part II. Functions of Several Variables*. American Mathematical Society, Providence, RI.

Siegel, C.L. (1955). Meromorphic funktionen anf kompakten mannigfaltigkeiten, Nachrichten der Akademie der Wissenschaften in Göttingen. *Math.-Phys. Klasse*, 4, 71–77.

Silverman, J.H. (2009). *The Arithmetic of Elliptic Curves*. Springer, New York.

Siu, Y.-T. (1983). Every $K3$ surface is Kähler. *Invent. Math.*, 73, 139–150.

Swinnerton-Dyer, H.P.F. (1974). *Analytic Theory of Abelian Varieties*. Cambridge University Press, Cambridge.

Ueno, K. (1975). *Classification Theory of Algebraic Varieties and Compact Complex Spaces*. Springer, Berlin, Heidelberg.

Vanhaecke, P. (2001). *Integrable Systems in the Realm of Algebraic Geometry*. Springer-Verlag, Berlin.

Zheng, F. (2000). *Complex Differential Geometry*. AMS, Providence, RI, and International Press of Boston, Cambridge, MA.

Index

A, B

Abel
  –Jacobi map, 166
  theorem, 230
Abelian
  differential, 227
  variety, 133, 139, 140
absolute convergence, 9
addition
  formulas, 163
  theorem, 230
algebraic curve, 43
ample
  bundle, 125
  divisor, 125
amplitude, 241
analytic
  function, 11
  subset, 134
  variety, 37, 42
Appel–Humbert theorem, 97
atlas, 40
autonomy factor, 96
Baker–Akhiezer function, 171
base
  locus, 87
  point, 87
Bertini theorem, 127
Betti number, 130
biholomorphic
  map, 39
  mapping, 28
boundary of Šilov, 3

C, D

Camassa–Holm equation, 171
canonical
  bundle, 97
  divisor, 88, 136
  fiber, 88
Cartan structure equation, 76
Cartier divisor, 80
Cauchy
  –Riemann
    conditions, 4
    equations, 4
  inequalities, 20
  integral formula, 7
Cech cohomology, 70
change of
  chart, 37, 39
    formulas, 39
Chern class, 76
Chow
  –Kodaira theorem, 111
  theorem, 134
coboundary operator, 69
cochain, 69
cocycle, 82
codimension, 55

coherent sheaf, 69
compact Riemann surface, 227
compatible charts, 39
complete
  linear system, 86
  Reinhardt domain, 14
complex, 69
  algebraic curve, 227
  manifold, 37
  projective line, 50
  projective space, 47, 49
  torus, 51, 89
connection, 74
  matrix, 74
constant rank theorem, 53
contractile, 91
convergence of a multiple series, 9
coordinate
  change, 39
  neighborhood, 38
covering map, 41
critical point, 31
curvature, 75
  form, 75
  matrix, 75
de Rham cohomology, 71
decomposition on forms, 108
degree
  of a divisor, 229
  of polarization, 90
  of the isogeny, 141
differential, 3
  form, 227
  form of type (p, q), 103
distinguished boundary, 2, 3
divisor(s), 229
  linearly equivalent, 229
Dolbeault
  cohomology, 74
    groups, 105
  complex, 105
  operators, 104
  theorem, 74
domain of convergence, 11
doubly periodic, 234
dual
  Abelian variety, 144
  isogeny, 157
  of the line bundle, 82

E, F

effective divisor, 80, 229
Eisenstein series, 236
elementary divisors, 98
elliptic
  curve, 227
  function, 234
  integral, 240
embedding, 41, 89
equivalent functions, 27
étale, 157
Euler–
  Arnold equations, 220
  Poincaré characteristic, 89, 136
even
  half period, 161
  section, 155
  theta function, 161
exceptional divisor, 120
external product, 112
first Chern class, 106
Flaschka transformations, 224
flow, 184
Fubini–Study
  form, 128
  metric, 129
function
  $\mathbb{C}$-antilinear, 2
  $\mathbb{C}$-differentiable, 3
  $\mathbb{C}$-linear, 2
fundamental parallelogram, 233

G, H

genus, 227
germ, 27
global section, 65, 80, 83
grassmannian, 50
Green operator, 108
half period, 161

harmonic
  (p, q)-forms, 108
  form, 107
Hartog's
  extension theorem, 22
  theorem, 25
Heisenberg algebra, 185
Hénon–Heiles system, 190
hermitian
  form, 90
  manifold, 109
  metric, 109
Hodge
  form, 109
  theorem, 108
holomorphic
  differential, 227
  function, 4, 5
  map, 39
  section, 83
homogeneous coordinates, 48
Hopf surface, 130
hyperelliptic curve, 227
hypersurface, 55, 79

I, J

identity
  principle, 21
  theorem for
    holomorphic functions, 21
    power series, 34
immersion, 41
implicit function theorem, 30
integral curve, 183
invertible sheaf, 69
irreducible hypersurface, 79
isogenic varieties, 141
isogeny, 141
Jacobi
  elliptic function, 241
  identity, 164
  inversion problem, 231
Jacobian
  matrix, 28
  variety, 135
jets of order 1, 53

K, L

Kähler
  form, 109
  manifold, 109
  metric, 109
Kirchhoff equations, 213
Kodaira
  –Nakano vanishing theorem, 111
  embedding theorem, 125, 126
Kummer surface, 131, 203
Landau–Lifshitz equation, 172
lattice, 89
Laurent series, 16
Lefschetz theorem
  for classes of type (1,1), 116
  on theta functions, 173
Legendre
  canonical forms, 240
  relation, 238
Leray theorem, 70
line bundle, 81, 82
linear system of divisors, 87
linearly equivalent divisors, 86
Liouville theorem, 21
local coordinates, 38
logarithmically convex open set, 15

M, N

Manakov condition, 222
maximum modulus principle, 23
meromorphic section, 83
Milnor spheres, 47
Mittag–Leffler problem, 70
module, 241
Moishezon theorem, 110
Montel theorem, 35
multi-index, 3
multiple
  power series, 10
  series, 9
Néron–Severi group, 96, 143
non-singular surface, 59
normal
  convergence, 10
  representation, 166

O, P

$\overline{\partial}$-laplacian, 108
odd
  half period, 161
  section, 155
  theta function, 161
one-parameter group of diffeomorphism, 184
open mapping theorem, 24
order, 80
Osgood lemma, 8
period matrix, 138, 228
Pfaffian, 98
Picard group, 71, 83, 142
Poincaré
  complete reducibility theorem, 157
  dual, 77
  reducibility theorem, 157
  residue, 228
  theorem, 158
polarization, 90, 140
  of the type $(\delta_1, ..., \delta_n)$, 178
polarized Abelian variety, 140
poly-radius, 2
polydisk, 2
positive
  divisor, 80, 229
  line bundle, 107, 111
principal
  divisor, 80, 229
  polarization, 90, 178
principle of analytic continuation, 21
product of manifolds, 40
projective
  manifold, 103, 127
  variety, 103
projectively normal
  divisor, 154
  embedding, 154
Prym variety, 146, 151

Q, R

quasi-periods, 160
Ramanan theorem, 178
rank, 31
  of the lattice, 89
real torus, 89
regular point, 31, 42, 56
Reinhardt domain, 14
Riemann
  –Hurwitz formula, 230
  –Roch theorem, 230
  bilinear relations, 139, 228
  constant vector, 167
  form, 90, 98
  fundamental theorem, 168
  matrix, 159
  sphere, 43
  surface, 43

S, T

Schwarz lemma, 26
section, 65, 83
semicharacter, 95
sesquilinear form, 90
sheaf, 65
simple
  Abelian variety, 157
  pendulum, 244
sine-Gordon equation, 172
singular
  locus, 56
  point, 31, 42, 56
skyscraper sheaf, 66
space
  of harmonic forms, 108
  tangent, 52
special linear group, 58
stereographic projection, 43, 50
submanifold, 55
submersion, 41
subvariety, 55
surface $K3$, 130
symmetric product, 168
symplectic basis, 166, 228
tangent
  bundle, 52
  space, 53
  vector, 53

tautological
  bundle, 85
  fiber, 97
tensor product, 82
theorem of the square, 158
theta
  divisor, 168
  function, 159
    with characteristics, 160
Toda lattice, 224
torus, 234
transition
  formulas, 39
  function(s), 39, 81
  matrix, 82
trivial line bundle, 83, 99
trivializations, 81

U, V, W

uniform convergence, 10
unimodular group, 58
vector
  bundle, 82
  field, 183
  tangent, 52
very ample
  bundle, 125
  divisor, 125
Weierstrass
  $\wp$ function, 235
  elliptic integral, 236
  function $\sigma(z)$, 182
  invariants, 236
  sigma function, 239
  theorem, 24
  zeta function, 237
Weil divisor, 79
Wirtinger operators, 4

Other titles from

ISTE

in

Mathematics and Statistics

2025

DUPUY Jean-François
Generalized Linear Models: Problems with Censored, Missing, and Zero-inflated Data

GAIVORONSKI Alexei A., KNOPOV Pavel S., NORKIN Vladimir I., ZASLAVSKYI Volodymyr A.
Stochastic Modeling and Optimization Methods for Critical Infrastructure Protection 1: Stochastic Modeling
Stochastic Modeling and Optimization Methods for Critical Infrastructure Protection 2: Methods and Tools

PICARD Philippe, LEFÈVRE Claude
Abel-Gontcharoff Pseudopolynomials and Stochastic Applications

2024

BOURLÈS Henri
System Theory – A Modern Approach 1: Linear Ordinary and Functional Differential Equations
(New Mathematical Methods, Systems and Applications Set – Volume 1)

MAKAROV Volodymyr, MAYKO Nataliya
Traditional Functional-Discrete Methods for the Problems of Mathematical Physics: New Aspects

2023

AIT BEN HASSI El Mustapha
The Theory of Distributions: Introduction

KOROLIOUK Dmitri, SAMOILENKO Igor
Asymptotic and Analytic Methods in Stochastic Evolutionary Systems

PARROCHIA Daniel
Graphs, Orders, Infinites and Philosophy

2022

CHAKRAVARTHY Srinivas R.
Introduction to Matrix-Analytic Methods in Queues 1: Analytical and Simulation Approach – Basics
Introduction to Matrix-Analytic Methods in Queues 2: Analytical and Simulation Approach – Queues and Simulation

LESFARI Ahmed
Integrable Systems

RADCHENKO Vadym M.
General Stochastic Measures: Integration, Path Properties and Equations

DE SAPORTA Benoîte, ZILI Mounir
Martingales and Financial Mathematics in Discrete Time

SIMON Jacques
Distributions
(Analysis for PDEs Set – Volume 3)

2021

KOROLIOUK Dmitri, SAMOILENKO Igor
Random Evolutionary Systems: Asymptotic Properties and Large Deviations

MOKLYACHUK Mikhail
Convex Optimization: Introductory Course

POGORUI Anatoliy, SWISHCHUK Anatoliy, RODRÍGUEZ-DAGNINO Ramón M.
Random Motions in Markov and Semi-Markov Random Environments 1: Homogeneous Random Motions and their Applications
Random Motions in Markov and Semi-Markov Random Environments 2: High-dimensional Random Motions and Financial Applications

PROVENZI Edoardo
From Euclidean to Hilbert Spaces: Introduction to Functional Analysis and its Applications

2020

BARBU Vlad Stefan, VERGNE Nicolas
Statistical Topics and Stochastic Models for Dependent Data with Applications

CHABANYUK Yaroslav, NIKITIN Anatolii, KHIMKA Uliana
Asymptotic Analyses for Complex Evolutionary Systems with Markov and Semi-Markov Switching Using Approximation Schemes

KOROLIOUK Dmitri
Dynamics of Statistical Experiments

MANOU-ABI Solym Mawaki, DABO-NIANG Sophie, SALONE Jean-Jacques
Mathematical Modeling of Random and Deterministic Phenomena

SIMON Jacques
Continuous Functions
(Analysis for PDEs Set – Volume 23)

2019

BANNA Oksana, MISHURA Yuliya, RALCHENKO Kostiantyn, SHKLYAR Sergiy
Fractional Brownian Motion: Approximations and Projections

GANA Kamel, BROC Guillaume
Structural Equation Modeling with lavaan

KUKUSH Alexander
Gaussian Measures in Hilbert Space: Construction and Properties

LUZ Maksym, MOKLYACHUK Mikhail
Estimation of Stochastic Processes with Stationary Increments and Cointegrated Sequences

MICHELITSCH Thomas, PÉREZ RIASCOS Alejandro, COLLET Bernard, NOWAKOWSKI Andrzej, NICOLLEAU Franck
Fractional Dynamics on Networks and Lattices

VOTSI Irene, LIMNIOS Nikolaos, PAPADIMITRIOU Eleftheria, TSAKLIDIS George
Earthquake Statistical Analysis through Multi-state Modeling
(Statistical Methods for Earthquakes Set – Volume 2)

2018

AZAÏS Romain, BOUGUET Florian
Statistical Inference for Piecewise-deterministic Markov Processes

IBRAHIMI Mohammed
Mergers & Acquisitions: Theory, Strategy, Finance

PARROCHIA Daniel
Mathematics and Philosophy

2017

CARONI Chysseis
First Hitting Time Regression Models: Lifetime Data Analysis Based on Underlying Stochastic Processes
(Mathematical Models and Methods in Reliability Set – Volume 4)

CELANT Giorgio, BRONIATOWSKI Michel
Interpolation and Extrapolation Optimal Designs 2: Finite Dimensional General Models

CONSOLE Rodolfo, MURRU Maura, FALCONE Giuseppe
Earthquake Occurrence: Short- and Long-term Models and their Validation
(Statistical Methods for Earthquakes Set – Volume 1)

D'AMICO Guglielmo, DI BIASE Giuseppe, JANSSEN Jacques, MANCA Raimondo
Semi-Markov Migration Models for Credit Risk
(Stochastic Models for Insurance Set – Volume 1)

GONZÁLEZ VELASCO Miguel, del PUERTO GARCÍA Inés, YANEV George P.
Controlled Branching Processes
(Branching Processes, Branching Random Walks and Branching Particle Fields Set – Volume 2)

HARLAMOV Boris
Stochastic Analysis of Risk and Management
(Stochastic Models in Survival Analysis and Reliability Set – Volume 2)

KERSTING Götz, VATUTIN Vladimir
Discrete Time Branching Processes in Random Environment
(Branching Processes, Branching Random Walks and Branching Particle Fields Set – Volume 1)

MISHURA Yuliya, SHEVCHENKO Georgiy
Theory and Statistical Applications of Stochastic Processes

NIKULIN Mikhail, CHIMITOVA Ekaterina
Chi-squared Goodness-of-fit Tests for Censored Data
(Stochastic Models in Survival Analysis and Reliability Set – Volume 3)

SIMON Jacques
Banach, Fréchet, Hilbert and Neumann Spaces
(Analysis for PDEs Set – Volume 1)

2016

CELANT Giorgio, BRONIATOWSKI Michel
Interpolation and Extrapolation Optimal Designs 1: Polynomial Regression and Approximation Theory

CHIASSERINI Carla Fabiana, GRIBAUDO Marco, MANINI Daniele
Analytical Modeling of Wireless Communication Systems
(Stochastic Models in Computer Science and Telecommunication Networks Set – Volume 1)

GOUDON Thierry
Mathematics for Modeling and Scientific Computing

KAHLE Waltraud, MERCIER Sophie, PAROISSIN Christian
Degradation Processes in Reliability
(Mathematial Models and Methods in Reliability Set – Volume 3)

KERN Michel
Numerical Methods for Inverse Problems

RYKOV Vladimir
Reliability of Engineering Systems and Technological Risks
(Stochastic Models in Survival Analysis and Reliability Set – Volume 1)

2015

DEVOLDER Pierre, JANSSEN Jacques, MANCA Raimondo
Basic Stochastic Processes

LE GAT Yves
Recurrent Event Modeling Based on the Yule Process
(Mathematical Models and Methods in Reliability Set – Volume 2)

DE SAPORTA Benoîte, DUFOUR François, ZHANG Huilong
Numerical Methods for Simulation and Optimization of Piecewise Deterministic Markov Processes

2014

COOKE Roger M., NIEBOER Daan, MISIEWICZ Jolanta
Fat-tailed Distributions: Data, Diagnostics and Dependence
(Mathematical Models and Methods in Reliability Set – Volume 1)

MACKEVIČIUS Vigirdas
Integral and Measure: From Rather Simple to Rather Complex

PASCHOS Vangelis Th
Combinatorial Optimization – 3-volume series – 2nd edition
Concepts of Combinatorial Optimization / Concepts and Fundamentals – Volume 1
Paradigms of Combinatorial Optimization – Volume 2
Applications of Combinatorial Optimization – Volume 3

2013

COUALLIER Vincent, GERVILLE-RÉACHE Léo, HUBER Catherine, LIMNIOS Nikolaos, MESBAH Mounir
Statistical Models and Methods for Reliability and Survival Analysis

JANSSEN Jacques, MANCA Oronzio, MANCA Raimondo
Applied Diffusion Processes from Engineering to Finance

SERICOLA Bruno
Markov Chains: Theory, Algorithms and Applications

2012

BOSQ Denis
Mathematical Statistics and Stochastic Processes

CHRISTENSEN Karl Bang, KREINER Svend, MESBAH Mounir
Rasch Models in Health

DECREUSEFOND Laurent, MOYAL Pascal
Stochastic Modeling and Analysis of Telecom Networks

DEVOLDER Pierre, JANSSEN Jacques, MANCA Raimondo
Stochastic Methods for Pension Funds

2011

MACKEVIČIUS Vigirdas
Introduction to Stochastic Analysis: Integrals and Differential Equations

MAHJOUB Ridha
Recent Progress in Combinatorial Optimization – ISCO2010

RAYNAUD Hervé, ARROW Kenneth
Managerial Logic

2010

BAGDONAVIČIUS Vilijandas, KRUOPIS Julius, NIKULIN Mikhail
Nonparametric Tests for Censored Data
Nonparametric Tests for Complete Data

IOSIFESCU Marius, LIMNIOS Nikoloas, OPRISAN Gheorghe
Introduction to Stochastic Models

VASSILIOU P.C.G.
Discrete-time Asset Pricing Models in Applied Stochastic Finance

2008

ANISIMOV Vladimir
Switching Processes in Queuing Models

FICHE Georges, HÉBUTERNE Gérard
Mathematics for Engineers

HUBER Catherine, LIMNIOS Nikolaos, MESBAH Mounir, NIKULIN Mikhail
Mathematical Methods in Survival Analysis, Reliability and Quality of Life

JANSSEN Jacques, MANCA Raimondo, VOLPE Ernesto
Mathematical Finance

2007

HARLAMOV Boris
Continuous Semi-Markov Processes

2006

CLERC Maurice
Particle Swarm Optimization

Printed and bound by CPI Group (UK) Ltd, Croydon, CR0 4YY
13/09/2026
14961014-0004